Reconstructing Evolution

Reconstructing Evolution

New Mathematical and Computational Advances

Edited by

OLIVIER GASCUEL AND MIKE STEEL

OXFORD
UNIVERSITY PRESS

OXFORD
UNIVERSITY PRESS

Great Clarendon Street, Oxford OX2 6DP

Oxford University Press is a department of the University of Oxford.
It furthers the University's objective of excellence in research, scholarship,
and education by publishing worldwide in

Oxford New York

Auckland Cape Town Dar es Salaam Hong Kong Karachi
Kuala Lumpur Madrid Melbourne Mexico City Nairobi
New Delhi Shanghai Taipei Toronto

With offices in

Argentina Austria Brazil Chile Czech Republic France Greece
Guatemala Hungary Italy Japan Poland Portugal Singapore
South Korea Switzerland Thailand Turkey Ukraine Vietnam

Published in the United States
by Oxford University Press Inc., New York

First published 2007

British Library Cataloguing in Publication Data
Data available

Library of Congress Cataloging in Publication Data
Data available

Typeset by Newgen Imaging Systems (P) Ltd., Chennai, India
Printed in Great Britain
on acid-free paper by
Biddles Ltd., King's Lynn, Norfolk

ISBN 978–0–19–920822–7

1 3 5 7 9 10 8 6 4 2

ACKNOWLEDGEMENTS

Many thanks to:

All the contributors, who have spent time, energy, and patience in writing and writing again their chapters, and have cross-reviewed other chapters with much care: Elizabeth S. Allman, Cécile Ané, Michäel G. B. Blum, Alexei Drummond, Oliver Eulenstein, Gregory Ewing, Joseph Felsenstein, David Fernãndez-Baca, Stefan Grünewald, Stéphane Guindon, Luke J. Harmon, Klaas Hartmann, Stephen B. Heard, Katharina Huber, Daniel Huson, Junhyong Kim, Michelle M. McMahon, Arne Ø. Mooers, Raul Piaggio-Talice, John A. Rhodes, Allen Rodrigo, Michael J. Sanderson, Charles Semple, Dennis H. J. Wong.

A number of distinguished anonymous referees, whose suggestions, recommendations, and corrections greatly helped to improve the contents of this volume.

Dietrich Radel (Christchurch) for his tireless work in reformatting the original submissions; Megan Foster (Christchurch) for proofreading some chapters; Jessica Churchman and Alison Jones (Oxford University Press) for encouragements and helpful advice.

The people from Institut Henri Poincaré (Paris) and LIRMM (Montpellier), who helped in organizing the 'Mathematics of Evolution and Phylogeny' conference in June 2005, from which this book arose: Céline Berger, Denis Bertrand, Samuel Blanquart, Isabelle Duc, Etienne Gouin-Lamourette, Julie Hussin, Sylvie Lhermitte, Corine Mélançon.

Olivier Gascuel and Mike Steel
Montpellier–Christchurch, November 2006.

INTRODUCTION

Olivier Gascuel and Mike Steel

It has become clear that the fundamentals of biology are much more complex than expected in the 1950s and 1960s following the discovery of the DNA double-strand and of the genetic code. The 'one gene, one protein, one function' hypothesis and the 'central dogma of molecular biology' have been profoundly revised and enriched. Now we know that alternative splicing [41] is frequent in eukaryotes and viruses. In this process, a single pre-messenger RNA transcribed from one gene can lead to different mature messenger RNA molecules (mRNA) and therefore to different proteins (up to tens of thousands [58]). Moreover, we understand the central role of post-RNA-translation modifications more clearly; these can extend the range of functions of a protein by attaching to it other biochemical functional groups, by changing the chemical nature of certain residues, or by modifying its sequence and/or structure. The discovery of micro RNAs [1] and interference RNAs [26] which appear to underlie the regulation of numerous biological functions have considerably augmented the repertoire of known non-coding genes. From these discoveries, it appears that one gene may correspond to a non-protein functional unit as well as to a number of proteins and biochemical functions. However, it is also clear that the gene content of an organism is only one factor, and that gene regulation could be at least as important in explaining the differences between species. For example, microarray-based studies [30] have shown that gene regulation in chimps and humans is significantly different, although their gene repertoire is almost identical. Moreover, species cannot be understood without considering their ecological environment and their interactions with other species. For example, we are just starting to explore the relationships between humans and their (bacterial and archaeal) intestinal flora, which involve numerous interactions and regulations between the host and symbiont genes [31]. These few examples show that biology is extraordinarily complex and constitutes a territory that is currently being explored more deeply and rapidly, but still has many uncharted regions.

Our vision of evolution has also changed considerably during the last few years. The mechanisms described above are likely to play an important role (e.g. alternative splicing could play a key role in the evolution of eukaryotic proteins [14, 62]). Moreover, molecular data have demonstrated that tree-like evolution as represented by Darwin (Fig. 1) is often a gross simplification of ancestry. Gene trees and species trees often differ, due to lineage sorting [18], or to lateral gene transfers [47]. Recent works have shown that gene transfers occurred (and

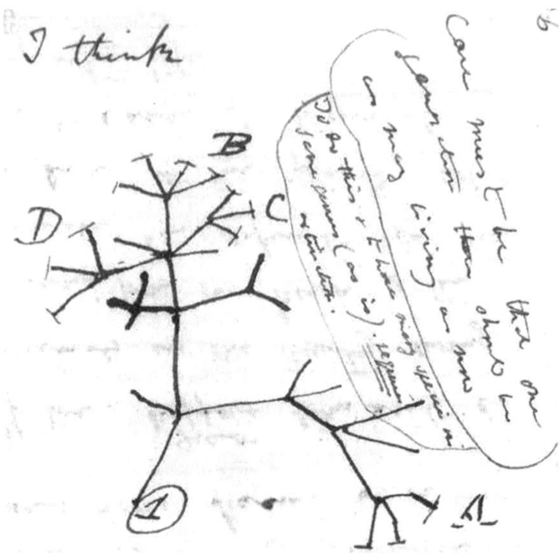

FIG. 1. Darwin's first sketch of an evolutionary tree (1837)

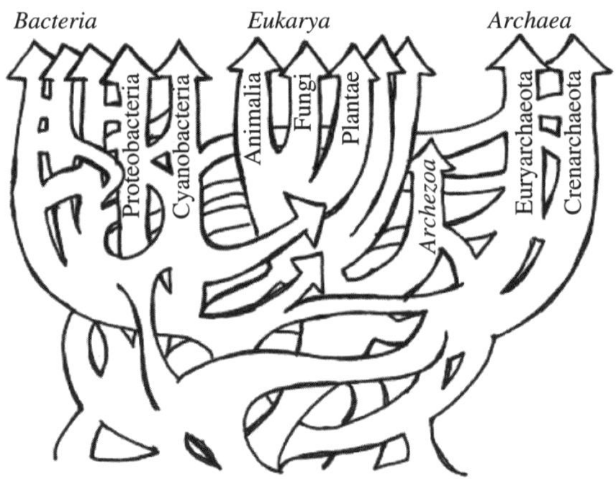

FIG. 2. Doolittle's network of life [20].

still occur) extensively in bacteria [40] and are not rare in eukaryotes [2, 6]. From Darwin's tree of Fig. 1, we are thus moving to a network view. Fig. 2 [20] is an artist's view of such a network, showing the reticulations that occurred in an organism's evolutionary history. It shows how a single species may have multiple ancestor species corresponding to its different parts. It may have one ancestor for its nuclear genome (several in case of major endosymbiotic events, e.g. in *Guillardia theta* [22] or *Plasmodium falciparum* [28]) and others for its organelles such as mitochondria and chloroplasts. In this emerging 'web of life' viruses play a pervasive role as they seem to have a central part in lateral transfer mechanisms [16, 27]. We are now at the point where the notion of species may be hard to define [21, 48], particularly for simple, primitive organisms such as bacteria and archaea, which appear as a genetic puzzle arising from multiple inheritance (transfers, hybridizations, endosymbiosis, etc.) rather than the result of a progressive and continuous evolution of a unique lineage.

Genomes have their own evolutionary dynamics and are subjected to various rearrangements (inversions, translocations, segmental or global duplications, etc., see [57] and Chapters 9 to 13 in [29]) which may be heavy even within a (relatively) short time period [23]. Relationships between these genome rearrangements and phenotypes are still unclear, and the variability of genomic configurations within any given species or along the course of time is just starting to be explored. Computer scientists have intensively investigated genome rearrangements, following the seminal works of D. Sankoff [56] and S. Hannenhali and P. Pevzner [32], but we are still at the early stages of understanding the biological implications of these rearrangements.

Genes may be seen as elementary building blocks, but sometimes they also have complex histories. They are subjected to duplications which tend to modify their function and to create new genes with new functions [49]. But genes are also subject to gene conversion, whereby multiple variable copies of a single gene become partially or fully homogenized. Genes may undergo recombinations and segmental transfers which make them mosaic-like; they are then composed of interspersed blocks of nucleotide sequence which have different evolutionary histories. Such mosaic genes are relatively frequent in bacteria [45], but have also been reported in eukaryotes [38]. All these events may have to be accounted for when reconstructing gene histories, which may be non-tree-like and resemble the scheme shown in Fig. 2. Yet most genes still seem to fit well with standard Darwinian tree scheme, although evolutionary forces are variable through time, and the structure and/or the function of the proteins may change. Detailed reconstruction of evolution thus necessitates the use of models which account for this variability, in order to be able to describe the precise history of the genes at the site level.

All life arises by evolution, via inheritance, mutation and selection. Even though evolutionary mechanisms are complex (as described above), and sometimes result in mosaic-like taxa with network-like histories, they reinforce the much cited assertion by T. Dobzhansky [19]: 'Nothing in biology makes sense, except in the light of evolution'. In particular, phylogenetics and the study of sequence evolution are fundamental for bioinformatics and the deciphering of genomes. One of the central goals in this field is to infer the function of proteins from genomic sequences. To this end, alignment methods are nowadays the most frequently used, based on the fact that homologous proteins most often have similar structure and function. To estimate (through alignment) the similarity between any sequence pair, we rely heavily on Markovian substitution models such as the famous Dayhoff [17] or JTT matrices [37]. Moreover, to obtain reliable functional predictions, we frequently distinguish between paraloguous and orthologous proteins (only the latter are likely to share the same function), which is a complex task requiring phylogenetic analysis of extensive sets of homologous proteins [59]. However, alignment typically gives functional indications for only $\sim$50% of the proteins in a newly sequenced genome. This limit encourages the development of new methods, a number of them being based on evolutionary analyses, such as phylogenomic profiling [24], gene cluster conservation [50], and phylogenetic footprinting [7]. Another non-sequence example of the pervasiveness of evolutionary approaches, is the elucidation and analysis of regulatory networks and metabolic pathways, which has become topical with the flood of microarray gene expression data. A deeper understanding of the structure and function of regulatory networks and metabolic pathways is emerging from comparative studies, phylogenetic analysis [46] and the search for conserved motifs [5].

Phylogenetics is also central to species-level studies. Most notably, several Tree of Life projects [60] are underway worldwide, aiming to establish the phylogenetic relationships between all living species. Massive sequencing approaches such as barcoding [9] and metagenomics [61, 15, 31] are becoming mainstream to the point where an organism's place in the Tree of Life will often become one of the first things we know about it. Phylogenies are becoming a preferred way to represent and measure biodiversity, to survey invasive species, and to assess conservation priorities [42]. Notably, interspecies phylogenies with divergence dates contain information about rates and distributions of species extinctions and about the nature of radiations after previous mass extinctions [8]. Comparative approaches have also been used to model extinction risk as a function of a species' biological characteristics [52], which could then be used as a basis for evaluating the status of species with an unknown extinction risk.

Phylogenetic analysis is also fundamental to modern epidemiology. Understanding how organisms, as well as their genes and gene products, are related to one another has become a powerful tool for identifying and classifying rapidly evolving pathogens, tracing the history of infections, and predicting outbreaks. Phylogenetic studies were crucial in identifying emerging viruses such as SARS [44], and in understanding the relationships between the virulence and the genetic evolution of HIV [53] and influenza [25].

Due to recent progress [43] in sequencing technologies, genomic data continue to grow exponentially. The genomic database Genbank has information on about 265,000 species and contains over 100 billion base pairs. Moreover, a number of species have been completely sequenced, e.g. $\sim$400 bacteria, but also 12 mammals (see Ensembl web site). Consequently, ever increasing numbers of phylogenetic studies are performed, as assessed by the citation numbers of the most famous phylogeny programs (e.g. above 14,000 for NJ and 3,000 for MrBayes, see Web of Science). However, due to the complexity of evolutionary processes, building phylogenetic trees is neither straightforward nor an end in itself, and new concepts and computational tools flourish—for example, for exploring phylogenetic networks, for studying evolution within populations, and for understanding evolution at the molecular level. This quantity of data provides us with extraordinary new possibilities to understand and reconstruct the past. For example, thanks to complete sequencing of both *Human* and *Tetraodon* (a fish), we have been able to reconstruct (in broad terms) the genome of a vertebrate ancestor [36]. As another example, the complete sequencing of *Paramecium tetraurelia* (an unicellular eukaryote) showed that most of the genes arose through at least three successive whole-genome duplications; moreover, phylogenetic analysis indicated that the most recent duplication coincides with an explosion of speciation events that gave rise to a number of sibling species [3].

But reconstructing evolution faces similar challenges to those that arise in other disciplines that deal with events that occurred in the past (e.g. astrophysics or earth history). We have no time machine, as imagined by H.G. Wells, evolution occurred just once, and there are few direct observations or experimental results on evolutionary processes. Most data are contemporary, and we rely on mathematical models to understand the past.

Pioneering work on the mathematical aspects of phylogenetics began during the 1960s and 1970s, and some of these early papers, particularly by D. Sankoff [54, 55] and P. Buneman [11, 12, 13] were enlightened predictors of the field to come in later decades. Statistical approaches, pioneered by A. Edwards and J. Felsenstein began by considering simple models of sequence site evolution. Typically these involved symmetric (and often two-state) Markov models in which each site evolves at a constant rate across the tree. This model is still studied for its mathematical properties (and it has been studied in related fields such as statistical physics and broadcasting theory). More recently, however, models have become increasingly sophisticated to account for the inherent complexity of evolution. They usually involve non-symmetric Markov processes which can vary across sites, and sometimes also across the tree (as with covarion-type processes). This has led to some debate as to what is the 'right' model for a phylogenetic study and an emerging pragmatism that there is no global model, rather each data set has its own characteristics that can suggest (and support) the most appropriate model [51].

Modelling of site substitutions has been primarily a statistical exercise, first studied within a likelihood framework, and more recently from the Bayesian (MCMC) perspective. Site substitution models also harbour a good deal of mathematical structure – for example, the Hadamard representation [33], as well as phylogenetic invariants. These invariants are algebraic identities first described in the mid 1980s, and which have been investigated with sporadic intensity ever since. Recent advances this century have stemmed from algebraic geometers and experts in commutative algebra, particularly B. Sturmfels and colleagues at UC Berkeley, together with E. Allman and J. Rhodes.

Site substitution is just one aspect of genomic evolution, and other genome rearrangement and insertion events are becoming increasingly important as phylogenetic markers. In the case of gene order, computer scientists during the 1990s devoted much effort to finding the smallest number of transformations of given types required to transform one gene sequence into another. At the same time, a group based around D. Sankoff investigated the properties of the more easily-computed breakpoint distance. In contrast to site sequence data, for gene order and for other rare genomic events, such as Short interspersed nuclear elements (SINEs), the state space is potentially very large, and this can be useful for methods that work well on data that exhibits low (or zero) homoplasy. The concept of reconstructing a tree from such compatible characters was investigated mathematically back in the 1970s and 1980s by G. Eastabrook, F. McMorris, C. Meacham, and others; it was resurrected in the early 1990s by T. Warnow and her colleagues as the 'perfect phylogeny problem' and has enjoyed further development due to the rich connection this problem has with chordal graph theory and closure operators. One recent result in this area is the theorem [34] that every fully-resolved phylogenetic tree can be uniquely specified by just *four* homoplasy-free characters, a finding that is surprising to many biologists (and some mathematicians!).

Although the reconstruction of evolutionary trees directly from character data is widespread, distance-based approaches are also popular due to their flexibility (distances can be easily computed and 'corrected'), and the computational efficiency of algorithms such as Neighbor-Joining. Mathematically, the idea of modelling distances on a tree seems to have first appeared in the 1960s in Russia after K. Zaretskii's pioneering work [63], and many of the classic results—the four-point condition, and the uniqueness of a tree representation—have since been rediscovered several times. A unified treatment was provided by A. Dress and H.-J. Bandelt in a series of papers between the late 1980s and early 1990s. One of the outcomes of their collaboration was the development of split decomposition theory [4] which provided, for the first time, a mathematically natural way to construct phylogenetic networks (rather than just trees) from distance data. This method is still used and it is implemented in the software package *SplitsTree* [35]. However the theory has also inspired more effective techniques for network reconstruction, including the now widely-used *Neighbor-Net* algorithm [10]. The turn of this century also saw mathematicians and computer scientists mount a series of attacks on the problem of reconstructing phylogenetic networks from different types of data—trees, characters, and distances. Supertree methods have also enjoyed a recent renaissance, as have methods for using phylogenetic trees to study processes of molecular evolution (such as selection and recombination), and to investigate processes of speciation and extinction.

This book aims to present these recent models, their biological relevance, their mathematical basis, their properties, and the algorithms for applying them to data. In addition, the book highlights some of the ways in which mathematics and computer science have been enriched by their interaction with evolutionary biology. These include results from the emerging field of 'phylogenetic combinatorics' which is developing a detailed theory for studying trees and networks, as well as some recent algebraic advances in the theory of phylogenetic invariants. The range of topics involves mathematics, statistics, and computer science, and in particular the subfields of combinatorics, graph theory, probability theory and Markov models, algebraic geometry, statistical inference, Monte Carlo methods, and continuous and discrete algorithms.

This book contains ten chapters, which are grouped into five main parts:

I. Evolution within populations

The first two chapters investigate within-species evolution of gene copies, under relatively short time scales, as opposed to standard phylogenetics which considers between-species evolution of genes and much larger time periods. Chapter 1, by J. Felsenstein, shows that the coalescent trees (coalescents for short), first proposed by J. F. C. Kingmann [39], allow us to think about evolution within and between populations, and to make the connection between phylogenies and population genetic analyses. Coalescents are essential in developing methods for making inferences about populations. The chapter reviews the properties of coalescents, and the likelihood-based and Bayesian inference methods which are based on them. Chapter 2, by A. Rodrigo and co-authors, deals with rapidly evolving species, typically viruses such as HIV. Because these species are evolving so rapidly, their sequences accumulate a significant number of substitutions over short time periods (~1% per year with HIV), and serial sampling gives us useful insights on their evolution. The chapter reviews the methods that have been developed to study these measurably evolving populations, e.g. for estimating the substitution rate and its time variations, the population size, or the migration rates.

II. Models of sequence evolution

The mathematical and statistical properties of models that describe the evolution of aligned DNA sequences have been intensively studied since the 1970s. Indeed this branch of molecular phylogenetics is arguably the most well-developed theoretically. But many questions still remain, as does the potential for further work. Early models concentrated on simple scenarios in which site substitution was described by a basic (usually symmetric) process running at a constant rate across the sequences. Increasingly sophisticated models have allowed for more complex (and realistic) processes that may vary across the sequence and throughout the tree. In Chapter 3, O. Gascuel and S. Guindon show how standard Markov models of DNA site substitution can be further extended to handle these complexities and to detect selection, and the authors illustrate the use of these models on data sets from plants and HIV-1. In Chapter 4, E. Allman and J. Rhodes describe the current state-of-the-art in phylogenetic invariants. These fundamental algebraic identities arise within site substitution models and they are becoming useful for answering basic questions such as whether one can estimate certain parameters (including the tree) when the models become sufficiently complex. They also look promising for the future development of more efficient ways to undertake maximum-likelihood analysis or the development of new statistical approaches to phylogenetic reconstruction.

III. Tree shape, speciation, and extinction

Phylogenetic trees relate contemporary species which have arisen from past speciation and extinction events. Depending on periods and places, evolution may be diversifying and induce high speciation levels (up to 'explosive radiation'), or may tend towards massive extinction, as is the case today due to increasing human impact. Phylogenetic trees retain signatures of the evolutionary conditions and mechanisms that gave rise to them, and are invaluable tools to represent biodiversity. Chapter 5, by A. Mooers and co-authors, reviews a variety of models designed to represent different hypotheses about diversification processes. These models range from the simple Yule model to more complex approaches that treat species as collections of individuals rather than simple lineages. The fit of these models to real data is discussed in the light of two widely-used measures of phylogenetic tree shape, that is, tree imbalance, which measures the variation in subgroup size, and a waiting-time index based on the root-to-tip distribution of speciation events. Chapter 6, by K. Hartmann and M. Steel, discusses 'phylogenetic diversity' which measures the biodiversity of a set species as being the length of the phylogenetic tree connecting them. Phylogenetic diversity has been widely used for prioritising taxa for conservation and is the basis of the 'Noah's ark problem' in biodiversity management. The chapter reviews some new and recent algorithmic, mathematical, and stochastic results concerning phylogenetic diversity, ranging from survival probabilities and diversity loss, to tree reconstruction.

IV. Trees from subtrees and characters

One of the challenges faced by attempts to reconstruct a 'Tree of Life' is that typically one has a great deal of partial information–for example, trees for certain collections of taxa may be obtained from different groups or different data, or fundamental partitions of taxa may be made on the basis of the presence or absence of various markers. How to combine these efficiently and effectively into a phylogeny is a complicated task, involving mathematical and computational questions. In Chapter 7, M. Sanderson and colleagues describe some new approaches for studying collections of trees, going beyond the current 'supertree' approach. Using graph-theoretic approaches, they describe ways to extract phylogenetic signal, cluster subsets of data, and identify 'groves' of phylogenetic trees. In Chapter 8, S. Grünewald and K. Huber use combinatorial techniques to investigate how trees can be reconstructed from multi-state characters (and subtrees). These characters can arise in several ways–either as primary data describing how taxa are partitioned by complex genomic characters, or from existing taxonomic classifications of groups that represents different divisions of life. The results are also relevant to supertree construction where overlapping taxon sets are combined into a larger parent tree.

V. From trees to networks

As we explained above, evolution is not always tree-like and network representations are required (see Fig. 2). Actually, there are several types of reticulation events (lateral transfer, recombination, hybridization, etc.) and even more types of phylogenetic networks. Chapter 9, by D. Huson, makes a clear distinction between the implicit network methods that aim to display (non-tree-like) phylogenetic signals, and the explicit networks aiming to model reticulate evolution. This chapter looks at split networks as a major class of implicit networks and discusses a number of approaches to produce split networks from sequences, evolutionary distances, and tree collections. This chapter also discusses explicit network methods for analysing hybridization and recombination. Chapter 10, by C. Semple, deals with the combinatorics of hybridisation networks and the problem of finding the smallest number of reticulation events that are required to explain conflicting phylogenetic signals. Here, the signals correspond to rooted phylogenetic trees—for example trees for genes collected within the species under consideration—and the chapter mostly deals with the case where we just have two conflicting trees. A number of mathematical and algorithmic properties are described, and these establish close connections between this problem, the rooted subtree prune and regraft distance, agreement forests, and recombination networks.

References

[1] Ambros, V. (2001). MicroRNAs: Tiny regulators with great potential. *Cell*, **107**, 823–826.

[2] Andersson, J. O. (2005). Lateral gene transfer in eukaryotes. *Cellular and Molecular Life Sciences*, **62**(11), 1182–1197.

[3] Aury, J. M. *et al.* (2006). Global trends of whole-genome duplications revealed by the ciliate Paramecium tetraurelia. *Nature*, **444**(7116), 171–178.

[4] Bandelt, H. -J. and Dress, A. W. M. (1992). A canonical decomposition theory for metrics on a finite set. *Advances in Mathematics*, **92**, 47–105.

[5] Berg, J. and Lässig, M. (2004). Local graph alignment and motif search in biological networks. *Proceedings of the National Academy of Science USA*, **101**(41), 14689–14694.

[6] Bergthorsson, U., Adams, K., Thomason, B., and Palmer, J. (2003). Widespread horizontal transfer of mitochondrial genes in flowering plants. *Nature*, **424**, 197–201.

[7] Blanchette, M., Schwikowski, B., and Tompa, M. (2002). Algorithms for phylogenetic footprinting. *Journal of Computational Biology*, **9**(2), 211–223.

[8] Bromham, L., Phillips, M. J., and Penny, D. (1999). Growing up with dinosaurs: molecular dates and the mammalian radiation. *Trends in Ecology and Evolution*, **14**(3), 113–118.

[9] Brownlee, C. (2004). DNA Bar Codes: Life under the scanner. *Science News*, **166**(23), 360–361. (see also: http://phe.rockefeller.edu/barcode/)

[10] Bryant, D. and Moulton, V. (2004). Neighbor-Net: an agglomerative method for the construction of phylogenetic networks. *Molecular Biology and Evolution*, **21**(2), 255–65.

[11] Buneman, P. (1971). The recovery of trees from measures of dissimilarity. In *Mathematics in the Archaeological and Historical Sciences* (ed. F. R.

Hodson, D. G. Kendall, and P. Tautu), pp.387–395. Edinburgh University Press, Edinburgh.

[12] Buneman, P. (1974*a*). A characterisation of rigid circuit graphs. *Discrete Mathematics*, **9**, 205–212.

[13] Buneman, P. (1974*b*). A note on the metric property of trees. *Journal of Combinatorial Theory, Series B*, **17**, 48–50.

[14] Chothia, C., Gough, J., Vogel, C., and Teichmann, S. A. (2003). Evolution of the protein repertoire. *Science*, **300**(5626), 1701–1703.

[15] Daniel, R. (2005). The metagenomics of soil. *Nature Reviews Microbiology*, **3**(6), 470–478.

[16] Daubin, V. and Ochman, H. (2004). Start-up entities in the origin of new genes. *Current Opinion in Genetics & Development*, **14**(6), 616–619.

[17] Dayhoff, M., Schwartz, R., and Orcutt, B. (1978). A model of evolutionary change in proteins. In *Atlas of Protein Sequence and Structure* (ed. M. Dayhoff), Volume 5, 345–352. National Biomedical Research Foundation, Washington, D. C.

[18] Degnan, J. H. and Rosenberg, N. A. (2006). Discordance of species trees with their most likely gene trees. *PLoS Genetics*, **2**, 762–768.

[19] Dobzhansky, T. (1973). Nothing in biology makes sense except in the light of evolution. *The American Biology Teacher*, **35**, 125–129.

[20] Doolittle, W. F. (1999). Phylogenetic classification and the universal tree. *Science*, **284**, 21246–2129.

[21] Doolittle, W. F. and Papke, R. T. (2006). Genomics and the bacterial species problem. *Genome Biology*, **7**(9), 116.

[22] Douglas, S., Zauner, S., Fraunholz, M., Beaton, M., Penny, S., Deng, L. T., Wu, X., Reith, M., Cavalier-Smith, T., and Maier, U. G. (2001). The highly reduced genome of an enslaved algal nucleus. *Nature*, **410**(6832), 1091–1096.

[23] Eichler, E. E. and Sankoff, D. (2003). Structural dynamics of eukaryotic chromosome evolution. *Science*, **301**(5634), 793–797.

[24] Eisenberg, D., Marcotte, E. M., Xenarios, I., and Yeates, T. O. (2000). Protein function in the post-genomic era. *Nature*, **405**(6788), 823–826.

[25] Ferguson, N. M., Galvani, A. P., and Bush, R. M. (2003). Ecological and immunological determinants of influenza evolution. *Nature*, **422**(6930), 428–433.

[26] Fire, A., Xu, S., Montgomery, M. K., Kostas, S. A., Driver, S. E., and Mello, C. C. (1998). Potent and specific genetic interference by double-stranded RNA in Caenorhabditis elegans. *Nature*, **391**, 806–811.

[27] Forterre, P. (2006) Three RNA cells for ribosomal lineages and three DNA viruses to replicate their genomes: a hypothesis for the origin of cellular domain. *Proceedings of the National Academy of Science USA*, **103**(10), 3669–3674.

[28] Gardner, M. J. *et al.* (2002). Genome sequence of the human malaria parasite Plasmodium falciparum. *Nature*, **419**(6906), 498–511.

[29] Gascuel, O. (ed) (2005). *Mathematics of Evolution & Phylogeny*, Oxford University Press, Oxford.

[30] Gilad, Y., Oshlack, A., Smyth, G. K., Speed, T. P., and White K. P. (2006). Expression profiling in primates reveals a rapid evolution of human transcription factors. *Nature*, **440**, 242–245.

[31] Gill, S. R., Pop, M., Deboy, R. T., Eckburg, P. B., Turnbaugh, P. J., Samuel, B. S., Gordon, J. I., Relman, D. A., Fraser-Liggett, C. M., and Nelson K. E. (2006). Metagenomic analysis of the human distal gut microbiome. *Science*, **312**(5778), 1355–1359.

[32] Hannenhalli, S. and Pevzner, P. A. (1999). Transforming cabbage into turnip: Polynomial algorithm for sorting signed permutations by reversals. *Journal of ACM*, **46**(1), 1–27.

[33] Hendy, M. D. (1989). The relationship between simple evolutionary tree models and observable sequence data. *Systematic Zoology*, **38**, 310–321.

[34] Huber, K., Moulton, V., and Steel, M. (2005). Four characters suffice to convexly define a phylogenetic tree. *SIAM Journal on Discrete Mathematics*, **18**(4), 835–843.

[35] Huson, D. H. and Bryant, D. (2006). Application of phylogenetic networks in evolutionary studies. *Molecular Biology and Evolution*, **23**, 254-267. Software available from www.splitstree.org.

[36] Jaillon, O. *et al.* (2004). Genome duplication in the teleost fish Tetraodon nigroviridis reveals the early vertebrate proto-karyotype. *Nature*, **431**(7011), 946–957.

[37] Jones, D., Taylor, W., and Thornton, J. (1992). The rapid generation of mutation data matrices from protein sequences. *Computer Applications in the Biosciences (CABIOS)*, **8**, 275–282.

[38] Keeling, P. J. and Palmer, J. D. (2001). Lateral transfer at the gene and subgenic levels in the evolution of eukaryotic enolase. *Proceedings of the National Academy of Science USA*, **98**(19), 10745–10750.

[39] Kingman, J. F. C. (1982). The coalescent. *Stochastic Processes and Their Applications*, **13**, 235-248.

[40] Lerat, E., Daubin, V., Ochman, H., and Moran N. A. (2005). Evolutionary origins of genomic repertoires in bacteria. *PLoS Biology*, **3**(5), e130.

[41] Lopez, A. J. (1998). Alternative splicing of pre-mRNA: developmental consequences and mechanisms of regulation. *Annual Review of Genetics*, **32**, 279–305.

[42] Mace, G. M., Gittleman, J. L., and Purvis, A. (2003). Preserving the tree of life. *Science*, **300**(5626), 1707–1709.

[43] Margulies, M. *et al.* (2005) Genome sequencing in microfabricated high-density picolitre reactors. *Nature*, **437**(7057), 376–80.

[44] Marra, M. A. *et al.* (2003). The Genome sequence of the SARS-associated coronavirus. *Science*, **300**(5624), 1399–1404.

[45] Maynard Smith, J., Dowson, C. G., and Spratt, B. G. (1991). Localized sex in bacteria. *Nature*, **349**, 29–31.

[46] Medina, M. (2005). Genomes, phylogeny, and evolutionary systems biology. *Proceedings of the National Academy of Science USA*, **102** (Suppl. 1), 6630–6635.

[47] Ochman, H., Lawrence, J. G., and Groisman E. A. (2000). Lateral gene transfer and the nature of bacterial innovation. *Nature*, **405**(6784), 299–304.

[48] Ochman, H., Lerat, E., and Daubin, V.(2005). Examining bacterial species under the specter of gene transfer and exchange. *Proceedings of the National Academy of Science USA*, **102**(Suppl 1), 6595–6599.

[49] Ohno, S. (1970). *Evolution by Gene Duplication.* Springer-Verlag, Berlin.

[50] Overbeek, R., Fonstein, M., D'Souza, M., Pusch, G. D., and Maltsev, N. (1999). The use of gene clusters to infer functional coupling. *Proceedings of the National Academy of Science USA*, **96**(6), 2896–2901.

[51] Posada, D. (2006). ModelTest Server: a web-based tool for the statistical selection of models of nucleotide substitution online. *Nucleic Acids Research*, **34**, W700-W703.

[52] Purvis, A., Gittleman, J. L., Cowlishaw, G., and Mace, G. M. (2000). Predicting extinction risk in declining species. *Proc. Royal Society of London, Series B Biological Sciences*, **267**(1456), 1947–1952.

[53] Ross, H. A. and Rodrigo, A. G. (2002). Immune-mediated positive selection drives human immunodeficiency virus type 1 molecular variation and predicts disease duration. *Journal of Virology*, **76**(22), 11715–11720.

[54] Sankoff, D. (1972). Reconstructing the history and geography of an evolutionary tree, *American Mathematical Monthly*, **79**, 596-603 (Correction: American Mathematical Monthly 79, p.1100).

[55] Sankoff, D. (1975) Minimal mutation trees of sequences. *SIAM Journal on Applied Mathematics*, **28**, 35–42.

[56] Sankoff, D. (1992). Edit distances for genome comparison based on non-local operations. In *Proc of 3rd Conference on Combinatorial Pattern Matching (CPM'92)* (ed. A. Apostolico, M. Crochemore, Z. Galil, and U. Manber), Volume 644 in Lecture Notes in Computer Science, 121–135, Springer-Verlag, Berlin.

[57] Sankoff, D. (2003). Rearrangements and chromosomal evolution. *Current Opinion in Genetics & Development*, **13**(6), 583–587.

[58] Schmucker, D., Clemens, J. C., Shu, H., Worby, C. A., Xiao, J., Muda, M., Dixon, J. E., and Zipursky S. L. (2000). Drosophila Dscam is an axon guidance receptor exhibiting extraordinary molecular diversity. *Cell*, **101**(6), 671–84.

[59] Tatusov, R. L., Koonin, E. V., and Lipman, D. J. (1997). A genomic perspective on protein families. *Science*, **278**(5338), 631–637.

[60] Tree of Life (2003). *Science*, special issue, **300**(5626), 1691–1709.

[61] Venter, J. C. *et al.* (2004). Environmental genome shotgun sequencing of the Sargasso Sea. *Science*, **304**(5667), 66–74.

[62] Xing, Y. and Lee C. (2005). Evidence of functional selection pressure for alternative splicing events that accelerate evolution of protein subsequences. *Proceedings of the National Academy of Science USA*, **102**(38), 13526–135231.

[63] Zarestkii, K. (1965). Reconstructing a tree from the distances between its leaves. *Uspehi Mathematicheskikh Nauk*, **20**, 90–92 (in Russian).

CONTENTS

LIST OF CONTRIBUTORS

Elizabeth S. Allman
Department of Mathematics and Statistics
University of Alaska Fairbanks, Fairbanks, AK USA
http://www.dms.uaf.edu/~eallman
e.allman@uaf.edu

Cécile Ané
Department of Statistics
University of Wisconsin-Madison, USA
http://www.stat.wisc.edu/~ane
ane@stat.wisc.edu

Michaël G. B. Blum
Laboratoire TIMC
Université Joseph Fourier & CNRS, Grenoble, France
http://sitemaker.umich.edu/michael.blum/home
michael.blum@imag.fr

Alexei Drummond
Bioinformatics Institute and Department of Computer Science
University of Auckland, New Zealand
alexei@cs.auckland.ac.nz

Oliver Eulenstein
Department of Computer Science
Iowa State University, USA
http://www.cs.iastate.edu/~oeulenst
oeulenst@cs.iastate.edu

Gregory Ewing
Bioinformatics Institute, and
Allan Wilson Centre for Molecular Ecology and Evolution
University of Auckland, New Zealand, and

Center for Integrative Bioinformatics Vienna (CIBIV)
Max F. Perutz Laboratories (MFPL), Austria
gregory.ewing@univie.ac.at

Joseph Felsenstein
Department of Genome Science and Department of Biology
University of Washington Seattle, Washington, U.S.A.
http://www.gs.washington.edu/faculty/felsenstein.htm
joe@gs.washington.edu

David Fernández-Baca
Department of Computer Science
Iowa State University, USA
http://www.cs.iastate.edu/~fernande
fernande@cs.iastate.edu

Olivier Gascuel
Centre National de la Recherche Scientifique
LIRMM (CNRS-UM2), Montpellier, France
http://www.lirmm.fr/~gascuel
gascuel@lirmm.fr

Stefan Grünewald
CAS-MPG Partner Institute for Computational Biology
Shanghai Institutes for Biological Sciences, Chinese Academy of Sciences
http://www.picb.ac.cn
stefan@picb.ac.cn

Stéphane Guindon
Centre National de la Recherche Scientifique
LIRMM (CNRS-UM2), Montpellier, France
http://www.lirmm.fr/~guindon/wordpress
guindon@lirmm.fr

Luke J. Harmon
Biodiversity Centre
University of British Columbia, Vancouver, Canada
http://www.zoology.ubc.ca/biodiversity/centre/harmon
harmon@zoology.ubc.ca

Klaas Hartmann
Biomathematics Research Centre
University of Canterbury, Christchurch, New Zealand
k.hartmann@math.canterbury.ac.nz

Stephen B. Heard
Department of Biology
University of New Brunswick, Fredericton, Canada
http://www.unb.ca/fredericton/science/biology/Faculty/Heard.html
sheard@unb.ca

Katharina T. Huber
School of Computing Sciences
University of East Anglia, United Kingdom
http://www.cmp.uea.ac.uk/people/kth
katharina.huber@cmp.uea.ac.uk

Daniel Huson
Center for Bioinformatics
University of Tübingen, Germany
http://www-ab.informatik.uni-tuebingen.de
huson@informatik.uni-tuebingen.de

Junhyong Kim
Department of Biology
University of Pennsylvania, USA
http://kim.bio.upenn.edu
junhyong@sas.upenn.edu

Michelle M. McMahon
Department of Plant Sciences
University of Arizona, USA
http://cals.arizona.edu/~mcmahonm
mcmahonm@email.arizona.edu

Arne Ø. Mooers
Biological Sciences
Simon Fraser University, Burnaby, Canada
http://www.sfu.ca/~amooers
amooers@sfu.ca

Raul Piaggio-Talice
Department of Computer Science
Iowa State University, USA
rpiaggio@iastate.edu

John A. Rhodes
Department of Mathematics and Statistics
University of Alaska Fairbanks, Fairbanks, AK USA
http://www.dms.uaf.edu/~jrhodes
j.rhodes@uaf.edu

Allen Rodrigo
Bioinformatics Institute, and
Allan Wilson Centre for Molecular Ecology and Evolution
University of Auckland, New Zealand
a.rodrigo@auckland.ac.nz

Michael J. Sanderson
Department of Ecology and Evolutionary Biology
University of Arizona, USA
http://ginger.ucdavis.edu
sanderm@email.arizona.edu

Charles Semple
Biomathematics Research Centre
Department of Mathematics and Statistics
University of Canterbury, Christchurch, New Zealand
http://www.math.canterbury.ac.nz/~cas83
c.semple@math.canterbury.ac.nz

Mike Steel
Biomathematics Research Centre
University of Canterbury, Christchurch, New Zealand
http://www.math.canterbury.ac.nz/bio
m.steel@math.canterbury.ac.nz

Dennis H. J. Wong
Department of Biology
University of New Brunswick, Fredericton, Canada
dhjwong@gmail.com

I

EVOLUTION IN POPULATIONS

1

TREES OF GENES IN POPULATIONS

Joseph Felsenstein

Abstract

Trees of ancestry of copies of genes form in populations as a result of the randomness of birth, death, and Mendelian reproduction. Considering them allows us to think about evolution within and between populations, to make the connection between phylogenies and population genetic analyses. These trees, known as coalescents, are essential to developing methods for making inferences about populations. This chapter reviews coalescents and the inference methods based on them. The review concentrates on the population processes, and also briefly treats the inference methods, concentrating on those that attempt a likelihood or Bayesian treatment.

1.1 Introduction

Molecular evolution represents phylogenies as branching diagrams composed of thin lines. At the tip we often find one molecular sequence, sometimes described as 'the yeast sequence' or 'the mouse sequence'. It is as if we were viewing the evolutionary tree from a great distance, so that each branch appears thin. If each of these thin lines truly contained only one copy of this gene's sequence, we would have a species that consisted only of a single individual, and a haploid one at that. But the lines are not lineages of single copies. Coming closer to them, we find that in reality the lines are thick—they are whole species, consisting of multiple populations, each of many individuals. To understand what molecular evolution looks like when we consider whole populations, we have to consider population-genetic phenomena in addition to the usual models of molecular evolution. The two fields of molecular evolution and population genetics (or evolutionary genetics) have grown up largely separately. However, they are connected, and with the availability of large population samples of sequences, their connections are increasing. We are well into a Great Encounter—the mathematics and statistics of population processes are becoming more and more important to molecular evolution, and multispecies comparisons are becoming more and more important to evolutionary genetics.

To explain how population-genetic models relate to molecular evolution between species, we have to start within species and model the ancestry of a population sample of n copies of a gene drawn from a single random-mating population. This ancestry is itself a tree, but not one whose forks are speciations. Instead they are simply events in which one parent copy gives rise to two or more offspring copies, a routine occurrence. The resulting trees have come to be called *coalescents.* They are sometimes called 'gene trees', but this is ambiguous terminology, as that same phrase is also used for trees of descent of genetic loci by gene duplication, an entirely different phenomenon.

The most standard model of theoretical population genetics is the Wright–Fisher model. In it, each of the $2N$ copies of a gene in a diploid population of constant size N in effect chooses its parent copy from among the $2N$ parent copies available. These choices are independent. Thus for two copies in a population, there is a chance $1/(2N)$ that they came from the same copy in the previous generation. If they do not, the process occurs again when we go back one more generation. In effect, we toss a coin for each generation back, with the probability of Heads equal to $1/(2N)$. The time to the first Heads is drawn from a geometric distribution with that probability of Heads. This much was known to Sewall Wright and R. A. Fisher in the early 1930s.

In 1982, the eminent probabilist J. F. C. Kingman, who has had a lifelong interest in population genetics, asked what the process of ancestry would look like if we traced back from a sample of n copies in a large population of N individuals. He defined an excellent approximation which he called the *n-coalescent* [29, 30]. In it, one goes back in continuous time rather than in discrete generations. The ancestry of the n copies remains distinct for a time T_n generations, where T_n is drawn from an exponential distribution:

$$T_n \sim \text{Exp}\,[4N/(n(n-1))]\,. \qquad (1.1)$$

At that time two lineages chosen at random join, so that there are now $n-1$ lineages. The process then starts again, going back farther in time, but with the value of n decremented, as an independent draw from the same distribution with that smaller value of n. This continues until there are only two lineages, whose common ancestor is drawn by this process with $n = 2$.

Note that in the Wright–Fisher model the ancestry of copies of a gene can be discussed without considering whether or not the copies have the same or different DNA sequences. For the moment, there is assumed to be no natural selection. The copies reproduce in ways that do not depend on their DNA sequences.

This is an approximation to the genealogy implied by the Wright–Fisher model. It allows only two lineages at a time to combine, while in the discrete-generations Wright–Fisher model, more than two lineages can combine simultaneously since a single individual can have multiple offspring. Kingman derives his model by taking a series of discrete-generations Wright–Fisher models, with the kth of these having $N = k$ and a new time scale in which one unit of time is k generations. He shows that the limit of the genealogical processes of these models is one in which the (rescaled) time back to coalescence when there are n copies is distributed as

$$\tau \sim \text{Exp}\,[4/(n(n-1))]\,, \qquad (1.2)$$

and he also shows that, in the limit, all coalescences are of only two copies. Returning to the original time scale, the limiting process approximates the genealogy specified by equation (1.1).

This sort of limit is well-known in theoretical population genetics—it is the one used to approximate gene frequency change by a diffusion process [12]. In effect, Kingman's n-coalescent is a diffusion approximation. Although diffusion processes approximate discrete changes of gene frequencies by a continuous diffusion process, they are extraordinarily accurate. One way that we can check this in the coalescent process case is to calculate whether coalescence will involve more than two lineages in the Wright–Fisher model. In the Wright–Fisher model, if we have n lineages and go back one generation, the probability that two copies coalesce while the others all do not will be $\binom{n}{2}$ times the probability that copies 1 and 2 coalesce and others do not, by the exchangeability of the process. As each copy chooses its ancestor independently, we need the probability that copy 2 chooses the same ancestor as copy 1, copy 3 chooses a different ancestor, copy 4 chooses an ancestor different from those two, copy 5 chooses an ancestor different from those three, and so on, so that the total probability of pairwise coalescence is

$$\binom{n}{2}\frac{1}{2N}\left(1-\frac{1}{2N}\right)\left(1-\frac{2}{2N}\right)\left(1-\frac{3}{2N}\right)\cdots\left(1-\frac{n-2}{2N}\right). \tag{1.3}$$

The probability that some of the copies coalesce is found by subtracting from 1 the probability that none coalesce, to get, by a straightforward argument:

$$1-\left(1-\frac{1}{2N}\right)\left(1-\frac{2}{2N}\right)\left(1-\frac{3}{2N}\right)\cdots\left(1-\frac{n-1}{2N}\right). \tag{1.4}$$

To first order, both of these expressions are equal, as both are

$$n(n-1)/(4N) \ + \ O(1/N^2) \tag{1.5}$$

which indicates that as N increases they become close, so that the probability that a coalescence involves more than two lineages becomes negligible. Taking the ratio of the expressions in equations (1.3) and (1.4), we can compute the fraction of coalescences that are coalescences of two lineages when there are 10 lineages for increasing values of N and get some sense of this (Fig. 1.1).

The fraction of two-way coalescences becomes high as the population size passes 100, which is the square of the number of lineages. We can also examine, for $N = 10,000$, the fraction of two-way coalescences with different numbers of lineages (Fig. 1.2).

These patterns can be summarized by saying that most coalescences will be two-way if $n^2 < N$. However it is not obvious that having a modest fraction of three- or four-way coalescences will invalidate inference methods that assume the coalescent, so the coalescent may be a good approximation even when this condition is violated.

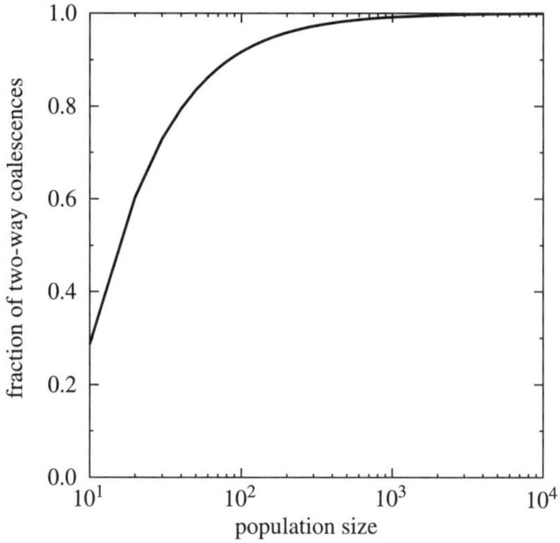

FIG. 1.1. The fraction of coalescences that are of two lineages, when there are 10 lineages, for different population sizes N.

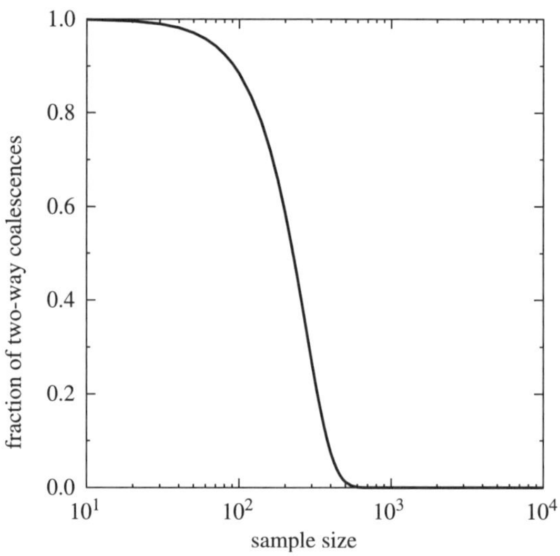

FIG. 1.2. The fraction of coalescences that are of two lineages, for different numbers of lineages, when population size $N = 10,000$.

The coalescent process predicts that the genealogy of copies in a population is a random branching tree. The coalescence times are individually exponentially distributed. The sum of their expectations is

$$\sum_{k=2}^{n} \frac{4N}{k(k-1)} = 4N \sum_{k=2}^{n} \left(\frac{1}{k-1} - \frac{1}{k} \right) = 4N \left(1 - \frac{1}{n} \right). \tag{1.6}$$

We might expect that the total time for coalescence of the ancestors of a sample from a population is proportional to the sample size (or even to its square), but this calculation shows that it is actually almost independent of sample size.

One simple modification of this result is to use Sewall Wright's N_e in place of N. This quantity, the 'effective population size' corrects for a variety of ways in which the mating system departs from a simple Wright–Fisher model. Formulas are available to calculate the appropriate corrections for separate sexes, unequal numbers of the two sexes, monogamy, overlapping generations, and variation of fertility from parent to parent. I will use N here, but the reader should keep in mind that N_e will usually be needed instead.

1.2 Effects of evolutionary forces on coalescent trees

1.2.1 *Population growth*

The above theory is for a single population of constant size. When population sizes grow or shrink, the rate of coalescence changes. For example, if the population size is N for the most recent 500 generations, but before that is $N/10$ for 100 generations, and before that again N, the effect of this bottleneck on the coalescent is straightforward. Going back 500 generations, we have the usual coalescent process with rate (for k lineages) of $k(k-1)/(4N)$. If we get back to the most recent end of the bottleneck period and have at that time ℓ lineages, the rate of coalescence back beyond that is $10\,\ell(\ell-1)/(4N)$. If when the farthest end of the bottleneck is reached we have m lineages, the rate beyond that point is $m(m-1)/(4N)$. Thus there will tend to be a burst of coalescence at the time of population bottlenecks, though there may not be many coalescent events in those bottlenecks unless the length of the bottleneck in generations approaches the population size at that time. A bottleneck of population size of 1000 individuals may not have much effect if it lasts for only 10 generations.

It was noticed by Kingman [29] that there is a simple way to treat population growth if we can integrate the reciprocal of the population size. It makes use of the fact that a smaller population causes proportionately more coalescence per unit time. For example, if the population size N grows exponentially at rate g, the population size t generations ago was $N(t) = N(0)\exp(-gt)$. The rate of coalescence of k lineages t units of time ago would then be $k(k-1)/(4N(t)) = \exp(gt)\,k(k-1)/(4N(0))$. A coalescent process that has such time-dependent rates can be defined and simulated. A simpler way is to note that coalescence occurs $\exp(gt)$ times faster t units of time ago, because the population is that factor smaller then. It is as if the clock were running $\exp(gt)$ times as fast. We can change the time scale going backwards, to one that accumulates $\exp(gt)$ as much time t units of time ago. It has this fictional time

$$\tau \;=\; \int_0^t e^{gu}\,du \;=\; \left(e^{gt}-1\right)/g. \tag{1.7}$$

On this fictional time scale, the coalescent process will have rates independent of time. The coalescent with an exponentially growing population is then simply the ordinary coalescent with population size $N(0)$, if we observe it on the fictional time scale τ. One can draw a random outcome of the coalescent process with exponential population growth by sampling the ordinary coalescent, considering the times of coalescence to be values of τ, and then computing the corresponding values of the actual time t by solving for t in equation (1.7) to get

$$t \;=\; \frac{1}{g}\,\ln(1+g\,\tau). \tag{1.8}$$

The effect of a positive growth rate g is to compress times in the past relative to the present. As Slatkin and Hudson [47] noted, the trees become closer to a 'star tree' in which all lineages simultaneously radiate from a single node. If the growth rate is negative, the times at the base of the tree are stretched (sometimes infinitely so).

1.2.2 *Migration*

When we have more than one population, a coalescent tree forms in each population, but lineages also move between populations. Going backwards in time, if m_{ij} is the probability that a lineage in population i came from population j in the preceding generation, there is an event with probability $m_{ij}\,dt$ in the previous small interval of time of length dt. For example, if there were 3 populations of size N_1, N_2, and N_3, and if currently they contain respectively k_1, k_2, and k_3 lineages, the events that can occur during a small interval of length dt, going backwards in time, include coalescences within each of the three populations and migrations. The former happen with rates $k_1(k_1-1)/(4N_1)$, $k_2(k_2-1)/(4N_2)$, and $k_3(k_3-1)/(4N_3)$ per unit time. In population 1 there is a total rate $k_1m_{12}+k_1m_{13}$ of migrations, and similarly for the other two populations. The total rate of events for p populations is then

$$\sum_{i=1}^{p} \frac{k_i(k_i-1)}{4N_i} \;+\; \sum_{i=1}^{p}\;\sum_{\substack{j\,=\,1\\ j\,\neq\, i}}^{p} k_i m_{ij}. \tag{1.9}$$

To draw a genealogy from the coalescent with migration, we proceed backwards in intervals. We draw the length of the interval from an exponential distribution whose mean is the reciprocal of the quantity in 1.9. We then decide whether the event is a coalescence or a migration, by drawing these in proportion to their total rates of occurrence, and then we decide in which population each event is and which lineage or lineages it involves.

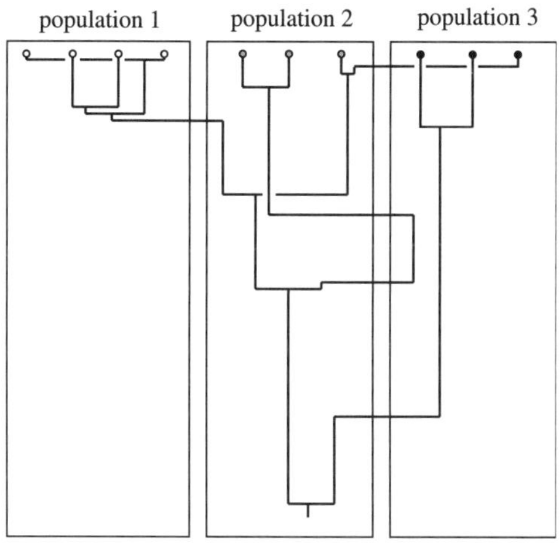

FIG. 1.3. A simulated coalescent with migration among adjacent populations with three populations of equal sizes and $4Nm = 1$ in each, going backwards from samples of 4, 3, and 3 lineages.

Figure 1.3 shows a randomly sampled coalescent from three populations of equal size N, who have adjacent symmetric migration with $4Nm_{ij} = 1$.

The coalescent process for migration was first investigated by Takahata [50] and (somewhat implicitly) by Hudson and Kaplan [27] and by Kaplan, Darden, and Hudson [28].

1.2.3 *Coalescents with recombination*

So far we have assumed that each copy of a gene is descended from a single copy in the preceding generation. This is true if there is no genetic recombination within the gene. If there is recombination possible, the copy could be descended from both copies in the parent. At any one site in the DNA sequence, the gene is descended from only one copy, and the coalescent at that site is the normal one. But when the sites are taken together, the genealogy is not a tree. When we approximate the genealogy of the sequence by a coalescent, recall that in effect we consider cases with large population size N, and small rates of such forces as migration. To obtain a coalescent approximation to a recombining genealogy, we also take the recombination rate per site per generation, r, to be small. This means that we will assume that there cannot be more than one recombination event in a sequence in one generation.

To model recombination, we assume that when a recombination event occurs in a sequence which has L sites, it does so at one of the $L - 1$ intervals between sites, chosen at random. The sequence before the point of recombination comes from one of the two parental copies, and the sequence after the point of recombination comes from the other parental copy. The two copies that are in the parent are themselves drawn at random from the population, so they go back in time along independent lineages that can coalesce with others, or even with each other. In tracking the ancestry of a population sample, we will want to have each lineage accompanied by a set S of sites. In the sample, the sets S are all $\{1, 2, \ldots, L\}$. As the lineages go back in time, they have the usual probabilities of coalescing and migrating. There are also recombination events occurring stochastically at rate $4Nr$ per interval between adjacent sites. When a recombination event occurs, if it occurs just after site ℓ it divides the set of sites into two subsets, $\{1, \ldots, \ell\}$ and $\{\ell + 1, \ldots, L\}$. The set of sites 'active' in the two parent haplotypes are then changed to $S \cap \{1, \ldots, \ell\}$ and $S \cap \{\ell + 1, \ldots, L\}$. When two lineages coalesce, the set of active sites is the union of the two sets of active sites, though the set of intervals available for recombination is from the leftmost site in that union to the rightmost site.

We can represent the genealogy by a graph called the *ancestral recombination graph* [20, 24]. Figure 1.4 shows an ancestral recombination graph with three tips, four coalescences (the shaded circles) and two recombination events (the white circles). Next to each line is the list of sites in that lineage (out of a total sequence length of 1000) that are 'active' in the sense of being ancestral to sites in the tip sequences. Note that one lineage has a disjoint list of active sites.

An alternative way of thinking of genealogies with recombination is to think of the genealogies at the different sites. At each site the genealogy is a simple coalescent. Neighbouring sites between which there has been no recombination

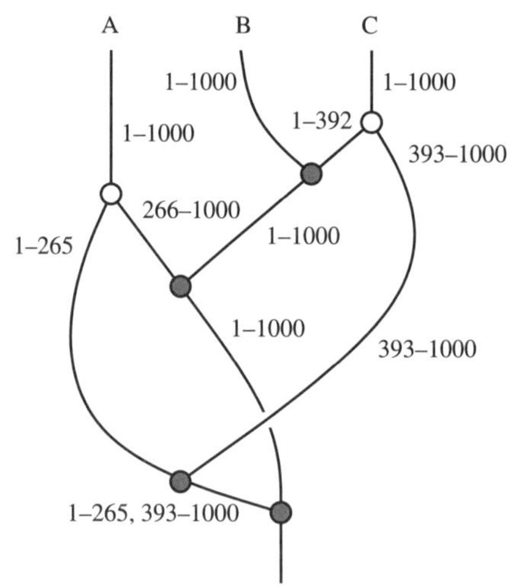

FIG. 1.4. An ancestral recombination graph for a sample of three sequences of 1000 bases. Next to each lineage are listed the sites in it that are ancestral to the tip sequences. Coalescent events are shown as shaded circles, recombination events as white circles.

have the same coalescent. In the example in Fig. 1.4 the first 265 sites have one coalescent tree, the next 127 sites another, and the final 608 sites a third. Wiuf and Hein [56] have defined a stochastic process that makes changes in the coalescent as one moves along a sequence in a way that correctly generates an ancestral recombination graph. Most computer simulation of ancestral recombination graphs uses the programme of Hudson [26] which generates the graph by moving backward in time and considering the sets of sites in different lineages.

It is helpful to have a sense of the rate at which the coalescent tree changes as one moves along the genome. How far must we go to have the tree be effectively independent? A simple calculation can be based on the distance we must move along the genome so that a lineage from a tip down to the root of the coalescent tree is expected to have one recombination event. The distance to the root is close to $4N$ generations. So we want to find how far along the genome we must go to have $4Nr = 1$. In a human meiosis there is about one recombination event per 10^8 bases. If the effective population size tens of thousands of years ago were 10^4, and the recombination rate were the same throughout the genome, this implies a short distance, 2500 bases. If the effective population size were higher, say 10^5, the distance is even shorter, only 250 bases!

You may wonder what justification I have for the rule $4Nr = 1$. In fact, the condition for similarity of trees is the same as the condition for there to be non-random association of alleles at loci. These associations are known as linkage disequilibrium. The coalescent tree at one site strongly affects the distribution of alleles in the sample. An allele that has arisen by mutation at that site tends to occur in the descendants of a single branch of the coalescent tree. If another site shares the same coalescent tree, one of its alleles will be strongly positively or negatively associated with the allele at the first site. Robertson and Hill [45] make a calculation closely similar to the above one, calculating the size of blocks of linkage disequilibrium.

Models can also be made of the effect of gene conversion on the coalescent, although as yet there has been little use of them.

1.2.4 *Natural selection*

It has been difficult to accomodate natural selection in coalescents, but recently there has been some progress in doing so. If there is no natural selection occurring, then the shape of the coalescent genealogy is not affected by which copies have which DNA sequence. In the presence of natural selection, there is such a dependence. If we have (say) five copies of one allele, and five of another, and if the first allele has higher fitness than the second, then most likely the first allele is spreading in the population. If so, it is more probable that two copies of it coalesce when we go back in time than two copies of the other allele. The result is that we cannot specify any coalescent without knowing more about the DNA sequence in the copies.

For many years this was thought to make it impossible to specify any coalescent process in the presence of natural selection. Krone and Neuhauser [31, 40] discovered a way to do so. It creates a coalescent by going back in time and having both coalescence events and also special forks that reflect a natural selection event. This produces a genealogy with loops in it, called *ancestral selection graph.* The genotype is then specified at the root of this genealogy, drawn from an appropriate population-genetic equilibrium distribution. Then genotypes are propagated up the genealogy, allowing for mutation events as well. When the top of a loop is reached, it is decided which side of that loop connects upward, depending on its genotype. Krone and Neuhauser's result is a breakthrough, though it does not specify a genealogy independent of the genotypes of the gene copies, as the other coalescent processes do.

Earlier treatments of natural selection [27, 28] could handle only cases of strong natural selection, which in effect divides the copies into subpopulations whose sizes are the consequence of the fitnesses.

1.3 Inference methods

Having understood the stochastic processes that produce treelike genealogies of gene copies, the next obvious step should have been to find a way to use these to compute likelihoods or carry out Bayesian inference of parameters. The central model framework for doing so is the neutral mutation theory of genetic variation, widely studied since the 1960s. Molecular sequences have been modelled as evolving under genetic drift and mutation, without natural selection. This model also serves as a null hypothesis against which to test for the presence of natural selection.

In a coalescent, mutation can be accomodated by allowing it to occur on the branches, modelled as happening in continuous time. This is the same model used in the inference of phylogenies. The difference is that in the coalescent case, the coalescent genealogy is not being estimated, but instead is part of the machinery of statistical inference of the population and genetic parameters.

The models of mutation used are the usual models of sequence mutation used with phylogenies. The presumption in most cases is that the mutations are selectively neutral, with no fitness differences. Two approximate models are also in wide use in the population genetics literature. One is the *infinite alleles model*, due to James F. Crow and Motoo Kimura [4]. In it there is a constant risk of mutation, at rate μ per locus, to a completely new allele. All alleles can be distinguished, but they give us no clue which ones are derived from which other ones. The same allele never arises twice. Mutations in DNA sequences behave approximately like this, as long as there are so many sites that the chance of the same site mutating again is small. However, in real DNA sequences, the sequence does give us information about which sequences are likely to be separated by one mutational event.

A closer approximation is the *infinite sites model* of Watterson [52]. It represents the gene by a line segment, and each mutation occurs at a random location chosen from (0,1). As such, no mutation ever recurs at the same exact location. It is assumed that we can see the line and the placement of the differences, but it is also usually assumed that we cannot know, at a site which has a variation, whether the presence or absence of the variation is the original state. Thus, if we see three copies that have their lists of variations present as $\{0.366, 0.8197\}, \{0.366\}$, and $\{0.684\}$, the variation counted as present at position 0.366 in the first two copies could also be considered as one that is absent in those copies but present in the third. The lists would then be $\{0.8197\}$, $\{\}$, and $\{0.366, 0.684\}$. If the variation at position 0.684 was considered absent in the third copy but present in the other two, the lists would be $\{0.684, 0.8197\}, \{0.684\}$, and $\{0.366\}$. These are all completely equivalent. As long as there is no recombination allowed within the locus, the exact locations on the line segment actually do not matter, and each mutational event in effect partitions the copies into two sets. The partitions are ordered and are compatible, in that when we intersect any two such partitions they form no more than three sets. We shall see the infinite sites model used in some of the inference methods below.

1.3.1 *Earlier inference methods*

It is a puzzling fact that little attention was paid to likelihood inference (and Bayesian inference) in population genetics until the 1990s. Some of this inattention may have been the result of the apparent intractibility of the problem. The only model for which a likelihood could be computed was Ewens's [9] model of a locus undergoing mutation and genetic drift under an infinite-alleles model of mutation. (One should mention also R. C. Griffiths for deriving a likelihood inference of population divergence time under that same model [18]). But one would have thought that the problem would at least have been posed as a major challenge for theoretical population geneticists. It was not.

This may be related to the high prestige in that field of closed-form solutions for distributions and changes of population composition, and the correspondingly low prestige of statistical and computational methods. For example, for a field with so much mathematically sophisticated theory, population geneticists maintain relatively few web sites and distribute relatively few computer programmes. They are far outclassed in this by systematists and molecular evolutionists, even though those fields are mathematically less sophisticated. Although likelihood and Bayesian inference methods became dominant in statistical inference from human pedigrees during this period, population geneticists working on evolution tended to ignore the likelihood paradigm and instead derived expectations and variances for particular statistics.

Many of those were heterozygosities which involved first and second moments of gene frequencies. These can be shown to lose statistical power compared to coalescent-based methods [13, 16]. Another widely-used statistic for the infinite-sites model, Watterson's number of segregating sites [52], is more powerful, but still less so than likelihood-based methods [13, 14, 16].

1.3.2 *The basic equation*

The first key to computation of the likelihood for a population sample of molecular sequences is that we can compute it straightforwardly once the coalescent tree is known. The likelihood models of phylogenetic inference allow the computation of Prob (D | T, $\mathcal{P}$), the probability of the sequences given the tree and the values of the relevant parameters. The second key is the realization that we do not know the tree T, but that the sequences do give us some information about it. The likelihood Prob (D | $\mathcal{P}$) is

$$\text{Prob}\,(\text{D}\,|\,\mathcal{P}) = \sum_{T} \text{Prob}\,(\text{D}, \text{T}\,|\,\mathcal{P}), \tag{1.10}$$

$$= \sum_{T} \text{Prob}\,(\text{D}\,|\,\text{T}, \mathcal{P})\ \text{Prob}\,(\text{T}\,|\,\mathcal{P}). \tag{1.11}$$

The summation is over all possible coalescent trees, and includes not only summation over tree topologies but integration over all possible combinations of coalescence times. The first term inside the summation in (1.11) is easily computed by the standard dynamic programming methods of phylogeny inference. The second is the density of the coalescent distribution.

1.3.3 *Rescaling times*

In the simplest case, of one population, the parameters in equation (1.11) are the population size, N, and the mutation rate per site, μ. In fact, they cannot be inferred separately. If we change the time scale of the branch lengths of the tree T so that they are given, not in generations, but in units of expected mutations per site, the expression for the likelihood now becomes a function of the product $4N\mu$ and the quantities μ and N do not appear separately. This makes intuitive sense—if we are computing the joint probability of a set of sequences observed at the present, there will be no difference between a tree with a given mutation rate μ and one which is twice as deep but has half the mutation rate. The depth of the tree is proportional to N, so that the likelihood is a function only of the product $N\mu$. It is a convenience to express the product as $\Theta = 4N\mu$.

In this simple case, the likelihood can then be written as

$$\text{Prob}\,(\text{D}\,|\,\Theta) = \sum_{\text{G}} \text{Prob}\,(\text{G}\,|\,\Theta)\ \text{Prob}\,(\text{D}\,|\,\text{G}) \tag{1.12}$$

since the branch lengths of the coalescent genealogy G are now expressed in mutational units.

The sum is of a product of two terms. The first is the coalescent density. If the ith coalescent interval on the tree G is u_i, measured in mutational units, then the coalescent density for n sequences is

$$f(G\,|\,\Theta) \;=\; \prod_{i=1}^{n-1} \left(e^{-\frac{(n-i+1)(n-i)}{\Theta}\, u_i}\ \frac{2}{\Theta} \right). \tag{1.13}$$

The density is easy to calculate once we know the u_i. Likewise the second term on the right-hand side of equation (1.12) is easy to compute, using the standard recursion for likelihoods on phylogenies. Although likelihood methods can be slow, this is not so much true for the computation of the likelihood for one tree, as we have one topology and are not optimizing the branch lengths.

1.3.4 *How many coalescent trees?*

This would seem to solve the problem, except for one matter. The summation is over all possible coalescent trees that could connect the sequences. Each tree is specified by a given sequence of pairs of lineages that coalesce, plus the times of these coalescences. With n lineages, the sequence of coalescence events is specified by choice of pairs of lineages to coalesce. The total number of possibilities is

$$\prod_{i=1}^{n-1} \binom{n-i+1}{2} = \frac{n!\,(n-1)!}{2^{n-1}}. \tag{1.14}$$

These different possibilities are called labelled histories—they are different trees in which we distinguish between the order of interior nodes in time. They were defined by Edwards [8]; the formula counting them is given in that paper.

The number of labelled histories rises rapidly, more rapidly than the number of tree topologies. For only 10 tip species, there are 2,571,912,000 histories. Worse yet, evaluating the likelihood involves integrating over all possible coalescence times. There are $n-1$ of these, so for 10 tips we must evaluate 2.571×10^9 integrals, each 9-dimensional. It would be a great economy if there were a closed-form formula for the integration, but there has been no progress toward that.

1.3.5 *Monte Carlo integration*

The integral in equation (1.12) can be thought of as the expectation of $\text{Prob}\,(\text{D} \mid \text{G})$ over the Kingman coalescent distribution for parameter value Θ. If we cannot do the integrals analytically, and cannot hope to do them all numerically, a natural alternative is Monte Carlo integration. Perhaps we can draw a large sample of coalescent genealogies from the Kingman density, compute $\text{Prob}\,(\text{D} \mid \text{G})$ for each, and average.

I have tried to implement this at least once, and the results were disastrous. For almost all of the possible genealogies G the value of $\text{Prob}\,(\text{D} \mid \text{G})$ is nearly zero; for a small minority it is much larger. The result is that the averages vary wildly from one sampling run to another, and no accurate estimate of the overall likelihood is obtained.

1.3.6 *Importance sampling*

It thus becomes essential to find some way of concentrating the sampling in the relevant regions. The correction that needs to be made for importance sampling has long been known. If we want to compute the expectation of function $h(x)$ over a distribution whose density function is $f(x)$, but we choose the samples from a distribution whose density function is $g(x)$, it is easy to see that

$$\begin{aligned}
\mathbb{E}_f\left[h(x)\right] &= \int_x f(x)h(x)\,dx, \\
&= \int_x \frac{f(x)}{g(x)}\,g(x)\,h(x)\,dx, \\
&= \mathbb{E}_g\left[\frac{f(x)}{g(x)}\,h(x)\right].
\end{aligned} \tag{1.15}$$

We correct for the importance sampling by averaging, not $h(x)$ but $(f(x)/g(x))$ $h(x)$. An intelligent choice of the density $g(x)$ can concentrate our sampling on coalescent trees that make a substantial contribution to the integral. The factor $f(x)/g(x)$ corrects for the excessive density of points in some areas of the space. If, for example, $g(x)$ concentrates twice as many sampling points around x as $f(x)$ would, the factor $f(x)/g(x)$ weights the samples to reflect the fact that each should be taken to represent half as much area in the space as it would if we sampled from the density $f(x)$.

Importance sampling makes numerical sampling approaches to likelihood inference or Bayesian inference with coalescents practical. Methods have been developed that draw independent samples, and also methods that draw correlated samples. I will call both of these 'sampling methods'. With the rise in popularity of Markov chain Monte Carlo (MCMC) methods as means of sampling from difficult distributions, it was inevitable that they would be applied to this task. Although the drawing of independent samples is a trivial case of a Markov chain, designation as MCMC methods is usually reserved for the correlated samplers.

1.3.7 *Independent sampling*

The pioneers in applying sampling methods for computing likelihood functions in coalescents were Griffiths and Tavaré [21]. For samples whose mutational process was the infinite sites model, Griffiths [19] had envisaged using a recursion (due to Golding [17]) to compute all possible sequences of mutational and coalescent events that could have led to the observed sample. This proved to be too difficult computationally for more than a few samples. Griffiths and Tavaré [21] proposed instead sampling paths through the recursion, and for each computing a functional that reflected the probabilities of events. Each such path is an independent sample, a very desirable property, as it thus completely avoids the problem of getting stuck in one region of the space.

At each stage, Griffiths and Tavaré consider the possible events that could happen (going backwards in time). If there is only one sequence that has a particular site in the mutant state, then it is possible that this event is a mutation. If there is more than one copy of a sequence, it is possible that this event is a coalescence of two of them. They sample these events proportional to their probability of occurrence, but not allowing those that would conflict with the data. Suppose that there was one sequence that carries a mutant allele at position 0.2, another with mutant alleles at positions 0.4 and 0.5, and a third with a mutant allele at position 0.2. With three sequences, we could have three possible coalescences, and there are four copies of the mutant that could have recently mutated (so that going backwards they unmutate). But as we have an infinite sites model, position 0.2 cannot unmutate in either of its positions (i.e. the most recent event cannot have been a mutation creating that mutant allele). Of the three possible coalescences, two of them could not have been the most recent event, as the genotypes of those pairs of sequences are different. In such a case, Griffiths and Tavaré sample from among the one allowable coalescence and two allowable mutations in proportion to their probabilities.

Griffiths and Tavaré go back in time, sampling possible events, until the sample coalesces to one sequence. They then compute a functional, which is simply the appropriate importance sampling weight. Their method can either be thought of as sampling paths through the recursion, or sampling sequences of past historical events. These are equivalent. The events define a genealogical tree with mutations indicated on it, but no time scale is needed.

There is one more subtlety. We can't actually know for any site that shows variation in our sample which of its two states is the original state and which the mutant. So Griffiths and Tavaré, in computing their importance sampling weights, use the probabilities of unrooted trees rather than of rooted trees, in effect summing up over all the ways that the ancestral state at the individual sites could be interpreted.

I have given a rather cursory description of their method here – a more detailed consideration of the way it fits into the framework of importance sampling is given by Felsenstein *et al.* [15].

This independent sampling (IS) method is attractive because it not only entirely avoids getting stuck in regions of tree space, but each sample is rapid. However, because the importance sampling is imprecise, it often needs large numbers of samples to be sure of sampling from the trees that contribute most of the probability. It also approximates the mutation process by an infinite sites model, which means that sites at which there are back mutations or parallel mutations must be removed from the data to avoid getting a likelihood of zero.

The original sampler allowed for either constant or exponentially growing populations. Bahlo and Griffiths [1] have extended the method to multiple populations with migration, and Griffiths and Marjoram [20] have extended it to sampling of ancestral recombination graphs.

The IS sampler can be extended to models of DNA sequences, but it then proves extremely slow owing to the high probability that mutations going backwards in time will lead to widely divergent sequences. This problem was addressed by Stephens and Donnelly [48], who have speeded up the IS sampler by a large factor in the DNA case by biasing the sampling of mutations in different sequences toward tracing back to a common ancestral sequence, and making the appropriate importance sampling correction. De Iorio and Griffiths [5] have derived an independent sampling method from consideration of the diffusion approximation. They show that this leads directly to Stephens and Donnelly's method, which thus can be seen to be a particular case of a more general approach. They also [6] extend their method to subdivided populations with migration among them. This approach can presumably be used as a general method for developing efficient independent sampling methods for other mixtures of evolutionary forces.

Fearnhead and Donnelly [10] have made another such correction that greatly speeds up independent sampling in the case of recombination, making it much more practical. They have presented simulation evidence that their independent sampler performs better than the correlated sampler described below.

1.3.8 *Correlated sampling*

A second approach by Kuhner *et al.* [34] comes from our lab. We sample our way through tree space by sampling coalescent genealogies. In the simple case of estimating Θ in a population of constant size, we used a trial value, the 'driving value' Θ_0, and wanted to achieve an importance sampling distribution whose density function was proportional to $\mathrm{Prob}\,(\mathrm{G}\,|\,\Theta_0)\ \mathrm{Prob}\,(\mathrm{D}\,|\,\mathrm{G})$. If Θ is close to Θ_0, this would be nearly an optimal choice. Using equations (1.12) and (1.15), if we are trying to compute the likelihood, it will be the average over sampled trees of

$$\mathrm{Prob}\,(\mathrm{G}\,|\,\Theta)\mathrm{Prob}\,(\mathrm{D}\,|\,\mathrm{G}) \Bigg/ \left(\frac{\mathrm{Prob}\,(\mathrm{G}\,|\,\Theta_0)\mathrm{Prob}\,(\mathrm{D}\,|\,\mathrm{G})}{\int_{\mathrm{G}} \mathrm{Prob}\,(\mathrm{G}\,|\,\Theta_0)\ \mathrm{Prob}\,(\mathrm{D}\,|\,\mathrm{G})} \right). \tag{1.16}$$

The denominator of the denominator is simply the likelihood at Θ_0, so after some cancellation this is

$$\frac{\mathrm{Prob}\,(\mathrm{G}\,|\,\Theta)}{\mathrm{Prob}\,(\mathrm{G}\,|\,\Theta_0)/\mathrm{L}(\Theta_0)}. \tag{1.17}$$

If we sample n genealogies $G_1, G_2, \ldots G_n$ in our Markov chain Monte Carlo run, and average this quantity, we find that $L(\Theta_0)$ can be factored out so that

$$\frac{L(\Theta)}{L(\Theta_0)} = \frac{1}{n}\sum_{i=1}^{n} \frac{\mathrm{Prob}\,(\mathrm{G_i}\,|\,\Theta)}{\mathrm{Prob}\,(\mathrm{G_i}\,|\,\Theta_0)}. \tag{1.18}$$

Thus the likelihood ratio between Θ and Θ_0 is estimated by the mean ratio of the Kingman coalescent densities for each tree at these two parameter values. The reader may wonder what happened to the data, which appears nowhere in equation (1.18). Its influence is felt entirely through the sampler that chooses the G_i.

1.3.8.1 *Tree proposals* To implement this sampler, we need a proposal mechanism and the usual Metropolis–Hastings acceptance-rejection method. Although we initially used a much more limited tree rearrangement method, the proposal mechanism we have found most useful (invented by Peter Beerli) is to choose a node in the coalescent tree (excluding the root), and then dissolve the connection between it and the node immediately ancestral to it. This lineage is then allowed to reconnect to the tree by a conditional coalescent. A conditional coalescent is a distribution whose density is proportional to the coalescent in all regions where it is not zero. We sample from this by having the lineage go back in time, having at any moment when there are k other lineages an instantaneous rate k/Θ_0 of coalescing with a random one of them. The lineage finally hooks itself back into the tree. This can result either in a small change of the time of the coalescent node or a major relocation of the lineage in the tree.

The Metropolis–Hastings sampler for this conditional coalescent proposal mechanism turns out to be to accept the new genealogy with probability

$$\min\left[1, \frac{\mathrm{Prob}\,(\mathrm{D}\,|\,\mathrm{G_{new}})}{\mathrm{Prob}\,(\mathrm{D}\,|\,\mathrm{G_{old}})}\right]. \tag{1.19}$$

The terms for the Kingman coalescent are cancelled by the Metropolis–Hastings correction for the biased proposal mechanism. This is convenient but not a large computational saving. The computations in 1.19 are still considerable, much more than for sampling a single event history in the independent sampler.

The sampler does considerably better if Θ_0 is close to the true Θ. In our programmes, we run an MCMC chain, infer a new value of Θ, and use that as Θ_0 for the next chain. In a typical run, we do this ten times, then use the resulting Θ as the basis for one longer chain to get an even more accurate Θ. This in turn is used for one final long chain to infer the likelihood ratio curve and the final estimate of Θ.

1.3.8.2 *Advantages and disadvantages* The correlated sampler has some obvious disadvantages. It could become stuck in one region of the tree space, and the calculations for each sample are much larger than for the independent sampler. However, there are advantages as well. If Θ_0 is close enough to Θ, the trees sampled are close to being an optimum sample of the trees proportional to their contribution to the likelihood. The independent sampler is less accurate, and that can lead it to need much larger numbers of samples than the correlated sampler. No clear conclusion has emerged about which method is superior.

1.3.8.3 *Extensions of the correlated sampler* Like the independent sampler, the correlated sampler has been applied to more complex cases. Kuhner *et al.* [35] have incorporated exponential population growth, Beerli and Felsenstein [2, 3] have incorporated migration among a number of populations, and Kuhner *et al.* [36] have incorporated recombination by having the sampler move in a space of ancestral recombination graphs.

One interesting discovery was made in the course of the work on exponential growth. It had been overlooked in previous coalescent studies. It was found [35] that the estimate of growth rate is strongly biased toward positive growth. If we estimate both Θ and the scaled growth rate g/μ, the maximum likelihood estimate of growth rate would usually be strongly positive even when true growth rate was 0. This behaviour is less alarming when it is considered that the interval of allowable growth rates is wide in these cases, and quite frequently contains 0 as well. The reality of this bias can be demonstrated in the case of a sample size of two sequences, when the integration can be done numerically without MCMC sampling. The bias is little reduced by adding more samples, but is strongly reduced by adding more loci. That allows us to rule out the possibility of a strong positive growth rate by occasionally finding loci with deep coalescences.

1.3.9 *Sampling from approximate distributions*

The computational difficulty of the sampling methods has led to the development of approximate methods that try to retain much of the statistical power of the exact samplers, while avoiding all or most of the sampling effort. This has been particularly tempting in the case of recombining coalescents, where the size and complexity of the ancestral recombination graph is daunting. Li and Stephens [37] have introduced the PAC (product of approximate conditionals) likelihood method for inferring the recombination from a sample of haplotypes. This approximates the coalescent distribution for the sample as the product of conditional distributions, each itself an approximation. The resulting calculation is far faster than any of the sampling approaches. It has become widely used. Hudson [25] and McVean *et al.* [39] have both used a different approximate method, one which approximates the distribution of haplotypes as the product of two-locus distributions. Fearnhead and Donnelly [11] give another approximate method based on using sampling methods on sub-regions and deriving an approximate likelihood from the results. Li and Stephens present simulations comparing these methods, finding that their method does best.

Those methods make an approximate computation of the likelihood of the full data. An alternative approach is to reduce the data to some appropriate summary statistics, and compute the likelihood for those reduced data. This was pioneered by Weiss and von Haeseler [53]. A more extensive consideration of methods for approximate inference that do not involve computing the full likelihood of the full data is given by Marjoram *et al.* [38]. While these methods enable much more rapid computation, the issue that must always be kept in mind is whether the summary statistics retain enough information.

1.3.10 *Ascertainment and SNPs*

The growth in the use of SNP (single nucleotide polymorphism) data has raised another issue, ascertainment bias. If sites are screened and only those found to be varying in some panel of genomes are included, we will find these sites to be much more variable in our sample than randomly sampled sites would be. If we included these sites without making any correction for the screening, the result would be an unrealistically high estimate of the mutation rate μ. That in turn would lead us to misestimate the rates of other parameters—for example, discrepancies in the picture of the tree from different sites that might actually be a sign of recombination would instead be too readily attributed to recurrent mutation.

Several papers have derived the corrections needed for the ascertainment of SNPs [6, 32, 42]. They treat various possible ways in which a SNP screening panel could be chosen. However, neither is able to treat the horrible reality. In some cases, ethical or legal concerns prevent the release of enough information about the panels to enable any sensible ascertainment correction to be made. The data are thus safe from being abused, and also safe from being used. Until recently, large-scale genomics projects acted as if they were blissfully unaware that analysis of their data required knowledge of how the screening was done. They either did not release the required information or, in some cases, they simply did not know it, or know that they had to know it. For some purposes (such as using the SNPs for linkage studies in pedigrees) this may not matter, but for all population analyses it matters a great deal. It is gradually beginning to be realized that an inability to correct the data for the way in which sites were chosen rules out many important uses of the data, making them largely a waste of money.

1.3.11 *Bayesian samplers*

I have so far discussed only likelihood inference. The spread in popularity of Bayesian inference has led it to be applied to coalescent-based inferences [7, 54, 55]. In Bayesian sampling one updates both the genealogy and the values of the parameters, sampling from these in proportion to their contribution to the posterior distribution of the parameter values. This can involve simultaneous updates of parameters and trees, or it can involve alternating updates of parameters and trees. The technology of sampling is very similar to the correlated sampler, but the use of the resulting sample is very different. In the likelihood-based methods, one uses the samples of the trees to compute a likelihood curve. In Bayesian methods one uses the sample of parameter values as a sample from the desired posterior, while ignoring the trees.

Bayesian samplers are attractive in their simplicity. They also have a tendency to avoid problems with driving values, as they sample broadly from the possible values of the parameters. When the objective is not Bayesian, these samplers can still be usefully employed and the posterior distribution of parameters ignored. One issue with posterior densities of parameters is that we need some means of interpolating density between the sampled parameter values. This leads to convolution of the extremely spiky posterior distribution with broader kernels that smooth out the density. All these are to some extent arbitrary.

As with likelihood methods, approximate calculations and use of summary statistics rather than the full data enable much faster computation. The Approximate Bayesian Computation (ABC) method of Tavaré and his coworkers [38, 44] takes advantage of this with, as is inevitable, the concomitant worries about whether one has chosen the best summary statistics.

1.3.12 *Future extensions*

I have barely skimmed the surface of the very active literature on coalescent-based inference. Coalescent methods are continually expanding. They will ultimately deal with all issues in evolutionary genetics. Some of the major extensions of the methods under way are:

Sequential sampling Coalescent methods have assumed that all samples are contemporary. If we can sample DNA from the past, some samples are at different levels in time in the tree. These need to be scaled using the mutation rate per generation (μ) and the generation time (T) to put them on the scale of branch length. In the simplest case [46], of the three quantities N, T, and μ, we can estimate two of them. This is an improvement over the case of contemporary tips, where we can only estimate one of these quantities, the product of N and μ. Sequential sampling is important in studies of ancient DNA, and is even more widely used in studies of rapidly evolving viruses such as HIV, where samples from the same patient over time must be considered to be at different levels of a tree. Sequential sampling methods are starting to be available in widely-distributed programmes [7]. For a more extensive treatment of sequential sampling coalescent methods see Chapter 2 by Rodrigo, Ewing, and Drummond in this book.

Uncertainty about haplotypes Data frequently come as diploid genotypes. The usual way of handling these has been to try to resolve haplotypes, then treat those reconstructed haplotypes as if they were observations. A more realistic treatment would be to sum the likelihoods for all possible haplotype resolutions, so that we incorporate our uncertainty about the haplotype resolution into our statistical analysis. This has been proposed by Kuhner and Felsenstein [33]. It requires extra rounds of MCMC sampling, as we sample from among all possible haplotype resolutions. The method is not available in most distributed programmes – when it is, it may replace most haplotype resolution calculations.

Multiple species It has been known since the work of Tajima [49] and Takahata and Nei [51] how to extend the coalescent to multiple related species. Each lineage in a tree of species will have a coalescent inside it, and such coalescents at different loci are independent of each other. If we arrive at a common ancestor, any gene copy lineages in each species that are not yet coalesced (going backwards in time) now join a common pool and are available to coalesce with each other. (It is best not to think of these matters forward in time, and thus not to use the confusing concept of 'lineage sorting'). Likelihood and Bayesian treatments of inferences about species trees from single and multiple loci have begun to appear [41, 43] and to be made available in computer programmes [7, 55].

Linkage disequilibrium mapping It is customary in genomics for researchers to debate which measure of linkage disequilibrium to use to characterize the joint distribution of variation at linked sites. The correct answer is 'none of them'. As we have seen, trees and D's are intimately related, and multiple-locus linkage disequilibrium describes the same phenomena as do trees of recombining haplotypes. While the two equivalent descriptions can be interconverted, it is the coalescent description that is easier to work with. For a fully powerful analysis of multiple linked sites, the correct way to compute the location score is to compute the likelihood for each possible location of the disease locus. A Bayesian approach might propose different locations for the disease locus, but it would accept or reject these based on these likelihoods. In either case one needs a full coalescent calculation. This point has been realized by all major researchers on recombining coalescents, but it has taken some time for linkage disequilibrium mapping methods based on coalescents to become available. That situation is about to change, and the discussion of methods in genomics will change with it.

Selection Inferring locations in the genome where there may have been selective sweeps or where there may be balanced polymorphisms is possible by likelihood or Bayesian methods. To do so, natural selection needs to be incorporated into the coalescent framework. This is perhaps the most interesting frontier of coalescent methods; it is under active exploration by a number of groups. As coalescent methods for detecting selection become widely available, they should replace the present summary-statistics methods.

Inferring the history As we sample past coalescent histories of our data, we can see historical events such as the times of particular coalescences. We could also imagine reconstructing when particular mutations occurred [22]. Knowing exactly what happened in the past has great appeal, and is always of interest to the popular science media. Taking a reasonable sample will usually show these inferences to be very noisy. In addition, they are not inferences of the parameters of the underlying models. As such, they are not maximum likelihood estimates, but rather maxmimum posterior probability estimates (in a Bayesian framework they have posterior probabilities just as do the parameters). The question arises: is reconstructing the exact history a trivial pursuit? The quantities which are needed in further analyses are usually the underlying parameter values rather than the exact times of particular events. However, the ages of mutations or the depths of particular coalescences can serve as indications of whether an allele is not neutral, or a population size not constant. The jury is not yet in on how interested we should be in these reconstructions of history.

1.4 Programmes

There are now many coalescent programmes available. As of the summer of 2006, some of the main ones I am aware of are:

LAMARC Likelihood-based inference including inference of migration, population growth, and recombination.
`http://evolution.gs.washington.edu/lamarc.html`

GENETREE Maximum likelihood estimation of mutation, migration, and population growth parameters and inference of times of coalescence and of mutation.
`http://www.stats.ox.ac.uk/~griff/software.html`

BEAST Bayesian estimation of population sizes and growth rates, allowing for sequential sampling. Allows a 'relaxed' molecular clock.
`http://evolve.zoo.ox.ac.uk/beast/`

BATWING (Bayesian Analysis of Trees With Internal Node Generation) Bayesian inference of mutation and population growth, with single or subdivided populations.
`http://www.mas.ncl.ac.uk/~nijw/`

msvar Bayesian inference of mutation rate and growth rate from microsatellite data for multiple loci in one population.
`http://www.rubic.rdg.ac.uk/~mab/software.html`

MDIV Likelihood inference of divergence time and migration rates for two populations.
`http://www.binf.ku.dk/~rasmus/webpage/mdiv.html`

MICSAT Likelihood inference for single-step microsatellite models.
`http://www.mas.ncl.ac.uk/~nijw/#micsat`

MISAT Likelihood inference of mutation rates for single- and multi-step models of microsatellite evolution in a single population.
`http://www.binf.ku.dk/~rasmus/webpage/misat.html`

IM (Isolation with Migration) Likelihood inference of divergence times and effective population sizes in a model with two diverged populations with subsequent migration between them.
`http://lifesci.rutgers.edu/~heylab/HeylabSoftware.htm#IM`

MCMCcoal Bayesian estimation of population sizes in a known tree of species.
`http://abacus.gene.ucl.ac.uk/software/MCMCcoal.html`

LDHAT Composite likelihood method for estimating recombination rates.
`http://www.stats.ox.ac.uk/~mcvean/LDhat/`

Hotspotter Product of Approximate Conditionals likelihood inference of recombination rates.
`http://www.biostat.umn.edu/~nali/SoftwareListing.html`

Recs Coalescent inference of recombination hotspots.
`http://www.maths.lancs.ac.uk/~fearnhea/software/Rec.html`

sequenceLD Approximate likelihood inference of recombination rate.
`http://www.maths.lancs.ac.uk/~fearnhea/software/Rec.html`

sequenceLDhot Approximate likelihood inference of recombination hotspots.
`http://www.maths.lancs.ac.uk/~fearnhea/`

popgen R package that includes neutral coalescent simulation of samples with recombination.
`http://www.stats.ox.ac.uk/mathgen/software.html`

CodonRecSim Simulation of sequence evolution under a codon model in a coalescent with recombination.
`http://www.binf.ku.dk/~rasmus/webpage/CodonRecSim.html`

SelSim Simulates samples under natural selection.
`http://www.stats.ox.ac.uk/mathgen/software.html`

hap and **dip** Simulate samples at a locus with natural selection.
`http://www.maths.lancs.ac.uk/~fearnhea/software/PS.html`

ms Simulates samples under a neutral coalescent with recombination and mutation.
`http://home.uchicago.edu/~rhudson1/source/mksamples.html`

SIMCOAL Simulates sequence evolution in a coalescent with migration.
`http://cmpg.unibe.ch/software/simcoal/`

Treevolve Simulates sequences evolving on a recombining coalescent with neutral mutation, population growth, and migration.
`http://evolve.zoo.ox.ac.uk/software.html?id=treevolve`

Mesquite Can simulate coalescents within species trees.
`http://mesquiteproject.org/Mesquite_Folder/docs/mesquite/popGen/PopGen.html#simulating`

I have not tried to describe which operating systems each programme requires. The programmes in this list are all free. I have omitted here a number of programmes that infer haplotypes rather than model parameters. By the time you read this, there will probably be many more programmes. Unfortunately, as yet there is no central list of coalescent programmes being maintained on the web.

1.5 The wave of the future

I have introduced the coalescent and some of the major approaches to inference that use it. I could not describe the full range of active work now going on, particularly with models of natural selection, models of recombination hot spots, and reconstruction of haplotypes from diploid data. We have passed the time when a single article could cover coalescent approaches. At least one major book on the coalescent has recently appeared [23]. It concentrates more on the population genetic phenomena than on inference methods.

To many researchers on evolutionary genetics and population genomics, coalescent inference methods may appear to be one of the major approaches, but only one. This perception will change, I hope rapidly. Coalescent inference methods are destined to replace most (perhaps all) other inference methods in these fields. They are currently limited by their computational burden, and by the difficulty of developing software to treat all cases. As those limitations are overcome, we will look back on the past decade as the period in which the major methods of analysis of population-level data developed, a period in which molecular evolution and population genetics began their ultimate merger. Students who now see coalescents as one interesting topic among many will ultimately understand that coalescents are the fundamental tool for analysing evolutionary data near the species level.

Acknowledgements

Work on this paper was supported by NIH grant R01 GM071639. I wish to thank the reviewers for many helpful comments, and for explaining to me what kind of book they would have written instead of this article.

References

[1] Bahlo, M. and Griffiths, R. C. (2000). Inference from gene trees in a subdivided population. *Theoretical Population Biology*, **57**, 79–95.

[2] Beerli, P. B. and Felsenstein, J. (1999). Maximum-likelihood estimation of migration rates and effective population numbers in two populations using a coalescent approach. *Genetics*, **152**, 763–773.

[3] Beerli, P. B. and Felsenstein, J. (2001). Maximum likelihood estimation of a migration matrix and effective population sizes in n subpopulations by using a coalescent approach. *Proceedings of the National Academy of Sciences, USA*, **98**, 4563–4568.

[4] Crow, J. F. and Kimura, M. (1964). The number of alleles that can be maintained in a finite population. *Genetics*, **49**, 725–738.

[5] De Iorio, M. and Griffiths, R. C. (2004). Importance sampling on coalescent histories. I. *Advances in Applied Probability*, **36**, 417–433.

[6] De Iorio, M. and Griffiths, R. C. (2004). Importance sampling on coalescent histories. II: Subdivided population models. *Advances in Applied Probability*, **36**, 434–444.

[7] Drummond, A. J., Nicholls, G. K., Rodrigo, A. G., and Solomon, W. (2002). Estimating mutation parameters, population history and genealogy simultaneously from temporally spaced sequence data. *Genetics*, **161**, 1307–1320.

[8] Edwards, A. W. F. (1970). Estimation of the branch points of a branching diffusion process. *Journal of the Royal Statistical Society, Series B*, **32**, 155–174.

[9] Ewens, W. J. (1972). The sampling theory of selectively neutral alleles. *Theoretical Population Biology*, **3**, 87–112.

[10] Fearnhead, P. and Donnelly, P. (2001). Estimating recombination rates from population genetic data. *Genetics*, **159**, 1299–1318.

[11] Fearnhead, P. and Donnelly, P. (2002). Approximate likelihood methods for estimating local recombination rates. *Journal of the Royal Statistical Society, series B*, **64**, 657–680.

[12] Feller, W. (1951). Diffusion processes in genetics. In *Proc. Second Berkeley Symposium on Mathematical Statistics and Probability* (ed. J. Neyman), pp. 227–246. University of California Press, Berkeley and Los Angeles.

[13] Felsenstein, J. (1992). Estimating effective population size from samples of sequences: inefficiency of pairwise and segregating sites as compared to phylogenetic estimates. *Genetical Research*, **59**, 139–147.

[14] Felsenstein, J. (2006). Accuracy of coalescent likelihood estimates: do we need more sites, more sequences, or more loci? *Molecular Biology and Evolution*, **23**, 691–700.

[15] Felsenstein, J., Kuhner, M. K., Yamato, J., and Beerli, P. (1999). Likelihoods on coalescents: a Monte Carlo sampling approach to inferring parameters from population samples of molecular data. In *Statistics in*

Molecular Biology and Genetics (ed. F. Seillier-Moiseiwitsch), IMS Lecture Notes-Monograph Series, volume 33, pp. 163–185. Institute of Mathematical Statistics and American Mathematical Society, Hayward, California.

[16] Fu, Y. X. and Li, W.-H. (1993). Estimating effective population size from samples of sequences: inefficiency of pairwise and segregating sites as compared to phylogenetic estimates. *Genetics*, **134**, 1261–1270.

[17] Golding, G. B. (1984). The sampling distribution of linkage disequilibrium. *Genetics*, **108**, 257–274.

[18] Griffiths, R. C. (1981). Lines of descent in the diffusion approximation of neutral Wright-Fisher models. *Theoretical Population Biology*, **17**, 37–50.

[19] Griffiths, R. C. (1989). Genealogical-tree probabilities in the infinitely-many-site model. *Journal of Mathematical Biology*, **27**, 667–680.

[20] Griffiths, R. C. and Marjoram, P. (1996). Ancestral inference from samples of DNA sequences with recombination. *Journal of Computational Biology*, **3**, 479–502.

[21] Griffiths, R. C. and Tavaré, S. (1994). Sampling theory for selectively neutral alleles in a varying environment. *Philosophical Transactions: Biological Sciences*, **344**, 403–410.

[22] Griffiths, R. C. and Tavaré, S. (1999). The ages of mutations in gene trees. *Annals of Applied Probability*, **9**, 567–590.

[23] Hein, J., Schierup, M. H., and Wiuf, C. (2005). *Gene Genealogies, Variation and Evolution. A Primer in Coalescent Theory.* Oxford University Press, Oxford.

[24] Hudson, R. R. (1983). Properties of a neutral allele model with intragenic recombination. *Theoretical Population Biology*, **23**, 183–201.

[25] Hudson, R. R. (2001). Two-locus sampling distributions and their application. *Genetics*, **159**, 1805–1817.

[26] Hudson, R. R. (2002). Generating samples under a Wright–Fisher neutral model of genetic variation. *Bioinformatics*, **18**, 337–338.

[27] Hudson, R. R. and Kaplan, N. L. (1988). The coalescent process in models with selection and recombination. *Genetics*, **120**, 831–840.

[28] Kaplan, N. L., Darden, T., and Hudson, R. R. (1988). The coalescent process in models with selection. *Genetics*, **120**, 819–829.

[29] Kingman, J. F. C. (1982). The coalescent. *Stochastic Processes and Their Applications*, **13**, 235–248.

[30] Kingman, J. F. C. (1982). Exchangeability and the evolution of large populations. In *Exchangeability in Probability and Statistics. Proceedings of the International Conference on Exchangeability in Probability and Statistics, Rome, 6th–9th April, 1981, in honour of Professor Bruno de Finetti* (ed. G. Koch and F. Spizzichino), pp. 97–112. North-Holland Elsevier, Amsterdam.

[31] Krone, S. M. and Neuhauser, C. (1997). Ancestral processes with selection. *Theoretical Population Biology*, **51**, 210–237.

[32] Kuhner, M. K., Beerli, P., Yamato, J., and Felsenstein, J. (2000). Usefulness of single nucleotide polymorphism data for estimating population parameters. *Genetics*, **156**, 439–447.

[33] Kuhner, M. K. and Felsenstein, J. (2000). Sampling among haplotype resolutions in a coalescent-based genealogy sampler. *Genetic Epidemiology*, **19 (Supplement 1)**, S15–S21.

[34] Kuhner, M. K., Yamato, J., and Felsenstein, J. (1995). Estimating effective population size and mutation rate from sequence data using Metropolis–Hastings sampling. *Genetics*, **140**, 1421–1430.

[35] Kuhner, M. K., Yamato, J., and Felsenstein, J. (1998). Maximum likelihood estimation of population growth rates based on the coalescent. *Genetics*, **149**, 429–434.

[36] Kuhner, M. K., Yamato, J., and Felsenstein, J. (2000). Maximum likelihood estimation of recombination rates from population data. *Genetics*, **156**, 1393–1401.

[37] Li, N. and Stephens, M. (2003). Modeling linkage disequilibrium and indentifying recombination hotspots using single-nucleotide polymorphism data. *Genetics*, **165**, 2213–2233 (Erratum, vol. 167, p. 1039, 2004).

[38] Marjoram, P., Molitor, J., Plagnol, V., and Tavaré, S. (2003). Markov chain Monte Carlo without likelihoods. *Proceedings of the National Academy of Sciences, USA*, **100**, 15324–15328.

[39] McVean, G., Awadalla, P., and Fearnhead, P. (2002). A coalescent-based method for detecting and estimating recombination from gene sequences. *Genetics*, **160**, 1231–1241.

[40] Neuhauser, C. and Krone, S. M. (1997). The genealogy of samples in models with selection. *Genetics*, **145**, 519–534.

[41] Nielsen, R. (1998). Maximum likelihood estimation of population divergence times and population phylogenies under the infinite sites model. *Theoretical Population Biology*, **53**, 143–151.

[42] Nielsen, R. (2000). Estimation of population parameters and recombination rates from single nucleotide polymorphisms. *Genetics*, **154**, 931–942.

[43] Nielsen, R. and Wakeley, J. (2001). Distinguishing migration from isolation: A Markov Chain Monte Carlo approach. *Genetics*, **158**, 885–896.

[44] Plagnol, V. and Tavaré, S. (2002). Approximate Bayesian Computation and MCMC. In *Monte Carlo and Quasi-Monte Carlo Methods 2000: Proceedings of a Conference held at Hong Kong Baptist University, Hong Kong SAR, China, Nov. 27-Dec.1, 2000* (ed. K. T. Fang, F. J. Hickernell, and H. Niederreiter), pp. 99–114. Springer-Verlag, London.

[45] Robertson, A. and Hill, W. G. (1983). Population and quantitative genetics of many linked loci in finite populations. *Proceedings of the Royal Society of London, Series B. Biological Sciences*, **219**, 253–264.

[46] Rodrigo, A. and Felsenstein, J. (1999). Coalescent approaches to HIV-1 population genetics. In *The Evolution of HIV* (ed. K. A. Crandall), pp. 233–272. Johns Hopkins University Press, Baltimore.

[47] Slatkin, M. and Hudson, R. R. (1991). Pairwise comparisons of mitochondrial DNA sequences in stable and exponentially growing populations. *Genetics*, **129**, 555–562.

[48] Stephens, M. and Donnelly, P. (2000). Inference in molecular population genetics. *Journal of the Royal Statistical Society. Series B*, **62**, 605–635.

[49] Tajima, F. (1983). Evolutionary relationship of DNA-sequences in finite populations. *Genetics*, **105**, 437–460.

[50] Takahata, N. (1988). The coalescent in two partially isolated diffusion populations. *Genetical Research*, **52**, 213–222.

[51] Takahata, N. and Nei, M. (1995). Gene genealogy and variance of interpopulational nucleotide differences. *Genetics*, **110**, 325–344.

[52] Watterson, G. A. (1975). On the number of segregating sites in genetical models without recombination. *Theoretical Population Biology*, **7**, 256–276.

[53] Weiss, G. and von Haeseler, A. (1998). Inference of population history using a likelihood approach. *Genetics*, **149**, 1539–1546.

[54] Wilson, I. J. and Balding, D. J. (1998). Genealogical inference from microsatellite data. *Genetics*, **50**, 499–510.

[55] Wilson, I. J., Weale, M. E., and Balding, D. J. (2003). Inferences from DNA data: population histories, evolutionary processes and forensic match probabilities. *Journal of the Royal Statistical Society: Series A (Statistics in Society)*, **166**, 155–201.

[56] Wiuf, C. and Hein, J. (1999). Recombination as a point process along sequences. *Theoretical Population Biology*, **55**, 248–289.

2

THE EVOLUTIONARY ANALYSIS OF MEASURABLY EVOLVING POPULATIONS USING SERIALLY SAMPLED GENE SEQUENCES

Allen Rodrigo, Gregory Ewing, and Alexei Drummond

Abstract

A population is said to evolve measurably if, when sequences are obtained over time, there is a significant accumulation of substitutions. Examples of Measurably Evolving Populations (MEPs) include rapidly evolving viruses, and populations from which it is possible to obtain ancient DNA sequences across long periods of geological time. In this chapter, we review the methods that have been developed to study the evolutionary genetics of MEPs. In particular, we describe (a) phylogenetic methods, including the reconstruction of serial sample phylogenies, and the estimation of substitution rate(s), and (b) coalescent methods to estimate population size and migration rates. We conclude with a discussion of where research in this area is heading, and some of the open questions that remain.

2.1 Introduction

When two neutrally-evolving homologous gene sequences are drawn randomly from an unfragmented haploid population of constant size, N, theory tells us that they have a common ancestor, on average, about N generations in the past. Theory also tells us that with a constant rate of substitution, μ, these two sequences will accumulate, on average, $N\mu$ substitutions each, so that between them one expects to see $2N\mu$ substitutions. These very simple statements about the times to common ancestry and numbers of substitutions lead to some quite powerful methods that allow us to work backwards from sequence data to derive estimates of population size, rates of growth or decline, migration, and selection. But what if each sequence was drawn at a different time? Now, the expected number of substitutions that separate the two is no longer a function of $N\mu$ alone, but also of the time between sampling, and the substitutions that accrue over this interval. Extend the thought experiment, and consider sampling two sequences first, and another two later. The expected number of substitutions between the pair of sequences sampled first ('early' sequences) or the pair of 'late' sequences will not be the same as that expected between an 'early' and a 'late' sequence. In fact, the expected difference between an 'early–late' pair and an 'early–early' pair will be equal to the product of the substitution rate and the time between early and late samples. This was pointed out by Shankarappa [52], Drummond and Rodrigo [4] and Fu [15]. If this expected difference is statistically different from zero for a reasonable sample size, we refer to such a population as a Measurably Evolving Population (MEP; [7]). The MEP is an empirical concept, obviously dependent on the size of the samples, the length of the sequences, the sampling interval, and the substitution rate. This should not detract from its utility because some populations obviously fit the definition better than others: as Drummond *et al.* note [7], 'although all populations evolve, only some evolve measurably'.

Population genetic studies that utilize molecular sequences, typically rely on samples of sequences that have been obtained contemporaneously (or *isochronously*). However, recently there has been increased interest in the analysis of samples that are gathered serially, each at a different time (i.e. *heterochronously*). Clearly, if it is our aim to derive estimates of the types of population parameters mentioned above, it may be inappropriate to treat these samples as contemporaneous. On the face of it, a plausible solution may be to treat each sample as an independent replicate from the same population, and derive estimates (or make inferences) using sequences obtained from each sampling occasion separately. However, this approach is potentially flawed as well, since the genealogies of the samples taken at different times may overlap extensively. At the very least, this correlation across samples biases the variances of estimates derived in this way. If the intent is to obtain estimates of how a parameter changes over time, treating each sample independently is analogous to, say, treating moving averages as independent. The latter are clearly not, and neither are serially sampled sequences, although there may be some exploratory benefits in such an exercise. In any case, the best approach would be to acknowledge the temporal dimension of the data and the correlations that are imposed by the overlap in genealogies.

There are two approaches one may adopt when analysing serially sampled sequences. The first is a 'phylogenetic' approach, in which the phylogeny of the sequences obtained is used as the foundation on which inferences are based. With this approach, a set of evolutionary relationships (i.e. a phylogenetic topology) is specified, and the only phylogenetic uncertainty that is usually admitted is the uncertainty in the branch lengths. This uncertainty exists because of the finite lengths of the sequences used in the analysis. Therefore, evolutionary parameters estimated using a phylogenetic approach are subject to variation only as a consequence of sequence length. With the phylogenetic approach, the fact that sequences are obtained randomly from the population is of no consequence – inferences are based on the phylogeny of these sequences only.

The second approach is to acknowledge that the sequences are a sample from a population, and that the phylogeny is a stochastic realization of an underlying evolutionary or demographic process acting on that population. This approach allows us to estimate the parameters associated with these processes. In this case the (intra-population) phylogeny of the sequences actually represents a genealogy. Uncertainty in the reconstruction of the genealogy of the sampled sequences is still a consequence of finite sequence lengths. However, the variances of the population parameter estimates are influenced both by sequence length and the limited number of coalescent events (and the intervals between these events) in the reconstructed genealogy.

Arguably, the most appropriate framework for making genealogy-based inferences of evolutionary parameters is based on the coalescent (see Chapter 1, and [14, 29, 30]). The coalescent is a mathematical description of the genealogy of a small sample of individuals from a large population. More specifically, it is a statistical description of the amounts of time independent genealogical lineages take to coalesce to common ancestors. Rodrigo and Felsenstein [47] showed how the coalescent can be extended to serially sampled sequences from MEPs.

Serially sampled populations allow us to estimate substitution rates, and other evolutionary parameters, directly. Another strength of serial sample inference is its ability to estimate changes in evolutionary parameters over time. This means that we can estimate the change in substitution rates, population size, migration rates, selection intensity—in fact, any evolutionary parameter one can think of—over the sampling intervals. And with MEPs, these evolutionary parameters certainly do change. Consider the Human Immunodeficiency Virus Type-1 (HIV-1), a retrovirus with an RNA genome that is reverse-transcribed to viral DNA, and is subsequently integrated into the host's chromosomal DNA. An individual infected with HIV-1 has a good chance of living for more than 10 years with the infection, particularly in the developed world. Over this period of time, the most variable portions of the envelope (*env*) gene of the virus (which encodes the Envelope protein) may diverge by 10% or more from the founding strain of the virus, the consequence of an error-prone reverse-transcription mechanism. To put this in context, if we mapped eukaryotic evolution with, say, 18S ribosomal RNA, which accumulates substitutions at the rate of around 1% per 50 million years [41, 46], a 10% divergence would correspond to 500 million years of evolution, over which we would have seen the radiation of the major animal and plant groups. In the same way, over the course of HIV-1 evolution within the host, the virus population changes: it grows in size, colonizes a variety of systems and tissues [43, 62], hides out in viral reservoirs [38], changes in response to the pressures of the host immune system [50], and recombines to form new and potentially more virulent forms [54]. Evolutionary analysis of MEPs should take account of these changes within the population, and provide tools to estimate their magnitude.

This chapter is organized as follows. In the next section, we describe a simple distance-based, least-squares (LS) method to estimate clock-constrained phylogenies of serially sampled sequences. This provides a useful introduction to the types of inferences that may be performed with serial sampled data. We then describe the maximum-likelihood (ML) estimation of single and multiple substitution rates. We move on to discuss the coalescent and its extension to serial samples. We describe some of the analyses that have been developed using the serial coalescent, including the estimation of migration rates and effective population size. We conclude with a look at where we think this research is heading.

2.2 Constructing phylogenetic trees from serially sampled data

If a researcher is interested in the phylogeny of sequences obtained at several timepoints, standard methods of phylogenetic reconstruction may be employed. Therefore, we may choose to build maximum parsimony, maximum-likelihood, or neighbor-joining trees with serially sampled sequences.

If, however, the researcher wishes to impose a molecular clock on the phylogeny, standard tree-building methods—even those that allow clock-like constraints—will not work. This is because with all standard tree-building methods that impose a molecular clock—e.g. UPGMA, or clock-constrained maximum-likelihood reconstruction—it is assumed that all sequences are sampled at the same time, and are therefore equally distant from the root of the tree. This is, of course, untrue for serially sampled sequences. Any clock-constrained tree-building method for serially sampled sequences must take account of the fact that sequences sampled at different times terminate at different distances from the root of the tree, with these distances proportional to the time that has elapsed. This last condition—that the distances are proportional to the time that has elapsed—is a particularly important one, because unlike 'standard' phylogenies of isochronous sequences, time is not confounded with substitution rate, but is separately measured as real time. This independent measure of time derives from the intervals between the times of sampling, and is typically measured in chronological units (e.g. days, months, years).

A further corollary of the fact that we have an independent measure of time (in chronological units) is that we are able to obtain an estimate of substitution rate in the same chronological units in which time is measured. As noted above, this means that with serially sampled sequences, we are able to decouple branch lengths of a phylogeny into time, t, and substitution rate, μ, instead of simply estimating the composite parameter, μt, as one does in any 'standard' phylogenetic analysis.

It is easiest to illustrate this using the LS approach that Drummond and Rodrigo [4] developed as a simple and rapid means of reconstructing serial phylogenies. Consider the following sampling scheme. A population is sampled several times over the course of a study period, and at each sampling time a number of sequences are obtained. In Fig. 2.1, for instance, six sequences have been sampled, two sequences at each of three timepoints. One method for reconstructing phylogenies when sequence evolution is clock-like is UPGMA (Unweighted Paired-Group Method with Arithmetic Means; see [49][55]). However, with UPGMA all tips on the tree terminate at the same time (i.e. the tree is ultrametric). Drummond and Rodrigo [4] developed an extension of UPGMA – serial sample UPGMA or sUPGMA – that allows the tips to terminate at different times, but constrains tips sampled at the same time to terminate at identical

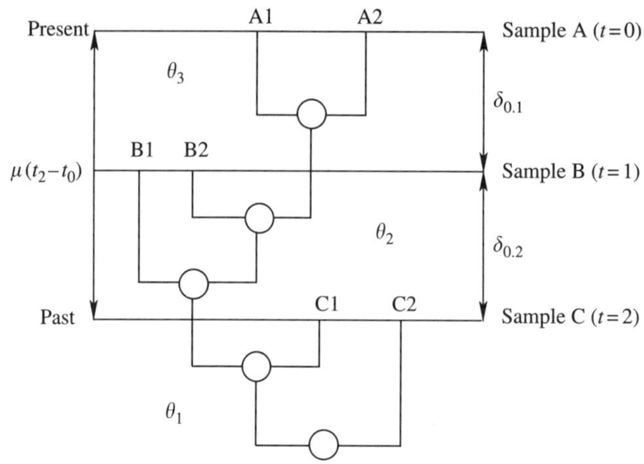

FIG. 2.1. Phylogeny of serially sampled sequences. In this example, six sequences are sampled, two at each of three times. Time is labelled sequentially from present to past. The δs measure the expected number of substitutions over a given interval. It is also possible to estimate a single rate of substitution, μ (see text). For each timepoint, θ estimates the intra-timepoint diversity. Note that there is no chronological information between the earliest timepoint ($t = 2$) and the root of the tree. Consequently, substitution rates cannot be directly estimated in that interval. Open circles indicate nodes for which heights from the root are also estimated. Chronological times can be associated with these heights if a uniform rate, μ, is estimated.

distances from the root. Serial sample UPGMA consists of four sequential steps, as follows:

- Estimation of the expected number of substitutions rate(s) in each interval.
- Correction of pairwise distances.
- Clustering using UPGMA.
- Trimming back branches.

Each step is developed in the following sections; particular emphasis is placed on the first section, where the logic of substitution rate estimation is best illustrated.

2.2.1 *Estimation of the expected number of substitutions in each interval, or a uniform substitution rate*

The first step estimates the expected number of substitutions per site that accumulates between sampling times. The expected distance between a pair of sequences, one from a later timepoint and the other from an earlier timepoint is [15, 52]:

$$\mathrm{E}[d(S_{early}, S_{late})] = \mathrm{E}[d(S^{(1)}_{early}, S^{(2)}_{early})] + \delta_{early,late}. \tag{2.1}$$

Equation (2.1) partitions the genetic distance, or number of substitutions, between early and late sequences into two parts: first, the expected distance between any two randomly sampled sequences in the earlier timepoint, and second, the distance that accrues over the interval between early and late timepoints (Fig. 2.1). The first term on the right hand side of equation (2.1) estimates the former distance, and $\delta_{early,late}$ measures the second of these distances. It is important to note that it is from $\delta_{early,late}$ that we are able to derive an estimate of μ, the substitution rate, separately from time: μ is simply estimated by $\delta_{early,late}$ divided by the time that has elapsed (in chronological units) between the early and late samples.

The problem becomes tricky when there are more than two timepoints, because now it becomes possible to calculate δ's for every possible pair of sampling times. Now, it may happen that, for any three timepoints A, B, C (where C is earlier than B, which is earlier than A), $\hat{\delta}_{CA} \neq \hat{\delta}_{CB} + \hat{\delta}_{BA}$ (where $\hat{\delta}$ is the estimated value) when, in fact, under any reasonable model the equivalent equality must be true. To overcome this problem, Drummond and Rodrigo [4] adopted a general LS approach to estimate δ, as follows. Consider a dataset of p samples with sample m obtained more recently than sample $m+1$ ($m \in \{1, \ldots p\}$). Let $d(m_i, n_j)$ be the evolutionary distance between the i'th sequence of the m'th sample and the j'th sequence of the n'th sample; by convention we will assume that $m \geq n$, i.e. we will only consider elements in the diagonal and lower triangular matrix of pairwise distances. Then for a haploid population with a constant intra-sample diversity, Θ, we can write the linear equation relating the distances to the parameters as:

$$d(m_i, n_j) = \Theta X_0 + X_{(2,1)m,n}\delta_{2,1} + X_{(3,2)m,n}\delta_{3,2} + \ldots + X_{(p,p-1)m,n}\delta_{p,p-1} ,$$

where $\delta_{k,k-1}$ is the expected number of substitutions that have accumulated between the k'th and $(k-1)$'th sample; $X_0 = 1$ and

$$X_{(k,k-1)m,n} = \begin{cases} 1 & \text{if } m \geq k \text{ and } n \leq k-1, \\ 0 & \text{otherwise.} \end{cases}$$

The solution for the vector of parameters $\boldsymbol{\beta} = \{\Theta, \delta_{2,1}, \ldots, \delta_{p,p-1}\}$ is obtained by the standard LS solution:

$$\boldsymbol{\beta} = (\boldsymbol{X}^{\mathrm{T}}\boldsymbol{X})^{-1}\boldsymbol{X}^{\mathrm{T}}\boldsymbol{d},$$

where $\boldsymbol{d}$ is a vector of pairwise distances.

This is simply a mathematical way of expressing the regression between the vector of pairwise genetic distances, and dummy variables ($\equiv X_{(k,k-1)m,n}$) that indicate the presence of one or more sampling intervals along the path between any two sequences. With this approach, the estimate of the δ's satisfies the condition that $\hat{\delta}_{CA} = \hat{\delta}_{CB} + \hat{\delta}_{BA}$. One additional constraint that we make to the δ's is to set any value of δ that has been estimated as a negative value to zero. Note also that the estimation process is easily extended to allow for multiple values of Θ, so that there is no need to assume that the intra-sample diversity is constant for all samples.

For the estimation approach above, it is not essential to know the actual sampling times, only the order in which the samples were drawn. Each of the δ's correspond to different expected amounts of substitutions that accumulate over the respective intervals. If the actual sampling times are known, then we may choose to divide each δ by the time between the appropriate sampling occasions, to derive interval-specific substitution rates. This may be interesting if we want to see whether there is any change in substitution rate as one moves along the tree.

If we wish to estimate a single substitution rate that spans all sampling intervals, an alternative approach to estimating δ is to estimate a single constant, μ, effectively the number of substitutions per unit time, and multiply this by the time interval between two sampling occasions. As above, we can estimate μ using a regression procedure. In this case,

$$d(m_i, n_j) = \Theta + \mu(t(i) - t(j)),$$

where $t(i)$ is the time at which the i'th sequence was obtained. Note that μ is not the substitution rate per generation, unless time is expressed in generation units. However, μ can be converted to the substitution rate per generation (i.e. number of substitutions per site per generation) if the generation time is known.

There are a few interesting aspects to the LS procedure outlined here. First, the interpretation of the Θs: each Θ effectively measures the intra-sample pairwise diversity per site for the earlier sample in a given sampling interval. Under standard population genetic theory, for a haploid population with constant size, the average pairwise diversity of a neutrally evolving locus is an estimate of twice the effective size, N, of the population multiplied by the substitution rate (either in units per site per generation for isochronous data, or units per site per unit of chronological time, for heterochronous data). Therefore, we may think of the Θs broadly as estimates of population size. If substitution rate is constant, then multiple Θs mean different population sizes, that remain constant over an interval and change in a stepwise manner from one interval to the next. The latter is clearly biologically unrealistic. Nonetheless, allowing multiple Θs provides a way of incorporating at least some of the variation in intra-sample diversity – perhaps a consequence of changing population size or migration – into the analysis.

Our use of Θs as measures of intra-sample diversity, however, means that we ignore the precise phylogenetic relationships of lineages within the sampling intervals. This, in turn, means that for any given phylogeny, we are ignoring patterns that could potentially improve estimation of substitution rates and, in fact, Drummond and Rodrigo [4] showed that the distributions of estimates of substitution rates using sUPGMA were unbiased, but tended to be positively skewed.

A second point worth noting is that there is no δ associated with the interval from the root of the tree to the time of the earliest sample (Fig. 2.1). This is quite important – it means that with any serial sample analysis, we really have no direct or empirical information on which we can base our estimates of substitution rate for the time period immediately prior to the first sample. Of course, if we fit a single μ, we can assume that this constant rate continues along the entire tree, including the lineages of sequences obtained in that earliest sample, but this is really an assumption on our part, and should be recognized as such. If we are prepared to make this assumption, then it is possible to date the nodes of the tree in real time, and that is certainly an advantage.

Finally, it may be obvious but it is probably worthwhile pointing out that our estimates of δ(s) or μ apply across all branches that span the sampling intervals. The approach described above, and indeed, all of the methods we describe in this chapter, do not fit lineage-specific substitution rates (although methods have now been developed that permit relaxed-clock models to be fitted – [5, 27, 59, 60]).

2.2.2 *Correction of pairwise distances*

Once the δs have been estimated, each pairwise distance d_{ij} in the distance matrix is transformed to a corrected distance, c_{ij} as follows:

$$c_{ij} = d_{ij} + \delta_{t(i),0} + \delta_{t(j),0},$$

where $t(i)$ and $t(j)$ are the time points from which the i'th and j'th sequences are obtained, and $\delta_{t(i),0}$ and $\delta_{t(i),0}$ are the δs associated with the divergence between $t(i)$ and $t(j)$ and the most recent sampling occasion (labelled '0'). What this does, in effect, is extend the distances of sequences sampled earlier to a value that approximates the expected divergences of sequences obtained most recently.

A similar correction can be employed if μ has been estimated:

$$c_{ij} = d_{ij} + \mu(t(i) - t(0)) + \mu(t(j) - t(0)).$$

2.2.3 *Clustering using UPGMA*

The corrected distances mean that all sequences may now be treated as though they were collected at the same time. Since we want to impose a molecular clock on these sequences, we can use UPGMA[49, 55] on the corrected distance matrix. We may, of course, choose to employ other methods, e.g. a clock-constrained Fitch–Margoliash LS approach (as implemented in the module KITSCH, part of Joe Felsenstein's PHYLIP suite).

2.2.4 *Trimming back branches*

For a given terminal lineage on the clock-constrained UPGMA tree, that extends to sequence i, $\delta_{t(i),0}$ (or $\mu(t(i) - 0)$) is subtracted from the branch length. This effectively trims the branch length so that it terminates at the appropriate sampling time. Therefore, the sUPGMA tree has the topology recovered by UPGMA (on corrected distances) with branch lengths that reflect the appropriate order of sampling and/or the sampling times.

2.2.5 *sUPGMA and serial sample miscellany*

sUPGMA gives us a rapid means of reconstructing serial phylogenies. Methodologically, it also illustrates some of the key issues related to working with serial samples. With sUPGMA trees, and in fact, with all of the types of serial-sample phylogenies or genealogies we deal with in this chapter, sequences sampled earlier are never allowed to be direct ancestors of sequences sampled later. There are three reasons for this. First, with simple theoretical populations that have discrete generations, where parents die as soon as offspring are produced, the act of sampling an individual means that the individual will never produce an offspring. Hence, it cannot be an ancestor of any other individual. Second, it is also important to distinguish between genes (as individual units), and the sequences these genes have (which may be identical). Two genes may have identical sequences, but both are nonetheless different individuals. Why include genes with identical sequences in a serial-sample phylogeny? The reason is that sequence identity tells us something about substitution rate—if, over a period of time, the same sequences persist in a population of genes, we can infer that the substitution rate per unit time is low. Third, and finally, there is also a statistical justification: if we are dealing with large populations, it is very unlikely that any lineage which proceeds along the phylogeny to an earlier sampling time will encounter one of its own ancestors.

The LS estimation of substitution rates implemented in the sUPGMA algorithm is very flexible. For instance, Drummond, Forsberg, and Rodrigo [3] implemented a LS method that permits μ to change along the tree at some *a priori* specified time. This 'Multiple Rates with Dated Tips' model (MRDT; as opposed to the 'Single Rate with Dated Tips' or SRDT model where μ remains constant [45]) allows for stepwise changes in μ with the changepoint occuring at any time, including times *between* sampling occasions. Readers are directed to [3] for a description of the method, which is relatively straightforward.

One of the drawbacks of using a LS approach in phylogenetic reconstruction is that there is no analytic way to calculate the variances of our estimates. This is because the distances that constitute the datapoints to which LS regression is applied are not independent. Consequently, randomization approaches are typically employed. For instance, we can bootstrap the sequence data, and obtain bootstrap confidence intervals for estimates of our parameters. Alternatively, we can apply parametric bootstrap methods, whereby we simulate sequence data of the same dimensionality (i.e. same number and length of sequences) using the estimated parameter values in the simulating model, and then re-estimate these parameters to determine the variation we can expect to see in our estimates [9, 13, 23, 25].

Of course, most things that can be done with least-squares can also be done with likelihood (with appropriate distributional assumptions). Over the years, maximum-likelihood estimation has become one of the cornerstones of phylogenetic reconstruction, so it is only natural that the estimation of substitution rates using serially sampled sequences has also been described in a likelihood-based context. These approaches are described in the next section. It is worthwhile noting, however, that the analyses that have been developed within the likelihood framework are directly equivalent to those mentioned above; what is of value with the likelihood approach is that there are standard statistical hypothesis tests, as well as measures of confidence, that can be applied. Of course, likelihood estimators also tend to have lower variance, but the ability to make inferences beyond simple point estimation makes likelihood-based methods more appealing than the LS approaches we have already mentioned.

2.3 Maximum-likelihood estimation of evolutionary rates

A number of researchers have developed and tested maximum-likelihood (ML) methods that accommodate the structure of serially sampled sequences [3, 45, 51]. It has generally been assumed that the tree topology is known and that each tip of the tree has associated with it a date of isolation that is known without error.

2.3.1 *Single rate dated tips*

Rambaut [45] considered the case when there is a single rate of evolution (i.e. a strict molecular clock), the rooted phylogenetic tree topology is given *a priori* and all sample isolation dates are known without error. The unknowns, as in Fig. 2.1, are then the times (in chronological units) of the internal nodes of the tree and the substitution rate (which can be thought of as a scaling parameter that reconciles the information in the sample dates with the genetic differences between the sequences). The substitution rate therefore scales the internal nodes from chronological units into units of expected number of substitutions per site. Given a rooted tree (in chronological units) and a rate of substitution, we can calculate the expected number of substitutions per site for each branch of that tree, as well as the likelihood of a given model of evolution [12]. The vector of internal node times, $t = \{t_1, \ldots, t_{n-1}\}$ along with the substitution rate (μ) and any parameters of the substitution model (for example, the transition–transversion ratio) are then optimized using any standard multi-dimensional optimization procedure. The result of such a numerical optimization procedure will be to find the parameter values that maximize the likelihood:

$$L(\mu) = \Pr(D|\boldsymbol{G}, \mu, Q), \tag{2.2}$$

where $\boldsymbol{G}$ is the given tree, D are the sequence data and Q is the model of substitution, including the instantaneous rate matrix and any associated parameters (e.g. the proportion of invariant sites, the shape parameter of a gamma distribution of rates). This model has been labelled the 'single rate dated tips' (SRDT) model [45].

2.3.2 *Multiple rates dated tips*

Molecular sequences accumulate substitutions over time, but the rate at which this occurs may not be constant through time, among lineages or among sites.

The rate of substitution depends on various biological processes such as the intensity of selection, changes in population size (when selection is present), and changes in life history characteristics such as, say, a shift in mean generation time. These effects can change substitution rate (1) over time, (2) in different lineages, and (3) at different positions along the sequence. Drummond *et al.* [3] considered models which held the rate of evolution constant across all lineages at any instant in time, but allowed the rate to vary at different periods of time in the evolutionary history of the sequences. In particular, they developed a model that allowed for stepwise changes in the overall substitution rate to occur at pre-specified times in the past. These times create a series of epochs, each of which has its own unique substitution rate for the entire population. This model of rate variation as a step function of time is appropriate when we model extrinsic factors that affect the whole population simultaneously and suddenly. In the context of virus evolution, for instance, the administration of anti-viral therapy may be accompanied by a sudden, and almost complete cessation of viral replication, and we may expect to see a precipitous decrease in substitution rate.

Because this 'Multiple Rates Dated Tips' (MRDT) model was developed within a likelihood framework, model testing and comparison is readily achieved with the SRDT model described above being a special case. For the MRDT, the likelihood of a set of substitution rates, $\boldsymbol{M} = \{\mu_1, \ldots, \mu_k\}$ (where k is the number of rates estimated) is identical to that in equation (2.2), except that in place of μ, we substitute $\boldsymbol{M}$ and add a vector of change-point times τ:

$$L(\boldsymbol{M}) = \Pr(D|\boldsymbol{G}, \boldsymbol{M}, Q, \tau).$$

The MLEs of the rates, $\hat{\mu}_i$ are jointly chosen such that $L(\boldsymbol{M})$ is maximized. As with estimates of substitution rates using sUPGMA, we constrain each estimated substitution rate to be greater than or equal to zero. When considering multiple substitution rates, confidence interval estimation is less straightforward than for a single rate. There are at least two ways of computing confidence intervals for multiple rates. First, multivariate upper and lower $(1-\alpha)$ confidence limits may be obtained by locating rates that correspond to log-likelihood values differing from the maximum-log-likelihood value by $\chi^2_{k,\alpha}/2$. If unbiased, these confidence intervals have an asymptotic $(1-\alpha)$ probability of enclosing the true $\boldsymbol{M}$ as sequence length tends to infinity. Second, a profile confidence likelihood interval may be obtained for each μ_i as follows. Over a range of μ_i, locate the upper and lower values of μ_i such that

$$-2|\ln L(\mu_1^*, \mu_2^*, \ldots, \mu_k^*) - \ln L(\hat{\mu}_1, \hat{\mu}_2, \ldots, \hat{\mu}_k)| = \chi^2_{1,\alpha},$$

where $\hat{\mu}_j$ is the MLE of the j'th rate, and μ_j^* is the maximum-likelihood estimate of the j'th rate when μ_i is fixed at a given value.

In the case where all elements of $\boldsymbol{M}$ are equal, the MRDT model collapses to the SRDT model of a uniform molecular clock. If all μ parameters are set to zero, the MRDT model reduces to the standard isochronous clock model [17, 45].

In fact, under the likelihood framework, one is able to test whether the MRDT model is a significantly better model for the data than the SRDT model. Since the SRDT model is simply a constrained MRDT model, the standard asymptotic likelihood ratio test may be applied. In this case, the test statistic,

$$\Delta = 2[\ln L(\boldsymbol{M}, \text{not all } \mu \in \boldsymbol{M} \text{ equal}) - \ln L(\boldsymbol{M}, \text{all } \mu \in \boldsymbol{M} \text{ equal})]$$

is asymptotically distributed as χ^2 with $k-1$ degrees of freedom under the null hypothesis that the two models are not significantly different, where k is the number of μ parameters specified *a priori* that are free to vary.

When testing whether a tree that uses the SRDT model is significantly more likely than one in which all tips terminate at the same distance from the root (i.e. the standard clock-like tree with isochronous data), the null and alternative hypotheses are of the form:

$$H_0 : \mu = 0 \text{ and } H_1 : \mu > 0,$$

respectively. The test is a one-tailed test. If α is the chosen level of significance, the null hypothesis can be rejected when

$$\Delta = 2\left(\ln L(\mu > 0) - \ln L(\mu = 0)\right) > \chi^2_{1,2\alpha}. \tag{2.3}$$

The same test can also be derived by treating the constraint that μ has to be greater than or equal to zero as a boundary-value problem [42].

Finally, one may test a fully unconstrained tree against one constructed using the MRDT model. In this case, the likelihood ratio statistic under the null hypothesis is asymptotically distributed as χ^2 with degrees of freedom equal to $2n-3-(n-k+1) = n-2+k$. This suite of tests suggests a natural hierarchy of hypotheses that one may choose to test – (1) an unconstrained tree vs. a MRDT-constrained tree; (2) a MRDT-constrained tree vs. a SRDT-constrained tree; and (3) a SRDT-constrained tree vs. a isochronous clock-constrained tree.

What influences the statistical power of these hypothesis tests? In essence, the statistical detection of an accumulation of substitutions over time requires that we reject the null hypothesis that the substitution rate is zero. To pre-empt any doubts about the validity of a zero substitution rate, readers are reminded that the substitution rate estimated is only the rate that subtends one or more sampling intervals. It is not the rate that extends from the earliest sampling interval to the root of the tree, for which there is no direct information independent of chronological time. Therefore, it is still possible to obtain a set of non-identical sequences at different timepoints, and hypothesize a zero substitution rate. The Likelihood Ratio Test (equation (2.3)) described above is used to test the null hypothesis that the substitution rate is zero. Three factors influence the power of this test, that is, our ability to correctly reject this null hypothesis given that the substitution rate is truly greater than zero over the sampling interval [7]: the intra-sample diversity, the length of the sampling interval, and the lengths of the sequences.

For a given non-zero substitution rate, increasing the length of the sampling interval increases power, as does a lower intra-sample diversity. These results are intuitively obvious: increasing the sampling interval increases the expected number of substitutions that can accumulate and therefore, under a Poisson model of evolution, reduces the probability of seeing no substitutions at all. By the same token, high intra-sample diversity is typically accompanied by high expected variances on the distances (or branch-lengths) between sequences from the same timepoint. If we return to equation (2.1), it should be obvious that as the intra-sample variance on distances increases, it becomes more difficult to detect the true inter-sample distance, $\delta_{early,late}$, with finite-length sequences, because $\delta_{early,late}$ contributes progressively smaller amounts to the total variance of distances between 'early' and 'late' sequences. Finally, as our sequences get longer, we have more opportunity to observe substitutions between sequences from different timepoints, and this – coupled with the reduction in variances in branch-lengths – also leads to an increase in power.

2.3.3 *A few last words about likelihood and serial samples*

There have been other extensions of the likelihood approach. For instance, Rodrigo *et al.* [48] extended the ML approach to estimate a common substitution rate when there are two or more populations from which serial samples are obtained. Such an approach is appropriate when, say, a number of hosts are infected with a rapidly evolving pathogen. The methods described may also be applied when several independent measurably evolving loci are sampled. When there are several populations or loci, it may be that these are grouped into a number of rate categories, each with its own substitution rate. Rodrigo *et al.* [48] showed how a joint ML estimate may be obtained for substitution rates and the proportion in each group when there are two or more groups. They applied these methods to a dataset published by Gunthard *et al.* [22], where six HIV-1 infected individuals were treated with Highly Active Antiretroviral Therapy (HAART), and HIV-1 sequences were obtained just before HAART began, and two years later. Gunthard *et al.* [22] reasoned that if HAART is successful in halting virus replication, substitutions would not accumulate over the sampling interval. However, not all patients would respond to HAART, and we expect to see two groups of patients, one with a rate $\mu > 0$ and another with $\mu = 0$. In this example, the aim was to jointly estimate p, the proportion of patients who did not respond to HAART, and $\mu > 0$, the substitution rate of the virus in those individuals. Readers are directed to [48] for details on the method but it is worth mentioning two interesting features of this analysis.

First, with HIV-1 sequences from multiple patients, it is possible to construct a joint phylogeny of HIV-1 populations across all patients. Therefore, estimates of μ and p can be derived in two different ways: (1) each HIV-1 population (in each patient) can be treated as an independent source of information (the 'Sub-Tree Likelihood' or STL method) or (2) a single phylogeny of all HIV-1 populations (across all individuals) can be built (the 'Whole-Tree Likelihood' or

WTL method). Rodrigo *et al.* [48] found, however, that there was almost no difference in the estimates of μ and p derived using STL or WTL.

A second interesting point is this: on the face of it, it would appear that p may be estimated simply by counting the number of individuals with viral μs that are statistically greater than 0, and dividing by the total sample size. However, this estimate fails to take account of the fact that, even for those patients whose samples of viral sequences fail to allow us to detect μs that are *statistically* different from 0, it is still possible for $\ln L(\mu > 0) > \ln L(\mu = 0)$. If, in fact, most individuals fall into this category, we would want our estimate of p to reflect the fact that the proportion of individuals for whom HAART has failed (to halt virus replication) may be quite high, even though we are not able to demonstrate this failure for each individual separately. By estimating p using all the data simultaneously, we allow these likelihoods to influence its value as well.

The maximum-likelihood method is expected to be more sensitive and accurate than distance-based methods. Furthermore, the maximum-likelihood framework provides much greater flexibility in model selection, by allowing standard model comparison approaches such as the likelihood ratio test (LRT) for nested models and the Akaike Information Criterion (AIC) for non-nested models. However, one concern with current ML implementations, such as TIPDATE [45], is that they assume that the topology is known without error. Of course, this is not usually the case, and with the ML methods described above, the uncertainty inherent in phylogenetic reconstruction does not contribute to the variances associated with the estimated evolutionary rates. A second problem with assuming a known tree topology is that, in practice, this topology is often obtained by running an unconstrained phylogenetic analysis (for example, by using PAUP* [57] or MrBayes [26] with standard settings). However, the maximum-likelihood tree topology under the SRDT or MRDT models may differ from the maximum-likelihood tree topology obtained using a standard unconstrained model [3].This may seem counter-intuitive at first. After all, if the SRDT model is the correct model, then an unconstrained ML search should recover the correct topology, because the SRDT tree is simply a special case of the unconstrained tree. However, because we typically deal with finite-length sequences, random error can mean that the unconstrained ML tree is not topologically identical to the SRDT tree. Consequently, using an ML topology from PAUP* (or a consensus tree from MrBayes) may bias substitution rate estimation. Obviously, the best approach is to simultaneously estimate both the appropriately-constrained ML tree and the substitution rate(s), but at the time of writing, software that does this has yet to be released.

On the other hand, if the tree itself is not of direct interest, then a method that takes into account the shared ancestry of the data without basing inference on a single reconstruction of ancestral relationships would be useful. Markov chain Monte Carlo (MCMC) methods provide exactly this opportunity, and these methods have been used widely within the population genetics literature, and in particular, with the coalescent.

2.4 The serial coalescent

The coalescent, with its focus on the genealogies of individuals sampled from large populations, is a very useful descriptor of the types of phylogenies (or more appropriately, genealogies) that MEPs generate, for example, genealogies of rapidly evolving viral genes sampled from a host or population of hosts, or mitochondrial genes from fossils and sub-fossils. Our motivation for developing coalescent methods for MEPs derives from research we have undertaken with rapidly evolving viral populations and ancient DNA. RNA viruses, for instance, typically have high rates of substitution because of low-fidelity replication mechanisms [28]. As an example, Shankarappa [52] showed that in the *env* gene of HIV-1, substitutions accumulate at a rate of approximately 1% per year. At the other end of the biological spectrum, eukaryote populations can also yield samples that show an appreciable accumulation of substitutions. Highly sensitive DNA amplification and sequencing methods now allow us to obtain sequences from sub-fossil bones [1, 34, 53], amber-preserved organisms [2], tissue remains [16, 31, 53, 58], and one or a few intact or degraded DNA molecules from fossils [18]. All these populations evolve at far slower rates than RNA viruses but the fact that DNA is obtained across very large time intervals means that these populations still qualify as MEPs.

In this section we introduce the serial coalescent, or *s*-coalescent [47], an extension to the Kingman coalescent [29, 30] incorporating serial samples. Chapter 1 provides a detailed account of the standard Kingman coalescent (i.e. with isochronous samples) and readers should refer to that chapter before proceeding with this section.

If we sample two extant haploid individuals, the probability that they have the same parent in the previous generation is $1/N$. If they are not siblings, the probability that the two individuals have a common ancestor two generations ago is $(1 - 1/N)/N$. This is the probability that they are not siblings $(1 - 1/N)$ multiplied by the probability that they do, however, have the same grandparent $(1/N)$. The probability that two individuals are the extant members of two lineages that coalesce in ρ generations is therefore $p(1-p)^{\rho-1}$, where $p = 1/N$. This is the probability that the two lineages in question do not coalesce for $\rho - 1$ generations $(= (1-p)^{\rho-1})$ multiplied by the probability of a coalescence in the ρ-th generation $(= p)$.

If, instead of sampling two extant individuals, we sample n individuals, the probability of observing a coalescence in a single generation is $n(n-1)/2N$ because there are $n(n-1)/2$ ways of selecting the two lineages that may coalesce. Conversely, the probability of not seeing any coalescence is $(1 - n(n-1)/2N)$. If we let $p = n(n-1)/2N$, then the probability of observing two lineages coalesce after ρ generations is again $p(1-p)^{\rho-1}$.

When N is large, the application of a diffusion approximation allows us to move from discrete time to continuous time. Applying this approximation means that we can obtain the probabilities of coalescent intervals, by treating these intervals as continuously-valued random variables drawn from exponential distributions with expectations equal to $2N/[n(n-1)]$. Consequently, for a given genealogy, $\boldsymbol{G}$, the conditional density of obtaining that genealogy, given a population size of N, and a sample of n individuals, is simply the product of as many of these probabilities as there are coalescent intervals on the genealogy:

$$f(\boldsymbol{G}|N) = \frac{1}{N^{n-1}} \exp\left(-\sum_{r=1}^{n-1} \frac{k_r(k_r-1)}{2N}\rho_r\right), \tag{2.4}$$

where k_r is the number of lineages remaining in interval ρ_r.

If N is unknown, and we intend to estimate it, there is a problem, because to do so requires that we know the length of the coalescent intervals (i.e. the ρs) in generations. But the genealogies that we have access to rarely have times (or branch-lengths) measured in generations. Instead, with standard phylogenetic methods, where the data are gene sequences, time is measured by the number of substitutions along a branch or along a coalescent interval. Consequently, with these standard approaches, it is only possible to measure time as it is scaled by substitution rate, μ. For this reason, with the standard coalescent, we replace N with the composite parameter $\theta = 2N\mu$ (for haploid populations) or $\theta = 4N\mu$ (for diploid populations). With serially sampled sequences, we are able to separate chronological time and substitution rate. Consequently, we use θ in a different way: in our formulation, θ is equal to the (effective) population size scaled by the number of chronological units per generation, t_g, i.e. $\theta = Nt_g$. Therefore, whereas with the standard coalescent, time is measured in substitutions, with the s-coalescent, time is measured in chronological units. This is the first important difference between the standard coalescent and the s-coalescent. Consequently, we can now rewrite equation (2.4) as:

$$f(\boldsymbol{G}|\theta) = \frac{1}{\theta^{n-1}} \exp\left(-\sum_{r=1}^{n-1} \frac{k_r(k_r-1)}{2\theta}\delta_r\right), \tag{2.5}$$

where $\delta_r = t_g\rho_r$ is a rescaling of time in chronological units. Readers should convince themselves that equation (2.5) is mathematically identical to equation (1.13) in the previous chapter.

Another significant difference between the standard coalescent and the s-coalescent is illustrated in Fig. 2.2. With genealogies of contemporaneously sampled sequences from a panmictic population without recombination, the only mathematically interesting events are the coalescent events between pairs of lineages, and the lengths of the intervals between these events. In contrast, with the s-coalescent, in addition to coalescent nodes, we also have events that correspond to the entry of new sequences into the genealogy (as one moves back in time from present to past). In Fig. 2.2, internal nodes represent coalescent events and leaf nodes represent the points at which new samples join the genealogy. Let δ_r be the time between nodes r (both coalescent nodes and leaf nodes) and $r+1$ in chronological time, with time increasing into the past. If node $r+1$ is a coalescent node, the probability density of the rth interval contributes a

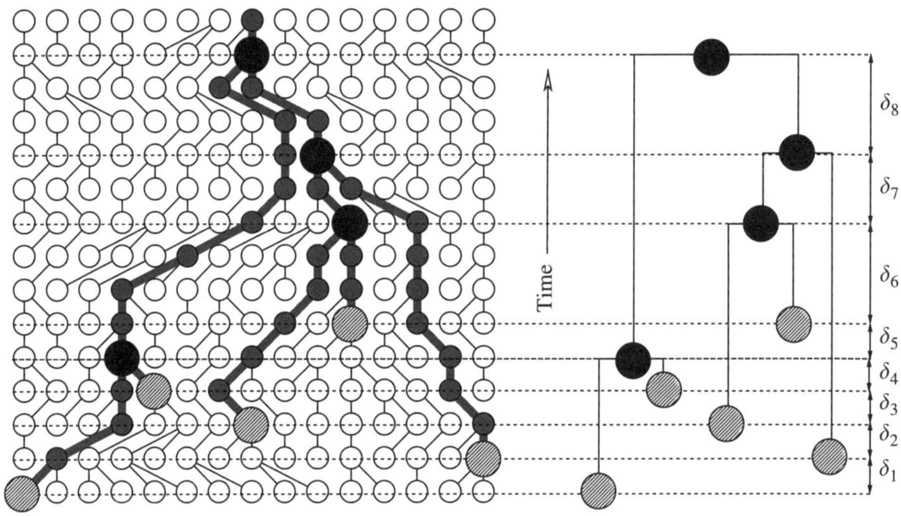

FIG. 2.2. Discrete-time population model for a haploid population sampled serially. Time is measured from present to past. Time intervals on the serial genealogy (right) are labelled as δs, and are measured between events that include both coalescent events (filled circles) and the entry of new sequences (hashed circles).

factor $\theta^{-1}\exp(-k_r(k_r-1)\delta_r/2\theta)$ to the overall coalescent density, where k_r is the number of lineages during interval r. This is, of course, the standard coalescent density for a single coalescent interval. If, however, the rth interval ends with the $r+1$-node being a leaf node, the contribution to the overall density is $\exp(-k_r(k_r-1)\delta_r/2\theta)$. This is simply the probability that no coalescent event has occurred in the interval δ_r; the probability of encountering a leaf node at the end of that interval is 1, because its entry is specified *a priori* as part of the sampling scheme. The coalescent density over the genealogy is then,

$$f(\boldsymbol{G}|\theta) = \frac{1}{\theta^{n-1}}\exp\left(-\sum_{r=1}^{m}\frac{k_r(k_r-1)}{2\theta}\delta_r\right), \tag{2.6}$$

where $m = 2n-2$; we arrive at this value of m because there are $n-1$ intervals that terminate in leaf nodes, and $n-1$ that terminate in coalescent nodes (Fig. 2.2). In fact, equation (2.6) looks similar to equation (2.5), except for the use of m instead of $(n-1)$ to index the sum within the exponential. Note that with the continuous-time approximation that the coalescent uses, it is assumed that in any instant of time, only a single event can occur. With the s-coalescent, we do not allow new sequences to join the genealogy at exactly the same moment of time that a coalescent event occurs. However, it is possible for several sequences to enter the genealogy simultaneously, as may happen when several sequences are obtained from the same sample. In this case, if there are d new sequences that join the genealogy at a single instant of time, we set $d-1$ of the δ_rs to 0. It follows that the standard coalescent, with n isochronously sampled sequences, is simply a special case of the s-coalescent because, although $m = 2n - 2$, the first $n - 1$ values of δ_r equal 0, leaving $n - 1$ non-zero δ_rs in equation (2.6).

There is a third difference between the standard coalescent and s-coalescent that our use of m points to: with the standard coalescent, the number of lineages decreases monotonically as time advances into the past. This is not so with the s-coalescent; instead, the number of lineages (i.e. the k_rs inside the exponential) can increase as new sequences join the genealogy. Whereas the fact that new sequences can join the genealogy at different times does not have profound effects on the mathematics of the coalescent, it has significant effects on our ability to make inferences with real data. Our ability to infer evolutionary and demographic parameters—population size, migration rates, recombination rates—are contingent on the the number of lineages that span each interval along the the coalescent. The smaller the number of lineages included in a given interval, the greater the variance of our estimate of the length of that interval, and consequently, the variances of any parameter estimates that may be unique to that interval. It is particularly difficult, therefore, to infer changes to these parameters over time because, with isochronous genealogies, the number of lineages decreases from n for the first coalescent interval, to 2 for the final coalescent interval. With serial samples, on the other hand, there is the opportunity to add lineages by incorporating historically derived sequences. This means that over the length of the genealogy, there can be high enough numbers of lineages and coalescent intervals, each providing an independent estimate of demographic parameters, so that our estimates are sufficiently reliable.

There is another interesting difference between the standard coalescent and the s-coalescent: with isochronous data, increasing the number of sequences sampled does not necessarily reduce the variance of our estimates, because under a standard coalescent process, most of these sequences will tend to join the genealogy towards the tips of the tree. In contrast, with serial samples, an investigator may be able to force sequences to join the tree at any stage he/she chooses. Consequently, with a judicious choice of sampling times—say, every N generations—an investigator can ensure that there is enough information across the tree to make reasonably efficient estimates of demographic parameters.

2.5 Estimating population size and substitution rates under the s-coalescent

In the above sections we have discussed reconstructing genealogies, estimating evolutionary rates and the s-coalescent. In this section we will discuss the simultaneous estimation of substitution rate, population size, and genealogy using the s-coalescent. By estimating all parameters simultaneously rather than individually, we take their interdependence into account properly. Further we wish to take into account the uncertainty that is present in the estimation of these parameters. This uncertainty comes from two sources: (1) the uncertainty that is inherent in our estimation of the underlying genealogy using molecular sequences of finite length, and (2) the uncertainty that is engendered by the fact that our sample of sequences, and the attendant genealogy, is just one stochastic realization of the coalescent process. It is also frequently the case that what is of interest is not the genealogy *per se*, but the historical processes that have acted on the population. The genealogy is therefore a 'nuisance' parameter. The approach that we have used, and which has become popular in recent years, is a Bayesian one, in which we estimate the joint and marginal posterior probability distributions of our parameters of interest, as a scaled proportion of their likelihood, and their prior probabilities [6]:

$$\Pr(\mu, \theta, \boldsymbol{G}|D) = z \Pr(D|\boldsymbol{G}, \mu)\, f(\boldsymbol{G}|\theta) \Pr(\mu, \theta)\,. \tag{2.7}$$

Here, D is the data, in this case the DNA sequences and sampling times at the tips of the genealogy, $\Pr(\mu, \theta)$ are the prior densities that quantify the uncertainty and our beliefs about the parameters in our model, and z is an unknown normalization constant. There is no general analytic solution for equation (2.7). Fortunately, a computational solution for difficult Bayesian problems has been well-characterized, and we may use Metropolis–Hastings Markov chain Monte Carlo to construct a distribution of the desired posterior probability [19, 24, 36].

Metropolis–Hastings Markov chain Monte Carlo (MHMCMC, or MCMC, for convenience), gives us a method to sample the joint posterior distribution without evaluating the normalization constant z [24, 36]. As the name suggests, an MCMC procedure generates a chain of parameter values, obtaining successive value(s) of one or more parameters by perturbing the present value(s) assigned to these parameters. The current parameters are altered in some random way to produce a proposed set of new parameter values. Then, with some well-defined probability, we either accept the new parameter values or discard them and keep the original parameter values for the next step in the chain. The chain must be able to sample all possible combinations of parameter values so it must be possible to move to any part of the parameter state space from any other part, not necessarily in a single step, but at least in a series of steps. In this chapter, we are not going to discuss the technical details of MCMC, nor are we going to discuss the problems of MCMC (e.g. problems associated with mixing, and non-stationarity of the chain), and potential solutions to these problems. This has already been covered in considerable detail elsewhere (see Chapter 1, and [6, 10, 19, 20, 21, 24, 33, 36, 63]), and readers are directed to these papers for a complete discussion of MCMC and its specific use in coalescent-based Bayesian inference. We do, however, want to comment briefly on the types of moves that we use in our s-coalescent-based Bayesian-MCMC analyses.

The state representation for our MCMC chain is $\psi = (\boldsymbol{G}, \theta, \mu)$. The genealogies $\boldsymbol{G}$ consist of edges and nodes together with node heights (i.e. the root-to-node distances). At each step the state is perturbed. We use the same types of moves for continuous-valued parameters—μ, or θ, for instance—as are routinely applied in other MCMC analyses. For example, a new value for $\theta' = u\theta$ may be generated with a random number u drawn from a suitable proposal distribution, usually uniformly on the interval $(\beta, 1/\beta)$ for $\beta > 1$. With coalescent-based MCMC, however, we also need moves that permit genealogies to change. One particularly effective move is the Wilson–Balding (WB) move [61] (as modified in ref. [6]) which is similar to Subtree Pruning and Regrafting (SPR), but tailored explicitly for the coalescent. With the WB-move, as with SPR, a random subtree is pruned from a genealogy, but the root-to-node distances of coalescent nodes (and leaf nodes, in the case of heterochronous data) on the pruned subtree and the residual genealogy are held constant. The pruned subtree is then regrafted onto *any* edge of the residual genealogy. When this happens, it is possible for the subtree to reattach to a node that is closer to the tips of the genealogy than the most distant coalescent node on the subtree, i.e. the subtree reattaches to a node which has a height greater than the minimum node-height on the subtree. This tree is illegal, and is rejected. When the WB-move results in legally regrafted trees, the standard MCMC acceptance ratio is used to accept or reject the state. The WB-move is particularly useful with heterochronous genealogies, because there is no need to constrain topology moves to respect the chronological sequence with which samples enter the genealogy—if a move results in an illegal tree, as when sequences sampled closer to the root are grafted on to edges closer to the tips, then it is simply rejected.

MCMC results in a chain of states, each of which varies slightly from the previous state $\{\psi, \psi', \ldots\} = \{(\boldsymbol{G}, \theta, \mu), (\boldsymbol{G}', \theta', \mu'), \ldots\}$. From this chain, we sample periodically, ideally choosing an optimal sampling frequency—one that delivers enough parameter estimates to construct meaningful distributions of posterior probabilities while at the same time maintaining as high a level of independence between successive samples as is practical.

In Fig. 2.3, we plot the marginal posterior distributions of substitution rate and θ, obtained from a MCMC analysis of a sample of 28 HIV-1 partial *env* sequences from two timepoints, seven months apart (with 15 sequences and 13 sequences, from the most recent and earlier timepoints, respectively). A coalescent model with population growth was applied (population growth rate was also estimated, but the recovered marginal distribution is not shown here). Uniform prior distributions on substitution rates, population size and population growth rates were used. The MCMC chain was run for two million generations and sampled every 500 generations. The results show clear modes for both substitution rate (0.000056 substitutions per site per day, or 2% per year), and θ (approximately 3500). In fact, these relatively well-defined marginal posterior distributions are not atypical of the types of results we obtain with serially sampled data.

In any Bayesian analysis, there is considerable focus on the appropriate choice of priors and, indeed, choosing priors for a particular analysis is not straightforward. Poorly specified priors can result in improper posterior distributions that cannot be normalized. Prior selection is far too vast a topic for proper treatment here, and readers are directed to [19] for a good introduction. We use priors to

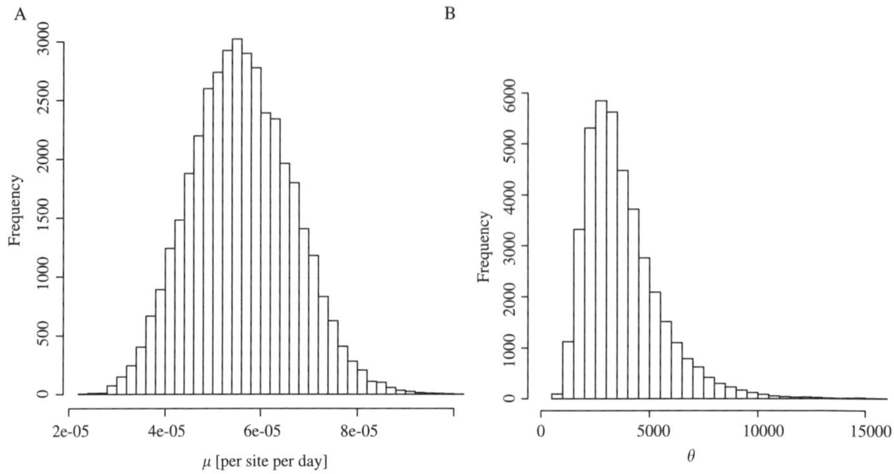

FIG. 2.3. Marginal posterior distributions of (A) substitution rate and (B) θ of serially sampled HIV-1 partial *env* sequences (see text for details).

specify our uncertainty about the values that parameters can take, and also to specify parts of the space of possible values where we are reasonably certain our parameters are unlikely to lie. In fact, there are usually reasonable bounds that one can impose on parameter space. In the case of inferences involving the coalescent, for instance, we know that the population size will be larger than zero but not infinitely large. We also have a fair idea that substitution rate is unlikely to be so large as to obliterate any phylogenetic information in the sequences. For both substitution rate and population size, we can define bounded intervals over which values of these parameters may vary. Bounded intervals are useful, because they ensure that the integral of the posterior density over the joint parameter space is finite (note that it is possible to have finite posterior density integrals with unbounded priors as well, but this is not generally true).

2.5.1 *Changing population sizes and skyline plots*

As Chapter 1 demonstrates, allowing the population size to vary as a function of time is reasonably easy to do in the coalescent. But what happens when we do not have an explicit model of change for population size? Here we describe an exploratory approach developed by Pybus *et al.* [44], and subsequently extended to serial samples by Drummond *et al.* [8].

In its simplest form, a skyline analysis takes a genealogy that has been constructed using standard phylogenetic methods, under the assumption of a molecular clock. Coalescent intervals are, therefore, known and are specified in substitution units. Consider the interval, t, between two coalescent events on a genealogy of contemporaneous sequences. One can obtain a simple estimate of θ

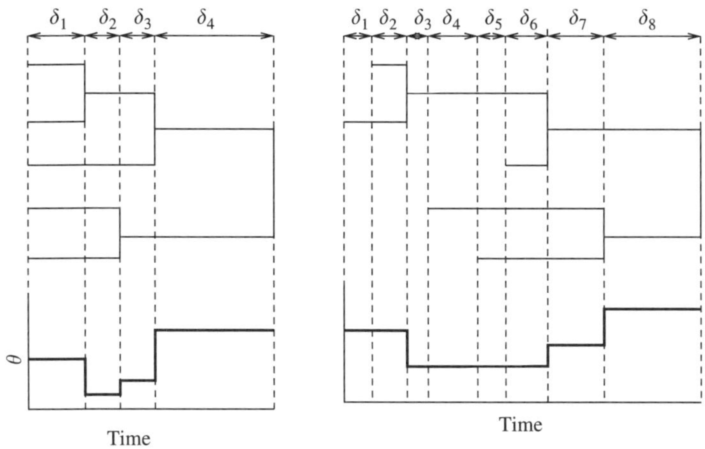

FIG. 2.4. Skyline plots for isochronous (left) and heterochronous genealogies. Note that with the heterochronous genealogy, the second coalescent interval from the left consists of several sub-intervals where new sequences enter the genealogy.

by finding the value that maximizes the conditional density

$$f(t|\theta) = \frac{1}{\theta}\exp\left(-\frac{k_r(k_r-1)t}{2\theta}\right). \tag{2.8}$$

In fact, the ML estimate that maximizes equation (2.8) is $\hat{\theta} = k_r(k_r-1)t/2$, and as one moves across all $n-1$ intervals of a (non-serial) genealogy, we obtain different values of θ for each interval. When these are plotted against the genealogy, we obtain a plot that resembles the skyline of a city (Fig. 2.4).

With serial samples, the intervals between certain pairs of adjacent coalescent events are interrupted by the addition of new leaves to the genealogy. If there are s such additions, then a single coalescent interval is partitioned into $s+1$ sub-intervals (Fig. 2.4). The probability density over the entire coalescent interval a is

$$f'(t_\delta|\theta_a) = \tfrac{1}{\theta_a}\exp\left(-\sum_{r=1}^{s+1}\tfrac{k_r(k_r-1)\delta_r}{2\theta_a}\right),$$

where $t_\delta = \sum_r \delta_r$. We can estimate a maximum-likelihood (ML) estimate of θ_a for just this interval. The ML estimator for θ_a is

$$\hat{\theta}_a = \tfrac{\sum_{r=1}^{s+1} k_r(k_r-1)\delta_r}{2}$$

and reduces to that given in [56] with isochronous data (i.e., when $s=0$). Repeating this over the whole genealogy gives a vector $\boldsymbol{\theta} = \{\theta_1, \ldots, \theta_{n-1}\}$ of estimates for all coalescent intervals. If it is assumed that the estimated values of θ_a are valid for the time interval of the corresponding coalescent event, we can plot the estimates of θ in the same way that we do with isochronous genealogies (Fig. 2.4).

Standard skyline plots are typically based on an *a priori* specified genealogy (fixed with respect to topology and branch-lengths) [44, 56], and fail to take account of the uncertainties in the genealogy and the times of coalescent events. Drummond *et al.* [8] have developed a Bayesian-MCMC skyline-plot analysis that incorporates uncertainties in topologies and interval lengths. The resulting plots are visually more appealing, and appear as smoothed population-size trajectories. Nonetheless, it is still important to realise that at the heart of this Bayesian-MCMC analysis is a stepwise model of population size change.

2.6 Estimating migration rates

The coalescent with a subdivided population (i.e. the structured coalescent) is a simple extension of the simple coalescent (see Chapter 1). When population subdivision is incorporated in the model, two types of events occur: coalescent events and migration events. Consider two distinct Fisher–Wright subpopulations A and B (Fig. 2.5). In each generation, each individual in one subpopulation has some probability of migrating to the other subpopulation. Since it is customary to treat migration as a stochastic process, we may model the intervals between migration events as exponentially-distributed waiting times. But these intervals between migration events add to the times between coalescent events, because lineages

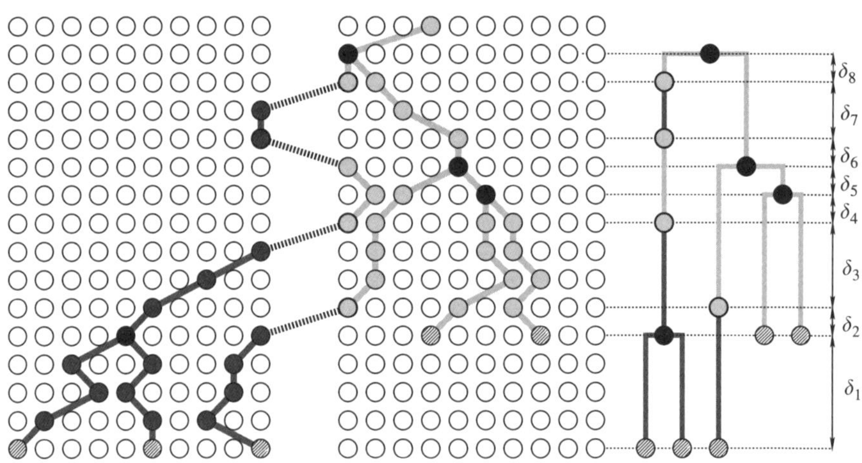

FIG. 2.5. A simplified view of Fisher–Wright subpopulations with migration. Migration events, shown as dashed lines between subpopulations, are explicitly placed on the genealogy (right), as bold circles. The δs signify intervals between migration nodes, coalescent nodes, and leaf nodes.

must be in the same deme or subpopulation to coalesce (Fig. 2.5). Consequently, the times between coalescent events tend to be longer than those obtained under a panmictic population model.

Extending the serial coalescent to include migration is relatively easy, as we now show. The island model of migration is a model of p subpopulations, or *demes*. For $j \in \mathcal{D}$, $\mathcal{D} = \{1, 2 \ldots p\}$, deme j is a panmictic population of N_j haploid individuals. Let λ_{ij} denote the *per capita* migration rate from deme i to j (time increases into the past, so in forward time the individual is moving from j to i).

A migration-coalescent genealogy $\boldsymbol{G_m}$ is a genealogy with explicit migration events as nodes. An example is given in Fig. 2.5. We can follow a lineage from the present backwards in time and we will encounter migration nodes as well as coalescent nodes. Every edge on the genealogy has a associated 'colour' or deme label, and thus every migration event represents a migration from deme i to deme j. Each coalescent event takes place within a particular deme. More formally, in the migration-genealogy there are n leaf nodes (with label set $\mathcal{L}$), $n-1$ coalescent nodes (label set $\mathcal{C}$), plus an indeterminate number, m, of migration nodes (label set $\mathcal{M}$). Let $\mathcal{A} = \mathcal{C} \cup \mathcal{M}$ denote the set of all ancestral (i.e. non-leaf) node labels and $V = \mathcal{L} \cup \mathcal{A}$ denote the set of all node labels. We let R be the root node label and V_{-R} be the set of all node labels excluding the root.

The demographic process realizing migration-coalescent genealogies is defined as follows. An ancestral lineage is associated with each sampled individual and carries a label indicating deme membership. As time increases into the past, each lineage in deme i migrates independently of all other lineages at rate λ_{ij} to deme j. Each pair of lineages in deme i coalesces at instantaneous rate $1/\theta_i$ where $\theta_i = N_i t_g$. Note that with our formulation, we allow asymmetric migration rates, and a different subpopulation size for each deme. A visual representation can be seen from Fig. 2.5; here the leaf deme membership is shown with either a grey line for one deme or a black line for the other deme. Migration nodes (events) are indicated when the lineage changes deme (line type).

We now write down the joint density $f(\boldsymbol{G_m}|\boldsymbol{\theta}, \boldsymbol{\lambda})$ for a migration-genealogy. As before the interval of time $\delta_r = t_{r+1} - t_r$ between consecutive nodes or events on the tree, and there are $m + 2n - 2$ such intervals on a tree ($m + 2n - 2$ decomposes into $n-1$ intervals terminating in leaf nodes, $n-1$ intervals terminating in coalescent nodes, and m intervals terminating in migration nodes; Fig. 2.5). Let k_{ir} denote the number of lineages in deme i in interval r. For $i \in \mathcal{D}$, let $\mathcal{D}_{-i}$ denote the set of demes omitting deme i. Every interval $(t_r, t_{r+1}]$ contributes a factor $\exp\left(-\sum_{i\in\mathcal{D}}\left[\frac{k_{ir}(k_{ir}-1)}{2\theta_i} + k_{ir}\sum_{j\in\mathcal{D}_{-i}}\lambda_{ij}\right]\delta_r\right)$ to the density, multiplied by a factor equal either to $1/\theta_i$, or λ_{ij}, or 1, depending on whether the event type at time t_{r+1} is a coalescent in deme i, or a $(i \to j)$-migration, or a leaf node, respectively (see also Chapter 1 equation (1.9)). Let m_{ij} denote the total number of $(i \to j)$ migrations and c_i denote the total number of coalescent events in deme i. The overall density is

$$f_m(\boldsymbol{G_m}|\boldsymbol{\theta},\boldsymbol{\lambda}) =$$

$$\prod_{i\in\mathcal{D}}\frac{1}{\theta_i^{c_i}}\prod_{j\in\mathcal{D}_{-i}}\lambda_{ij}^{m_{ij}}\exp\left(-\sum_{r\in V_{-R}}\sum_{i\in\mathcal{D}}\left[\frac{k_{ir}(k_{ir}-1)}{2\theta_i}+k_{ir}\sum_{j\in\mathcal{D}_{-i}}\lambda_{ij}\right]\delta_r\right). \quad (2.9)$$

With the isochronous migration-coalescent, the set V_{-R} that indexes the first summation of equation (2.9) is replaced by $\mathcal{A}_{-R}$.

In Section 2.5.1 we showed how skyline plots can be used to explore changes in population size over time. In fact, it is possible—and indeed, it is one of the strengths of serial sample methods—to formally model changes in population demographic structure over time. It is relatively straightforward to extend the coalescent to permit a pre-defined number of intervals and interval boundaries (= 'change-points'), over which demographic models change in some abrupt manner. Changes can be modelled for any set of parameters, including migration rates. It is also possible to model changes to the number of demes over the entire genealogy [11].

We will not discuss the technicalities of these analyses here and readers are directed to Ewing and Rodrigo [11] for details. Nonetheless, it is worthwhile spending a little time thinking about the uses to which such analyses may be applied. With HIV-1, for instance, it is known that as disease progresses, barriers between systemic compartments in the host (e.g. the blood–brain barrier) may become more permeable [35], so that there is a change in migration rates between these compartments over time. In other instances, we may want to allow changes to the number of demes. Colonization of new geographical areas adds to the number of populated demes over time. Similarly, glaciation may disrupt a continuous population for a period of time, before permitting the restoration of contact. In both of these instances, we can explicitly model the changes to the numbers of demes and, if unknown, estimate the times when these events occurred.

Of course, there is nothing in the theory of the standard coalescent that prohibits its use in modelling changes to population demographies. The difficulty, as we have noted before, is that as one moves back in time, the number of lineages diminishes quickly, and it becomes much harder to obtain good estimates of population parameters. Again, with the serial coalescent, the addition of new sequences (chosen appropriately, of course), improves estimation considerably.

2.7 Where to next?

In the introduction, we stated that a Measurably Evolving Population was a population in which a statistically significant accumulation of substitutions are detectable over a period of time. Clearly this definition is an operational one, that is, it defines a type of population on the basis of experimental design, the length of time between samples, the amount of sequence information we are able to collect, and the statistical properties of our estimators. In fact, as we also noted above, many populations that continue to accumulate substitutions will not be classified as MEPs because substitutions accumulate too slowly and/or there are no ancient samples from which sequences may be obtained. But it may also be that we are not able to statistically detect any accumulation of substitutions in serial samples from these populations.

An obvious question arises: why not treat all sequences collected at different times as non-isochronous data, and apply serial sample methods routinely? Why only use serial sample methods when we have rejected the zero-substitution rate hypothesis? It appears that when samples are too close in time, and very little substitution has occurred, there is some probability of inferring a higher-than-expected substitution rate just because some branches of later samples are longer by chance alone. This affects not only the estimate of μ, but may bias downstream analyses as well. We have seen the effects of this with sUPGMA: Drummond and Rodrigo [4] performed simulations to test the efficiency of sUPGMA over UPGMA, and found that at very low substitution rates, UPGMA outperformed sUPGMA at reconstructing the true topology. The problem can be alleviated if the number of sequences at each sampling time is increased. This makes intuitive sense, because increasing the number of sequences per time-point decreases the likelihood of detecting elevated substitution rates by chance alone.

The development of methods to analyse Measurably Evolving Populations has progressed steadily in the last few years. The foundations have been laid and what remains to be done is reasonably obvious. For one thing, we have not discussed selection and recombination, and how these evolutionary processes may be modelled with serially sampled sequences. We have not done so because these methods have not yet been published. Nonetheless, it is only a matter of time before these processes are included in models for MEPs – they already exist for isochronous data [32, 33, 37, 39, 40], and as we have seen, extensions to serially sampled data are typically straightforward.

Perhaps the Holy Grail of any evolutionary genetic analysis is to allow all processes that may influence the genetic diversity of a population to be modelled simultaneously. In this respect, we would want to build a unified model that would allow us to jointly estimate population size, rates of historical growth or decline in population numbers, migration and recombination rates, and the intensity of selection. Once again, we are not there yet, either with isochronous data or serially sampled data. It is almost certain, however, that when such models are built, a great deal of data will be required to disentangle the effects of each of these processes.

How much data do we need? How do we apportion the data we collect between sampling times and subpopulations? How efficient are our estimation procedures? These questions remain unanswered at present, and this is another area where work is needed. With MEPs, we need more studies on optimal experimental designs, particularly as our models become more complex. The strength of MEP analysis—its ability to detect changes in the evolutionary processes that shape the genetic diversity of a population—is also its liability, because it adds another level of complexity to our analyses.

The flexibility of Bayesian approaches provides a ready means to test the power of our estimation procedures. They also provide an avenue to determine which models fit our data best. Model averaging – where the model is a parameter that can take different 'values' within a Bayesian MCMC analysis – is an attractive possibility because it frees us from having to decide *a priori* which is the best demographic or evolutionary model to apply. We can envisage a model averaging procedure applied to population subdivision, for instance, when the number of demes is unknown. However there is the added non-trivial task of assigning priors to models. For a small and finite set of models, we may choose to set a uniform prior on each of our models, but this may not work when the model space is large.

The road ahead is easily visible, although there are likely to be potholes and pitfalls. Additionally, we are constantly forced to confront the challenges that real data present. There is no better way to foil a good model than with data. For now, therefore, our models for MEPs are simply stepping stones to reality.

Acknowledgements

Our methodological research on MEPs has been greatly aided by interactions with a number of people: Joe Felsenstein, Jim Mullins and members of his lab, Geoff Nicholls, Andrew Rambaut, Oliver Pybus, Roald Forsberg, Matthew Goode, and Wiremu Solomon. We also thank Joe Felsenstein, another anonymous reviewer, and Olivier Gascuel for comments that helped us improve this chapter considerably. This research has been supported by grants from the Allan Wilson Centre for Molecular Ecology and Evolution, the US Public Health Service, and the New Zealand Government. We would also like to thank Jayne Ewing for assistance in manuscript preparation.

References

[1] Cooper, A., Mourer-Chauvire, C., Chambers, G. K., Von Haeseler, A., Wilson, A. C., and Paabo, S. (1992). Independent origins of New Zealand moas and kiwis. *Proceedings of the National Academy of Sciences, USA*, **89**(18), 8741–8744.

[2] DeSalle, R., Barcia, M., and Wray, C. (1993). PCR jumping in clones of 30-million-year-old DNA fragments from amber preserved termites (*Mastotermes electrodominicus*). *Experientia*, **49**(10), 906–909.

[3] Drummond, A., Forsberg, R., and Rodrigo, A. G. (2001). The inference of stepwise changes in substitution rates using serial sequence samples. *Molecular Biology and Evolution*, **18**(7), 1365–1371.

[4] Drummond, A. and Rodrigo, A. (2000). Reconstruction genealogies of serial samples under the assumption of a molecular clock using serial-sample UPGMA. *Molecular Biology and Evolution*, **17**(12), 1807–1815.

[5] Drummond, A. J., Ho, S. Y. W., Phillips, M. J., and Rambaut, A. (2006). Relaxed phylogenetics and dating with confidence. *PLoS Biology*, **4**(5), e88.

[6] Drummond, A. J., Nicholls, G. K., Rodrigo, A. G., and Solomon, W. (2002). Estimating mutation parameters, population history and genealogy simultaneously from temporally spaced sequence data. *Genetics*, **161**, 1307–1320.

[7] Drummond, A. J., Pybus, O. G., Rambaut, A., Forsberg, R., and Rodrigo, A. G. (2003). Measurably evolving populations. *Trends in Ecology and Evolution*, **18**(9), 481–488.

[8] Drummond, A. J., Rambaut, A., Shapiro, B., and Pybus, O. G. (2005). Bayesian coalescent inference of past population dynamics from molecular sequences. *Molecular Biology and Evolution*, **22**(5), 1185–1192.

[9] Efron, B., Halloran, E., and Holmes, S. (1996). Bootstrap confidence levels for phylogenetic trees. *Proceedings of the National Academy of Sciences, USA*, **93**, 7085–7090.

[10] Ewing, G. B., Nicholls, G. K., and Rodrigo, A. G. (2004). Using temporally spaced sequences to simultaneously estimate migration rates, mutation rate and population sizes in measurably evolving populations. *Genetics*, **168**, 2407–2420.

[11] Ewing, G. B. and Rodrigo, A. G. (2006). Coalescent-based estimation of population parameters when the number of demes changes over time. *Molecular Biology and Evolution*, **23**, 988–996.

[12] Felsenstein, J. (1981). Evolutionary trees from DNA sequences: a maximum likelihood approach. *Journal of Molecular Evolution*, **17**, 368–376.

[13] Felsenstein, J. (1985). Confidence limits on phylogenies: an approach using the bootstrap. *Evolution*, **39**, 783–791.

[14] Felsenstein, J. (2003). *Inferring Phylogenies*, Chapter 26. Coalescent trees. Sunderland, Sinauer Associates.

[15] Fu, Y. X. (2001). Estimating mutation rate and generation time from longitudinal samples of DNA sequences. *Molecular Biology and Evolution*, **18**(4), 620–626.

[16] Gibbons, A. (2005). Ancient DNA - new methods yield mammoth samples. *Science*, **310**(5756), 1889–1889.

[17] Goldman, N. (1993). Simple diagnostic statistical tests of models for DNA substitution. *Journal of Molecular Evolution*, **37**(6), 950–661.

[18] Golenberg, E. M. (1991). Amplification and analysis of miocene plant fossil DNA. *Philosophical Transactions of the Royal Society, London, Series B*, **333**(1268), 419–427.

[19] Green, P. J. (1995). Reversible jump Markov Chain Monte Carlo computation and Bayesian model determination. *Biometrika*, **82**, 711–732.

[20] Green, P. J. (2003). *Highly Structured Stochastic Systems*, Chapter Trans-dimensional Markov chain Monte Carlo. Oxford University Press, Oxford.

[21] Griffiths, R. C. and Tavare, S. (1994). Ancestral inference in population genetics. *Statistical Science*, **9**, 307–319.

[22] Gunthard, H. F., Leigh-Brown, A. J., D'Aquila, R. T., Johnson, V. A., Kuritzkes, D. R., Richman, D. D., and Wong, J. K. (1999). Higher selection pressure from antiretroviral drugs in vivo results in increased evolutionary distance in HIV-1 pol. *Virology*, **259**(1), 154–165.

[23] Hall, P. and Martin, M. A. (1988). On bootstrap resampling and iteration. *Biometrika*, **75**, 661–671.

[24] Hastings, W. K. (1970). Monte Carlo sampling methods using Markov chains and their applications. *Biometrika*, **57**, 97–109.

[25] Huelsenbeck, J. P., Hillis, D. M., and Jones, R. (1995). Parametric bootstrapping in molecular phylogenetics: Applications and perfomance. In *Molecular Zoology: Strategies and Protocols* (ed. J. Ferraris and S. Palumbi). Wiley, New York.

[26] Huelsenbeck, J. P. and Ronquist, F. (2001). MrBayes. *Bioinformatics*, **17**, 754–755.

[27] Huelsenbeck, J. P., Ronquist, F., Nielsen, R., and Bollback, J. P. (2000). A compound possion process for relaxing the molecular clock. *Genetics*, **154**, 1879–1862.

[28] Jenkins, G. M., Rambaut, A., Pybus, O. G., and Holmes, E. C. (2002). Rates of molecular evolution in RNA viruses: a quantitative phylogenetic analysis. *Journal of Molecular Evolution*, **54**, 156–165.

[29] Kingman, J. F. C. (1982). The coalescent. *Stochastic Processes and their Applications*, **13**, 235–248.

[30] Kingman, J. F. C. (1982). On the genealogy of large populations. *Journal of Applied Probability*, **19A**, 27–43.

[31] Krings, M., Stone, A., Schmitz, R. W., Krainitzki, H., Stoneking, M., and Paabo, S. (1997). Neandertal DNA sequences and the origin of modern humans. *Cell*, **90**(1), 19–30.

[32] Krone, S. M. and Neuhauser, C. (1997). Ancestral processes with selection. *Theoretical Population Biology*, **51**(3), 210–237.

[33] Kuhner, M. K., Yamato, J., and Felsenstein, J. (2000). Maximum likelihood estimation of recombination rates from population data. *Genetics*, **156**, 1393–1401.

[34] Lambert, D. M., Ritchie, P. A., Millar, C. D., Holland, B., Drummond, A. J., and Baroni, C. (2002). Rates of evolution in ancient DNA from Adelie penguins. *Science*, **295**, 2270–2273.

[35] Langford, T. D., Letendre, S. L., Larrea, G. J., and Masliah, E. (2003). Changing patterns in the neuropathogenesis of HIV during the HAART era. *Brain Pathology*, **13**(2), 195–210.

[36] Metropolis, N., Rosenbluth, A., Rosenbluth, M., Teller, A., and Teller, E. (1953). Equations of state calculations by fast computing machines. *Journal of Chemical Physics*, **21**, 1087–1091.

[37] Neuhauser, C. and Krone, S. M. (1997). The genealogy of samples in models with selection. *Genetics*, **145**, 519–534.

[38] Nickle, D. C., Jensen, M. A., Shriner, D., Brodie, S. J., Frenkel, L. M., Mittler, J. E., and Mullins, J. I. (2003). Evolutionary indicators of human immunodeficiency virus type 1 reservoirs and compartments. *Journal of Virology*, **77**, 5540–5546.

[39] Nielsen, R. and Yang, Z. H. (1998). Likelihood models for detecting positively selected amino acid sites and applications to the HIV-1 envelope gene. *Genetics*, **148**(3), 929–936.

[40] Nielsen, R. and Yang, Z. H. (2003). Estimating the distribution of selection coefficients from phylogenetic data with applications to mitochondrial and viral DNA. *Molecular Biology and Evolution*, **20**(8), 1231–1239.

[41] Ochman, H. and Wilson, A. C. (1987). Evolution in bacteria: evidence for a universal substitution rate in cellular genomes. *Journal of Molecular Evolution*, **26**, 74–86.

[42] Ota, R., Waddell, P. J., Hasegawa, M., Shimodaira, H., and Kishino, H. (2000). Appropriate likelihood ratio tests and marginal distributions for evolutionary tree models with constraints on parameters. *Molecular Biology and Evolution*, **17**, 798–803.

[43] Poss, M., Rodrigo, A. G., Gosink, J. J., Learn, G. H., de Vange, P. D., Martin, H. L., Bwayo, J., Kreiss, J. K., and Overbaugh, J. (1998). Evolution of envelope sequences from the genital tract and peripheral blood of women infected with clade A human immunodeficiency virus type 1. *Journal of Virology*, **72**(10), 8240–8251.

[44] Pybus, O. G., Rambaut, A., and Harvey, P. H. (2000). An integrated framework for the inference of viral population history from reconstructed genealogies. *Genetics*, **155**, 1429–1437.

[45] Rambaut, A. (2000). Estimating the rate of molecular evolution: incorporating non-contemporaneous sequences into maximum likelihood phylogenies. *Bioinformatics*, **16**(4), 395–399.

[46] Rodrigo, A. G., Borges, K. M., and Bergquist, P. L. (1994). Pulsed-field gel electrophoresis of genomic digests of thermus strains and its implications for taxonomic and evolutionary studies. *International Journal of Systematic Bacteriology*, **44**, 547–552.

[47] Rodrigo, A. G. and Felsenstein, J. (1999). Coalescent approaches to HIV-1 population genetics. In *The Evolution of HIV* (ed. K. A. Crandall), pp. 233–272. Johns Hopkins University Press, Baltimore.

[48] Rodrigo, A. G., Goode, M., Forsberg, R., Ross, H., and Drummond, A. (2003). Inferring evolutionary rates using serially sampled sequences from several populations. *Molecular Biology and Evolution*, **20**, 2010–2018.

[49] Rohlf, F. J. (1962). *A Numerical Taxonomic Study of the Genus Aedes (Diptera: Culicidae) with Emphasis on the Congruence of Larval and Adult Classifications.* Ph. D. thesis, Department of Entomology, University of Kansas.

[50] Ross, H. A. and Rodrigo, A. G. (2002). Immune-mediated positive selection drives human immunodeficiency virus type 1 molecular variation and predicts disease duration. *Journal of Virology*, **76**(22), 11715–11720.

[51] Seo, T. K., Thorne, J. L., Hasegawa, M., and Kishino, H. (2002). A viral sampling design for testing the molecular clock and for estimating evolutionary rates and divergence times. *Bioinformatics*, **18**, 115–123.

[52] Shankarappa, R., Margolick, J. B., Gange, S. J., Rodrigo, A. G., Upchurch, D., Farzadegan, H., Gupta, P., Learn, C. R. Rinaldoand G. H., He, X, Huang, X.-L., and Mullins, J. I. (1999). Consistent viral evolutionary changes associated with the progression of HIV-1 infection. *Journal of Virology*, **78**, 10489–10502.

[53] Shapiro, B., Drummond, A. J., Rambaut, A., Wilson, M. C., Matheus, P. E., Sher, A. V., Pybus, O. G., Gilbert, M. T. P., Barnes, I., Binladen, J., Willerslev, E., Hansen, A. J., Baryshnikov, G. F., Burns, J. A., Davydov, S., Driver, J. C., Froese, D. G., Harington, C. R., Keddie, G., Kosintsev, P., Kunz, M. L., Martin, L. D., Stephenson, R. O., Storer, J., Tedford, R., Zimov, S., and Cooper, A. (2004). Rise and fall of the beringian steppe bison. *Science*, **306**, 1561–1565.

[54] Shriner, D., Shankarappa, R., Jensen, M. A., Nickle, D. C., Mittler, J. E., Margolick, J. B., and Mullins, J. I. (2004). Influence of random genetic drift on human immunodeficiency virus type I *env* evolution during chronic infection. *Genetics*, **166**(3), 1155–1164.

[55] Sneath, P. H. A. (1962). *Microbial Classifications*, Chapter The construction of taxonomic groups, pp. 289–332. Cambridge University Press, Cambridge.

[56] Strimmer, K. and Pybus, O. G. (2001). Exploring the demographic history of DNA sequences using the generalized skyline plot. *Molecular Biology and Evolution*, **18**(12), 2298–2305.

[57] Swofford, D. L. (1999). *PAUP*. Phylogenetic Analysis Using Parsimony (*And Other Methods)* Sinauer Associates, Sunderland.

[58] Thomas, W. K. and Paabo, S. (1993). DNA-sequences from old tissue remains. *Methods in Enzymology*, **224**, 406–419.

[59] Thorne, J. L. and Kishino, H. (2002). Divergence time and evolutionary rate estimation with multilocus data. *Systematic Biology*, **51**(5), 689–702.

[60] Thorne, J. L., Kishino, H., and Painter., I. S. (1998). Estimating the rate of evolution of the rate of molecular evolution. *Molecular Biology and Evolution*, **15**(12), 1647–1657.

[61] Wilson, I. J. and Balding, D. J. (1998). Genealogical inference from microsatellite data. *Genetics*, **150**, 499–510.

[62] Wong, J. K., Cignacio, C., Torriani, F., Havlir, D., Fitch, N. J., and Richman, D. D. (1997). In vivo compartmentalization of human immunodeficiency virus: evidence from the examination of pol sequences from autopsy tissues. *Journal of Virology*, **71**(3), 2059–2071.

[63] Yang, Z. (2005). Bayesian inference in molecular phylogenetics. In *Mathematics of Evolution and Phylogeny* (ed. O. Gascuel). Oxford University Press, Oxford.

II

MODELS OF SEQUENCE EVOLUTION

3

MODELLING THE VARIABILITY OF EVOLUTIONARY PROCESSES

Olivier Gascuel and Stephane Guindon

Abstract

The evolutionary processes that act at the molecular level are highly variable. For example, the substitution rates and the natural selection regimes vary extensively during the course of evolution and across sequence sites. This chapter describes the mathematical tools and concepts to describe and understand these variations. We show how the standard Markov models of sequence evolution are extended through mixture models to account for variability among sites, and how the mixture approach is further generalized by Markov-modulated Markov models (MMM) to incorporate variability among lineages. We illustrate these models using data sets from plants and human immunodeficiency virus type 1 (HIV-1). Both data sets are processed under the 3-component mixture codon-based model of Nielsen and Yang [62] and its MMM extension [28]. We show that these models allow us to get insight into important biological features such as positively selected sites at the surface of the envelope protein of HIV-1 and site-specific changes within selection regimes correlated to duplication events in plant genes.

3.1 Introduction

From a historical perspective, the first goal of statistical phylogenetics was to construct more accurate species phylogenies by comparing nucleotide or protein sequences. It is now quite clear that the most important advances brought by this research area do not only involve taxonomy. Indeed, statistical phylogenetics provides an adequate framework to improve our understanding of the evolutionary processes that act at the molecular level. The first probabilistic models of evolution assumed that these processes were the same across different regions of the sequences and/or at different stages of evolution. However, simple observation of nucleotide or amino acid sequences suggests a very different picture. For instance, some regions seem to evolve quickly while other barely change. It is also quite clear that different sequences accumulate substitutions at distinct rates.

This chapter introduces models that are well suited to test such hypotheses in a statistical framework. More specifically, we focus on modelling the heterogeneity of the molecular evolution processes. The remaining part of this section provides an overview of the different biological sources of variability. The mathematical tools that are used to account for distinct sources of heterogeneity are then described. We next present the models in action by analysing two reference data sets. We show how these models can be used to infer relevant features of molecular evolution.

3.1.1 *Among-site heterogeneity*

Structural and functional constraints vary across sites of a protein. For instance, the sickle-cell disease is caused by a single mutation at the sixth position in the haemoglobin chain. Hence, this position is likely to evolve slowly as most mutations occurring at this particular site have low probabilities of being transmitted to offspring. Less specifically, β-sheets, α-helices, and coils (the three main secondary structures in proteins) are not subject to identical structural constraints, and residues near the core of the molecule evolve under different processes than those exposed at the surface (e.g. Goldman *et al.* [23]).

As amino acid sequences derive from nucleotide sequences, the structural and functional constraints that shape protein evolution also affect substitution processes at the nucleotide level. The structure of the genetic code is also an essential source of heterogeneity which acts on coding DNA. Indeed, two thirds of the nucleotide changes at the third codon position do not modify the amino acid translated (synonymous changes), while the changes that occur at the second position systematically alter the amino acid (non-synonymous changes). Only a small proportion (10%) of changes at the first codon position are synonymous. Therefore, the three codon positions obviously evolve under distinct evolutionary constraints that are superimposed on the constraints already existing at the protein level.

To sum up, there is a negative correlation between the rate at which mutations occur and are fixed in the population (i.e. the substitution rate) and the strength of structural and/or functional constraints acting on the positions at which the mutations occur. A simple approach to model this evolutionary feature is to allow the substitution rates to vary across amino acid or nucleotide sites of the sequence. However, estimating a rate for each site of the alignment is not achievable (see Chapter 4 in this book). Thus, we assume that site rates are unknown but comply with a probabilistic distribution whose parameters or shape (in the non-parametric settings) are estimated from the data [22, 40, 59, 88, 91, 92]. The next section (3.2) describes the mathematical tools and assumptions that are used for this purpose.

The structure of the genetic code is also responsible for other evolutionary patterns which are more complex than 'simple' variations in rates across codon positions. Indeed, synonymous and non-synonymous mutations have variable probabilities of being fixed in the population, depending on the selective forces that act on the corresponding amino acids. Non-synonymous changes often modify the structure of the peptide and alter its function. In this case, natural selection gets rid of proteins that carry these changes. However, amino acid changes sometimes offer the protein the opportunity to get adapted to a changing environment, and such modifications may correspond to major adaptive events. Hence, identifying regions of a protein at which the ratio between the rates of non-synonymous and synonymous substitutions is larger than 1.0 provides valuable information about the underlying evolutionary forces. Section (3.2) describes codon-based models in the line initiated by Goldman and Yang [24] that aim at estimating this ratio (or ω ratio). We will see that this approach is highly relevant from a biological perspective (Section 3.4).

3.1.2 *Mixing among-site and time-dependent variability*

Pioneering work by Zuckerkandl and Pauling [99] and Sarich and Wilson [72] has shown that the rate at which substitutions accumulate in proteins is constant over long periods of time. This observation suggests that proteins could be used as molecular clocks. Hence, given a phylogenetic tree and a calibration point, it would be possible to date past evolutionary events. Unfortunately, this is not the only molecular clock. Rather there are several clocks that do not tick at a steady rate. Nowadays, the most accurate molecular dating methods more or less relax the molecular clock constraint and rely instead on statistical models that describe the variations of substitution rates across lineages (see [32, 35, 70, 71, 85] for instance). Such methods separate elapsed time along branches and substitution rates. In most cases, however, dating evolutionary events is not the first goal. Indeed, most phylogenetic methods do not aim at estimating substitution rates. Rather they estimate expected numbers of substitutions along each edge, i.e. the product of a substitution rate by a duration, and therefore produce non-ultrametric (i.e. non-clocklike) trees.

A common assumption is that the expected amount of substitutions that accumulated on a given branch is the same at every site of the alignment. The substitution rate on a given branch is therefore supposed to be constant across sites. However, biological evidence suggests that some sites evolve quickly in some lineages and slowly in other clades, while different patterns are observed at other sites. For instance, Lockhart *et al.* [53] exhibited such an evolutionary pattern among 16S rDNA and tufA sequences from nonphotosynthetic prokaryotes and oxygenic photosynthetic prokaryotes and eukaryotes. Gaucher *et al.* [21] clearly demonstrated the existence of a link between functional differences across lineages and site-specific variations of substitution rates in elongation factors Tu (EF-Tu) and 1α (EF-1α). Lopez *et al.* [54] showed that $\sim$95% of the variable positions in cytochrome b (a protein that is often used to decipher deep evolutionary events) are 'heterotachous', i.e. rate variations are distinct at different sites and different branches.

The substitution rate is not the only evolutionary parameter that displays complex patterns of variation in time and across sites. Indeed, several studies have shown that the selection regimes vary extensively across lineages [39, 57, 63, 69]. For instance, in a pioneering work Messier and Stewart [57] have shown that adaptive episodes (i.e. positive selection) during the evolution of primate lysozymes were most probably followed by episodes of negative selection. Hence, these observations combined with those presented in the previous section show the necessity to account for both the variability of processes across sites and across lineages in a unified statistical framework.

The next section describes suitable models for this purpose. Indeed, these models treat the changes of substitution rate or ω ratio as a random process. The rate at which these events occur is estimated from the data. We will explain the mathematical properties of these models and show how they are used to decipher relevant evolutionary features.

3.2 Mathematical tools and concepts

This section describes the basic tools and concepts to model the evolution of homologous sequences. We start with the basis and assumptions of the standard Markovian models, acting at the DNA, RNA, and protein levels. We then explain how these simple models can be used through mixture models to account for among-site variation of rates or evolutionary modes. Finally, we describe Markov-modulated Markov models (often called 'covarion-like' or 'heterotachous' models even though these terms do not properly describe their features and aims). These models provide a unified framework to account for both among-site and time-variability, and can be seen as natural extensions of the mixture-based approaches.

Figures 3.1 and 3.2 display the main features of the DNA and codon models that are discussed in this chapter. These figures also display the section numbers where these models are described and applied to illustrative data sets, which should help readers to navigate in the chapter. Protein models are not shown in these figures (their relationships are quite simple), but described in sections 3.2.2, 3.2.5, and 3.4.1.

3.2.1 *Markovian models of sequence evolution: the basis and assumptions*

Standard approaches apply to aligned sequences. Firstly, it is necessary to use a multiple alignment tool (e.g. CLUSTAL [84]) to extract homologous sites. The data set then comprises a set of sites, where each site contains the character (nucleotide, amino acid, codon) of each of the sequences at a given position. The alphabet (set of possible characters) is denoted as X, and we shall typically use x and y to denote characters. It is assumed that all sequence characters within a single site derive from a unique character in the ancestral sequence. Each site is then viewed as a statistical unit that contains information on the evolutionary process which led to the contemporary sequences. We shall typically use χ_i to denote the character of sequence χ at site i.

The first assumption is that sites evolve independently. Thus, the probability that sequence χ evolves to sequence χ' equals the product, over all sites i, of the probability that χ_i evolves to χ'_i. This simplifying assumption is almost essential for tractability (though it can be slightly relaxed, e.g. [18, 67, 93]). With this

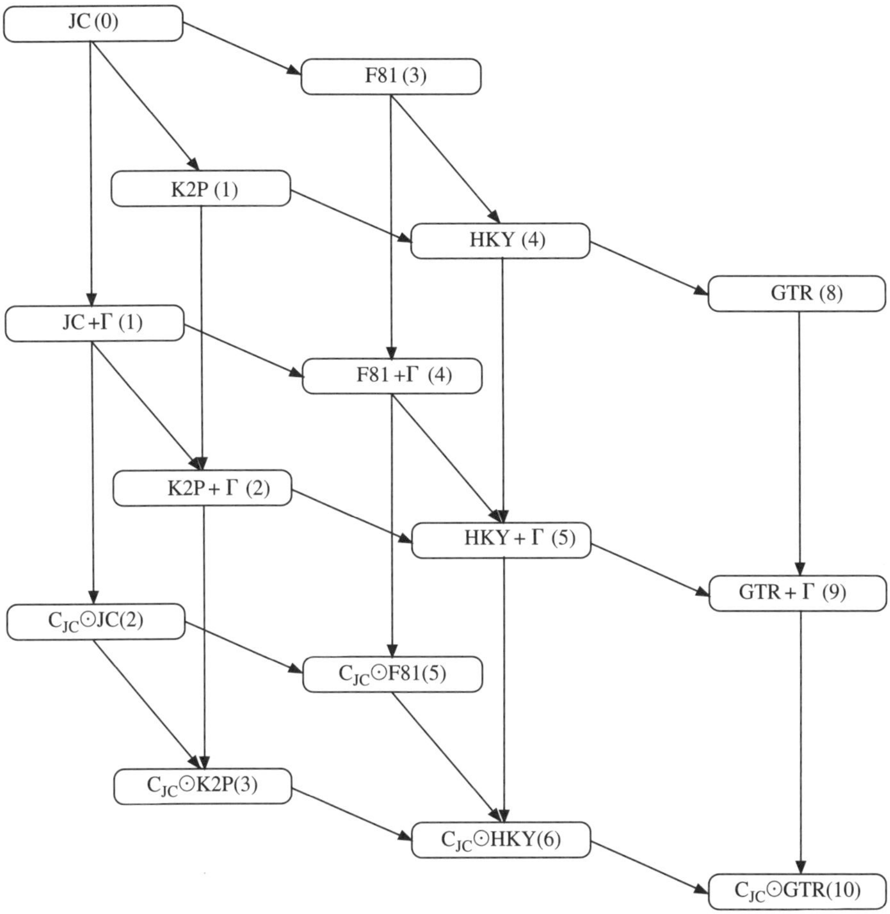

FIG. 3.1. **DNA models.** Arrows display the nested relationships. The parameter number of each model is indicated within parenthesis. Standard models (JC, K2P, F81, HKY and GTR) are described in section 3.2.2 and applied to illustrative data sets in section 3.4.1. Those simple models are extended using a gamma-based ($+\Gamma$) mixture approach to account for among-site variability of rates (section 3.2.5 and 3.4.1). In turn, the covarion-like approach of Galtier [19] extends gamma-based models to account for both among-site and time-variability of rates (section 3.2.7); changes of rate category are modelled thanks to a JC-like model ($\mathbf{C}_{\text{JC}}$) and the compound models incorporating both rate and nucleotide changes are denoted as $\mathbf{C}_{\text{JC}}\odot$M, where M is any of the standard nucleotide substitution models.

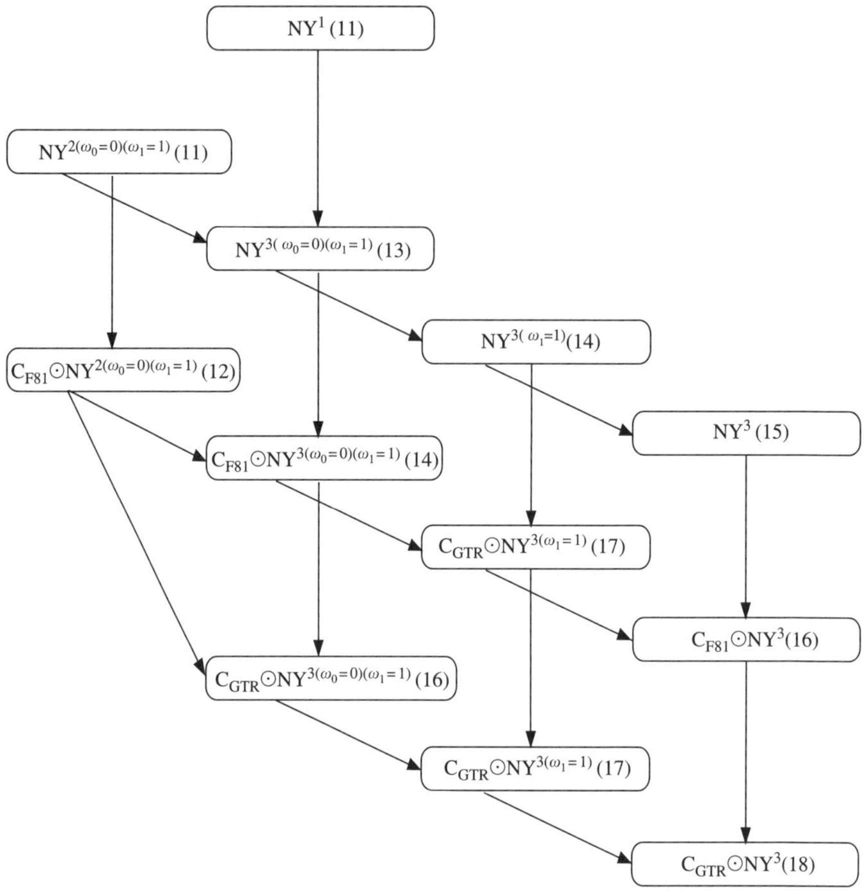

FIG. 3.2. **Codon models.** Arrows display the nested relationships. The parameter number of each model is indicated within parenthesis; 10 parameters are common to all models: 9 nucleotide frequencies (defining codon frequencies) and the transition/transversion ratio (κ). NY^1 belongs to the standard models we describe in section 3.2.2 and apply to data in 3.4.1. NY^1 is extended to NY^2 and NY^3 models to account for heterogeneity of selection regimes among sites, thanks to a mixture approach (sections 3.2.5, 3.4.1, and 3.4.2). Mixtures are in turn extended via Markov-modulated Markov models (denoted as $\mathbf{C}_X \odot NY^z$) to account for time-variability of selection regimes (sections 3.2.7, 3.4.3 and 3.4.4). Changes of selection regime are modelled using a F81-like model ($\mathbf{C}_{F81}$, equal rates of regime changes but unequal regime frequencies) or a GTR-like model ($\mathbf{C}_{GTR}$, unequal rates of regime changes and regime frequencies). Note that $\mathbf{C}_{F81}\odot NY^{2(\omega_0=0)(\omega_1=1)}$ and $\mathbf{C}_{GTR}\odot NY^{2(\omega_0=0)(\omega_1=1)}$ are identical as $\mathbf{C}_{F81}$ and $\mathbf{C}_{GTR}$ are identical when the number of states (selection regimes here) is equal to 2.

assumption made, we spend most of the chapter focusing on the evolution of an individual site.

The second assumption is about the Markovian nature of site evolution. We assume that evolution has no memory and is time-continuous, and we also commonly assume that it is time-homogeneous. Thus, any model can be characterized by a generator, or instantaneous rate matrix, which is denoted as $\mathbf{Q}$ and remains constant during evolution. The set of states corresponds to the characters in the studied sequences. Q_{xy} $(x \neq y)$ corresponds to the rate of substitutions from x to y, and the diagonal terms Q_{xx} are such that the row sums are all zero. Let $\mathbf{P}(t) = (P_{xy}(t))$ be the matrix of substitution probabilities, where $P_{xy}(t)$ is the probability of observing a substitution from x in one sequence to y in another sequence when the elapsed time separating both sequences is t. Note that multiple, hidden substitutions are possible and that $P_{xy}(t)$ sums over all possibilities $(1, 2, \ldots, \infty$ substitutions) and describes the final observable states in the two sequences at hand. The following relation holds:

$$\mathbf{P}(dt) = \mathbf{Q}dt + \mathbf{I},$$

where $\mathbf{I}$ is the identity matrix and dt represents an infinitesimal period of time. This equation basically states that the probability of changing from x to y $(x \neq y)$ in time dt is proportional to dt and to the corresponding coefficient in $\mathbf{Q}$. Furthermore, we have:

$$\mathbf{P}(t) = e^{\mathbf{Q}t}, \tag{3.1}$$

where the right term denotes the matrix exponential, which is computed via diagonalization of $\mathbf{Q}$ (see Bryant *et al.* [12] for more).

The third common assumption is that the evolutionary process is stationary. We can define the stationary distribution of the process, which is unique and corresponds to the *a priori* probability of the characters. This stationary distribution is denoted as $\mathbf{\Pi} = (\pi_x)$, where π_x is the *a priori* probability of character x. We have:

$$\mathbf{\Pi} = \lim_{t \to \infty} [\mathbf{\Pi}_{t=0}\mathbf{P}(t)],$$

where $\mathbf{\Pi}_{t=0}$ represents any starting distribution on X. This implies:

$$\mathbf{\Pi P}(t) = \mathbf{\Pi}, \ \forall t \geq 0,$$

or its equivalent:

$$\mathbf{\Pi Q} = \mathbf{0}.$$

It is assumed (stationary assumption) that the studied sequences comply with $\mathbf{\Pi}$: with infinitely long sequences, the character distribution within each sequence should be equal to $\mathbf{\Pi}$. However, as sequences are of limited length, it is expected that their character distribution slightly differs from $\mathbf{\Pi}$.

Finally, the process is assumed to be time-reversible, that is:

$$\pi_x P_{xy}(t) = \pi_y P_{yx}(t).$$

This assumption is complementary to the stationary assumption and implies the absence of time direction. The combination of these two assumptions implies that character distribution should be the same in the extant sequences and in the ancestral ones (with infinitely long sequences). Moreover, with a large number of sequences and sufficiently long times, the distribution of characters within each site should also be equal to $\mathbf{\Pi}$. However, the time scale being considered in phylogenetics is much shorter, characters within a single site tend to be strongly correlated to the ancestral character value and their distribution usually departs from $\mathbf{\Pi}$. Time reversibility is commonly used to rewrite the process generator as $\mathbf{Q} = (Q_{xy})$, with:

$$
\begin{aligned}
Q_{xy} &= \pi_y R_{x \leftrightarrow y} \text{ for } x \neq y \\
Q_{xx} &= -\sum_{y \neq x} Q_{xy}.
\end{aligned}
\tag{3.2}
$$

The R rates are symmetric and this writing of $\mathbf{Q}$ makes the stationary distribution $\mathbf{\Pi}$ explicit.

Up to now, we did not discuss time and time scale. In molecular phylogenetics, time is measured in number of substitutions per site, rather than years. Indeed, the rate of evolution can change markedly between different genes, different parts of the same genes, and even different periods of the past. Thus, we normalize the $\mathbf{Q}$ generator so that a time unit ($t = 1.0$) corresponds to 1 expected substitution per site. The normalized form of $\mathbf{Q}$ is then equal to $\frac{1}{\mu}(Q_{xy})$, where the normalization term is defined by:

$$
\mu = -\sum_{x} \pi_x Q_{xx}. \tag{3.3}
$$

3.2.2 *Neyman (two-state, DNA), GTR (DNA), WAG (protein), and NY*[1] *(codon) models*

To illustrate the formal presentation shown above, we now detail four models, starting from the simple two-state model of Neyman [61]. This model can be used in two different ways: (1) to analyse DNA data, in which case the two states are Purine (R, i.e. A or G) versus Pyrimidine (Y, i.e. C or T); (2) to express that sites can be in two different configurations, 'On' (i.e. free to mutate) or 'Off' (i.e. remaining invariant). We shall see (Section 3.2.7) that the 'On/Off' version is useful to account for heterogeneity of mutation rates over time and across sites. The normalized $\mathbf{Q}$ generator of Neyman model is given by:

$$
\begin{aligned}
\mathbf{Q}_{Neyman} &= \frac{1}{2\pi_{\mathrm{R}}\pi_{\mathrm{Y}}R_{\mathrm{R}\leftrightarrow\mathrm{Y}}} \begin{pmatrix} -\pi_Y R_{\mathrm{R}\leftrightarrow\mathrm{Y}} & \pi_Y R_{\mathrm{R}\leftrightarrow\mathrm{Y}} \\ \pi_R R_{\mathrm{R}\leftrightarrow\mathrm{Y}} & -\pi_R R_{\mathrm{R}\leftrightarrow\mathrm{Y}} \end{pmatrix} \\
&= \frac{1}{2} \begin{pmatrix} -\pi_{\mathrm{R}}^{-1} & \pi_{\mathrm{R}}^{-1} \\ \pi_{\mathrm{Y}}^{-1} & -\pi_{\mathrm{Y}}^{-1} \end{pmatrix},
\end{aligned}
\tag{3.4}
$$

where the stationary probabilities are subject to equality $\pi_{\mathrm{R}} + \pi_{\mathrm{Y}} = 1$. This model has just one free parameter, which can be easily estimated from the data.

The GTR (General Time Reversible) model [49, 83] applies to DNA. This is a four state model (A, C, G, and T), which is defined by:

$$\mathbf{Q}_{GTR} = \frac{1}{\mu} \begin{pmatrix} - & \pi_{\mathrm{C}} R_{\mathrm{A}\leftrightarrow\mathrm{C}} & \pi_{\mathrm{G}} R_{\mathrm{A}\leftrightarrow\mathrm{G}} & \pi_{\mathrm{T}} R_{\mathrm{A}\leftrightarrow\mathrm{T}} \\ \pi_{\mathrm{A}} R_{\mathrm{A}\leftrightarrow\mathrm{C}} & - & \pi_{\mathrm{G}} R_{\mathrm{C}\leftrightarrow\mathrm{G}} & \pi_{\mathrm{T}} R_{\mathrm{C}\leftrightarrow\mathrm{T}} \\ \pi_{\mathrm{A}} R_{\mathrm{A}\leftrightarrow\mathrm{G}} & \pi_{\mathrm{C}} R_{\mathrm{C}\leftrightarrow\mathrm{G}} & - & \pi_{\mathrm{T}} R_{\mathrm{G}\leftrightarrow\mathrm{T}} \\ \pi_{\mathrm{A}} R_{\mathrm{A}\leftrightarrow\mathrm{T}} & \pi_{\mathrm{C}} R_{\mathrm{C}\leftrightarrow\mathrm{T}} & \pi_{\mathrm{G}} R_{\mathrm{G}\leftrightarrow\mathrm{T}} & - \end{pmatrix}, \qquad (3.5)$$

where the diagonal terms are such that the row sums are all zero, and where μ is obtained from equation (3.3). This model is the most general DNA model assuming time reversibility. It has 10 parameters that are subject to 2 constraints ($\sum \pi_x = 1$ and $-\sum \pi_x Q_{xx} = 1$), and therefore 8 degrees of freedom. The parameters can be estimated from usual single gene data sets; but when the number of sites is low, the estimates of the (8 free) parameters are not reliable. We use simpler models in that case (see Fig. 3.1). The HKY model [29] assumes only two types of substitutions: transitions that conserve the Purine/Pyrimidine status (i.e. $R_{\mathrm{A}\leftrightarrow\mathrm{G}} = R_{\mathrm{C}\leftrightarrow\mathrm{T}} = \alpha$), and transversions that transform a Purine into a Pyrimidine, or the converse (i.e. $R_{\mathrm{A}\leftrightarrow\mathrm{C}} = R_{\mathrm{A}\leftrightarrow\mathrm{T}} = R_{\mathrm{G}\leftrightarrow\mathrm{C}} = R_{\mathrm{G}\leftrightarrow\mathrm{T}} = \beta$). The ratio $\kappa = \alpha/\beta$ (for a slightly different definition, see PAML [94] manual) is estimated from the data. Transitions occur much more frequently than transversions that correspond to strong modifications of the biochemical properties of nucleotides. κ generally varies in the $[0, 20]$ range and 4.0 is commonly used as default. The HKY model has 4 free parameters (3 stationary probabilities and the κ ratio). The Kimura [44] (K2P) model further simplifies HKY by assuming that the stationary probabilities are equal, requiring a single parameter (κ) to be estimated from the data. Note that the Kimura model is often called the 'Kimura 2 parameter' model (hence the 'K2P' abbreviation); the extra parameter (regarding previous explanations) corresponds to the time elapsed between the two sequences being considered. Felsenstein's [16] model (F81) simplifies HKY in a different way: it assumes $\kappa = 1$, but does not assume equal nucleotide frequencies. Finally, the Jukes and Cantor [42] (JC) model assumes both $\kappa = 1$ and equal nucleotide equilibrium frequencies. This is the simplest possible model and it does not require any parameter to be estimated from the data.

The WAG model [89] applies to proteins and expresses the substitution rates of the 20 amino acids. This is a refinement, dedicated to phylogenetic analysis, of the well known PAM1 [13], JTT [41] and Blosum62 [31] models, whose main concern is protein sequence alignment. These four models are homogeneous, stationary, and time-reversible. PAM1 was built from pairs of sequences that display 85% of sequence identity. Strictly speaking, PAM1 gives the probability of change between two amino acids that are separated by 0.01 substitutions on expectation [47]. In practice, however, PAM1 is considered as an instantaneous rate matrix. The JTT model is similar to PAM1, the only difference lying in the set of sequences that was used to estimate the change probabilities. Blosum62 corresponds to the default option implemented in the famous software BLAST [2] for identifying pairs of homologous sequences. It was designed to provide relevant mismatch scores for protein alignments, and was estimated using blocks of (non-independent) sites. WAG is a refinement of the previous models as it was established from the analysis of several protein families in a phylogenetic framework. Indeed, for each of these families a tree was reconstructed. The symmetrical parameters of the amino acid substitution rate matrix (the so-called WAG matrix) was then estimated from the whole set of protein families using an approximate maximum likelihood approach. The **R** symmetric (triangular) matrix of WAG is equal to:

Ala	Arg	Asn	Asp	Cys	Gln	Glu	Gly	His	Ile	Leu	Lys	Met	Phe	Pro	Ser	Thr	Trp	Tyr	Val
-																			
0.55	-																		
0.51	0.64	-																	
0.74	0.15	5.43	-																
1.03	0.53	0.27	0.03	-															
0.91	3.04	1.54	0.62	0.10	-														
1.58	0.44	0.95	6.17	0.02	5.47	-													
1.42	0.58	1.13	0.87	0.31	0.33	0.57	-												
0.32	2.14	3.96	0.93	0.25	4.29	0.57	0.25	-											
0.19	0.19	0.55	0.04	0.17	0.11	0.13	0.03	0.14	-										
0.40	0.50	0.13	0.08	0.38	0.87	0.15	0.06	0.50	3.17	-									
0.91	5.35	3.01	0.48	0.07	3.89	2.58	0.37	0.89	0.32	0.26	-								
0.89	0.68	0.20	0.10	0.39	1.55	0.32	0.17	0.40	4.26	4.85	0.93	-							
0.21	0.10	0.10	0.05	0.40	0.10	0.08	0.05	0.68	1.06	2.12	0.09	1.19	-						
1.44	0.68	0.20	0.42	0.11	0.93	0.68	0.24	0.70	0.10	0.42	0.56	0.17	0.16	-					
3.37	1.22	3.97	1.07	1.41	1.03	0.70	1.34	0.74	0.32	0.34	0.97	0.49	0.55	1.61	-				
2.12	0.55	2.03	0.37	0.51	0.86	0.82	0.23	0.47	1.46	0.33	1.39	1.52	0.17	0.80	4.38	-			
0.11	1.16	0.07	0.13	0.72	0.22	0.16	0.34	0.26	0.21	0.67	0.14	0.52	1.53	0.14	0.52	0.11	-		
0.24	0.38	1.09	0.33	0.54	0.23	0.20	0.10	3.87	0.42	0.40	0.13	0.43	6.45	0.22	0.79	0.29	2.49	-	
2.01	0.25	0.20	0.15	1.00	0.30	0.59	0.19	0.12	7.82	1.80	0.31	2.06	0.65	0.31	0.23	1.39	0.37	0.31	-
8.66	4.40	3.91	5.70	1.93	3.67	5.81	8.33	2.44	4.85	8.62	6.20	1.95	3.84	4.58	6.95	6.10	1.44	3.53	7.09

Those values were rounded, and the last line corresponds to standard amino acid percentages. The normalized **Q** generator is obtained by multiplying every column of **R** by the corresponding amino acid equilibrium frequency (π_y in equation (3.2)), then normalizing the resulting matrix (equation (3.3)). For example, $Q_{\text{Ile}\rightarrow\text{Val}} = \pi_{\text{Val}} \times R_{\text{Ile}\leftrightarrow\text{Val}} \times \mu^{-1} = 0.0709 \times 7.82 \times 1.241 = 0.688$, indicating that amino acids Ile and Val are likely to mutate one into the other (both are aliphatic and very similar). In the same way, we obtain $Q_{\text{Ala}\rightarrow\text{Trp}} = 0.00196$. This is a low substitution rate that is explained by the fact that Ala is tiny, while Trp is large, aromatic, and rare. The WAG model involves (20×19 / 2) free parameters to define **R**, plus 19 independent amino acid probabilities. Thus, it cannot be estimated from a single protein data set; the values of **R** and **Π** shown above were obtained by Whelan *et al.* [89] from a large database containing a number of alignments and thousands of sequences. An option (generally called 'F', available in some software) involves estimating **Π** from the analysed data set, which adds 19 free parameters in comparison to the standard option based on original **Π** (and **R**) values.

The Yang *et al.* [96] 'one-ratio' model is used to analyse genes at the codon level, with a focus on purifying/neutral/positive selection. This is a simplified version of the Nielsen and Yang [62] 'positive selection' model, which is itself inspired by Goldman and Yang [24] model. For the sake of homogeneity, we denote the 'one-ratio' model as NY (or NY^1, Fig. 3.2). The states are the 61 non-stop codons, as substitution of any codon into a stop codon is very likely to be deleterious. Moreover, simultaneous substitutions of nucleotides at a given codon are not allowed. This model distinguishes between synonymous substitutions which do not modify the corresponding amino acid, and non-synonymous substitutions that have an impact at the amino acid level and are less likely to occur (unless sites are under positive selection). For $x \neq y$, the $\mathbf{R}$ matrix is defined by:

$$R_{x \leftrightarrow y} = \begin{cases} 0 & : \text{if } x \text{ and } y \text{ differ at more than one position} \\ 1 & : \text{synonymous transversion} \\ \kappa & : \text{synonymous transition} \\ \omega & : \text{nonsynonymous transversion} \\ \kappa\omega & : \text{nonsynonymous transition} \end{cases} \tag{3.6}$$

κ is the transition/transversion ratio, just as α/β in the Kimura [44] model; ω is the non-synonymous/synonymous rate ratio. When ω is less than 1.0 the selection is purifying (changes in amino acids are deleterious); when ω is larger than 1.0, selection is positive (changes in amino acids are advantageous); when ω is equal to 1.0, evolution is neutral. Clearly, the among-site average value of ω is expected to be less than 1.0, but we shall see that proper use of a more general version of this model can be used to detect regions in proteins evolving under positive selection. The $\mathbf{Q}$ generator is obtained as in previous models, using equations (3.2) and (3.3). The $\mathbf{\Pi}$ distribution is usually deduced from the nucleotide frequencies at each of the three coding positions, which makes a total of 9 free parameters (3 nucleotide frequencies at each coding position). Therefore, besides branch lengths, this model requires estimating 11 free parameters (9 nucleotide frequencies, κ and ω).

3.2.3 *Trees and likelihood calculations*

Up to now we have discussed the evolution from one sequence to another. Phylogenetic studies involve sets of homologous sequences which are assumed to descend from a common ancestor through a tree-like scheme. In this tree, the leaves are labelled by the extant sequences and the internal nodes represent the ancestral sequences. The Markovian models explained in the previous section describe sequence evolution from one tree node to another, along a tree branch whose length is measured in expected number of substitutions per site. Usually trees are not clock-like, meaning that the distance from root to tips varies. Trees with this property reflect the fact that evolutionary rates are not constant among lineages. However, some models (not described here—see Introduction) are based on clock-like trees with explicit acceleration (or deceleration) events (e.g. [6, 35, 70, 85]). These models are typically used to estimate species divergence times.

Assume that the sequences evolve according to one of the standard models presented above. Moreover, assume that this model is identical throughout the whole tree and for all sites, as well as for the values of the model parameters (κ, ω, $\mathbf{\Pi}$, etc.). Let a be the tree root. The likelihood of the extant sequences (D), given the selected substitution model (M) and tree (T, which includes branch lengths), is defined by:

$$L(T, M; D) = \prod_i \left[\sum_x \pi_x L_i^a(x, T, M; D) \right] ; \tag{3.7}$$

the product runs over all sites in the alignment (which are assumed to evolve independently), and the sum is over all possible characters; $L_i^a(x, T, M; D)$ is the probability of the data at site i given that state x is observed at the i-th site of the sequence at node a. Let v be any tree node (vertex) and ν be the sequence attached to v. We use the notation $L_i^v(x, T, M; D)$ to express the (so-called partial) likelihood of observing the characters at position i in the extant sequences descending from v, given $\nu_i = x$, T and M. For short, we also use the simplified notation $L_i^v(x)$, as T, M, and D are the same for all sites and nodes. Partial likelihoods are defined recursively [16]. Let l and r be the right and left descendants (if any) of v, respectively, and t_{vw} be the length of branch (v, w). We have:

$$L_i^v(x) = \begin{cases} 1 \text{ if } v \text{ is a leaf and } \nu_i = x, \\ 0 \text{ if } v \text{ is a leaf and } \nu_i \neq x, \\ \left[\sum_{x'} P_{xx'}(t_{vl}) L_i^l(x')\right] \left[\sum_{x'} P_{xx'}(t_{vr}) L_i^r(x')\right] \text{ else.} \end{cases} \tag{3.8}$$

The likelihood of tree T with substitution model M, given sequence data D, is then obtained using equations (3.7) and (3.8); equation (3.1) is also used to compute the substitution probabilities $P_{xy}(t)$. Felsenstein [16] showed that when M is time-reversible (Section 3.2.1), the 'pulley principle' applies and the tree likelihood is the same for all root locations. Thus, trees are basically unrooted, even if one commonly selects a root (e.g. the first taxon) to compute the likelihood.

In the following, we explain how the standard models (and the corresponding likelihood calculations) are extended to account for among-site and among-lineage variability of evolutionary processes.

3.2.4 *Accounting for among-site variability using mixture models*

Sites in amino-acid or nucleotide sequences are subject to different functional and structural constraints, as explained in the Introduction. Therefore, we expect to find among-site variability in the rates and modes of evolution. In this section, we will assume (as do most of the practical solutions which account for among-site variability) that sites belong to categories, each one defining an evolutionary mode which is assumed to be the same for all the sites belonging to the category. The number of categories is fixed *a priori*, the set of categories is denoted as Θ, and θ denotes an element of Θ. Moreover, M_θ denotes the evolutionary model corresponding to category θ, and $\mathcal{M}_\Theta = \{M_\theta\}$ is the set of models.

Basically, two situations may occur: (1) the category of each site is known, or (2) site categories are unknown. Typically, codon positions are known (case 1), while precise structural configurations and functional roles of the sites are unknown (case 2). With proteins, we could have structural and functional information on the sites, but this information is incomplete and the way to use it in phylogenetic reconstruction is still unclear, so we generally deal with case 2. Finally, we could hypothetically predict the site categories using the data set being analysed, and use the predictions in likelihood calculations; but this would involve estimating one parameter per site, which is not possible, both for practical and theoretical reasons (see Chapter 4 in this book).

Assuming case (1), let θ_i be the (known) category of site i, and $\{\theta_i\}$ represent this *a priori* knowledge for all the sites. The tree likelihood becomes:

$$L(T, \{\theta_i\}, \mathcal{M}_\Theta; D) = \prod_i \left[\sum_x \pi_x L_i^a(x, T, M_{\theta_i}; D) \right],$$

that is, we simply extend equation (3.7) by accounting for the known evolutionary model corresponding to each site. Equation (3.8) is extended in the same way. Partial likelihoods now depend on the site category and are denoted as $L_i^v(x, T, M_{\theta_i}; D)$ or $L_i^v(x, \theta_i)$ for short, that is, the likelihood of site i of the extant sequences descending from v, when $\nu_i = x$ and when i belongs to θ_i. At the statistical level, the change (from the standard model) is not so simple: by multiplying the number of categories, we multiply the number of parameters to be estimated from the data. This approach, often called 'separate analysis', should then be used with caution. For example, using two categories (i.e. first and second codon position *versus* third codon position) to analyse coding DNA is achievable in most cases. But analysing concatenated genes may become tricky: we could be tempted to use one category per gene, or two categories (per gene) to account for third codon position, but this would involve a huge number of parameters. Genes are then usually clustered depending on their origin and role (e.g. mitochondrial, nuclear, protein coding, RNA coding, etc.). An alternative is to use a mixture model approach (thus abandoning the knowledge we have on each gene), as we shall now explain.

Assume case (2), where the site categories are unknown. Let π_θ be the *a priori* probability of category θ, and $\mathbf{\Pi}_\Theta = (\pi_\theta)$ the category probability distribution. To express the tree likelihood we use the total probability theorem, that is:

$$L(T, \mathbf{\Pi}_\Theta, \mathcal{M}_\Theta; D) = \prod_i \left[\sum_\theta \pi_\theta \sum_x \pi_x L_i^a(x, T, M_\theta; D) \right]. \tag{3.9}$$

In other words, each category is envisaged for each site and the corresponding likelihood is weighted by the category probability. Equation (3.8) is extended in the same way and becomes:

$$L_i^v(x,\theta) = \begin{cases} 1, \text{ if } v \text{ is a leaf and } \nu_i = x, \\ 0, \text{ if } v \text{ is a leaf and } \nu_i \neq x, \\ \left[\sum_{x'} P_{xx'}^{\theta}(t_{vl}) L_i^l(x',\theta)\right] \left[\sum_{x'} P_{xx'}^{\theta}(t_{vr}) L_i^r(x',\theta)\right], \text{ else,} \end{cases} \tag{3.10}$$

where $P_{xx'}^{\theta}(t)$ denotes the probability in model M_θ to observe a substitution from x to x' in time t. Note that in equations (3.9) and (3.10), we envisage every category for every site but preclude any change of category for a single site through time (which is the subject of section 3.2.6).

3.2.5 *Gamma-based rate across sites models and NY^3 (codon) models*

We shall now apply two mixtures to describe among-site variability. The first one is used to account for rate variability, both with DNA (Fig. 3.1) and protein sequences. The substitution model is the same for all categories, but categories evolve at different rates. In the simplest (and most widely used) version of Yang [91, 92], each category has the same probability, i.e. $\pi_\theta = 1/|\Theta|$, and the rates within categories are defined by a gamma distribution with parameter γ. Moreover, the (relative) rate expectation is set to 1 so as to conserve the same branch length scaling for all γ values. When γ is large (i.e. $\gg 1$) the rate distribution has a low variance, which implies that sites evolve at similar rates. When γ is small (i.e. in the $[0,1]$ range), the distribution is exponential-like with high variance. For example, with four categories and $\gamma = 0.75$, the (relative) rates within each category are (approximately) 2.580, 0.943, 0.387, and 0.086. This means that in the fastest category (2.580), sites evolve about 30 times faster than in the slowest category (0.086). This γ value (0.75) is typical of real data, which shows that site rates are highly variable. To account for this model in likelihood calculations, we simply use $P_{xx'}^{\theta}(t) = P_{xx'}(r_\theta \times t)$, where r_θ is the rate of category θ, and where $P_{xx'}(r_\theta \times t)$ is computed using equation (3.1) based on the substitution model that is shared by all categories. In other words, assuming θ we compute the tree likelihood as usual, but multiplying all the branch lengths by r_θ. This simple model has been refined in several ways. Most notably: Gu *et al.* [25] extended Yang's [91, 92] model by adding an invariant category to account for sites showing the same character across the different sequences; Susko *et al.* [80] and Felsenstein [17] refined the discretization of the gamma distribution by using rate categories with unequal *a priori* probabilities; Susko *et al.* [80] also proposed a non-parametric approach to estimate the rate distribution.

Our second example involves the codon model (NY) which is described in Section 3.2.2 (see also Fig. 3.2). Nielsen and Yang [62] and Yang *et al.* [96] extended this model with mixtures, to account for the variability of selection regimes across sites. Their aim was to test whether certain sites (e.g. sites that play a role in defining the 3D structure or the biochemical function of the protein) are subject to negative selection pressure, while other sites (e.g. in coils) evolve neutrally and, finally, that certain sites (e.g. located in the epitope regions of viral proteins) are subject to positive selection. The basic mixture model is then based on three categories, denoted as 0, 1, and 2. Within each category, sites evolve under the NY model, but with different ω values; typically $\omega_0 \approx 0.0$ (negative selection), $\omega_1 \approx 1.0$ (neutral evolution), and $\omega_2 > 1.0$ (positive selection). However, we shall see (Section 3.4) that ω values estimated from real data may depart significantly from this ideal scheme. Category prior probabilities are denoted as π_0, π_1, and π_2. Besides branch lengths, equilibrium distribution of codons, and transition/transversion ratio, which are common to all categories, this model thus involves 5 free parameters (3 ωs, 2 πs). This model is called M3 by Yang *et al.* [96], but we call it NY^3 for consistency with the rest of the chapter. Moreover, Yang *et al.* [96] envisage three restrictions to this model for exploring alternatives between the full NY^3 and the simple NY, which is denoted from now on as NY^1 for the sake of consistency. These restrictions are as follows:

- $\text{NY}^{3(\omega_1=1)}$ is the same as NY^3 but ω_1 is fixed to 1.0 which corresponds to a strictly neutral process of evolution. This model has one free parameter less than NY^3. It is similar to the model called M2a by Yang *et al.* [97] which adds the constraints $\omega_0 < 1.0$ and $\omega_2 > 1.0$.
- $\text{NY}^{3(\omega_1=1)(\omega_0=0)}$ further simplifies $\text{NY}^{3(\omega_1=1)}$ by fixing $\omega_0 = 0.0$. The $\omega_0 = 0.0$ class models sites at which non-synonymous changes are prohibited. This model is called M2 by Yang *et al.* [96] and has one free parameter less than $\text{NY}^{3(\omega_1=1)}$.
- $\text{NY}^{2(\omega_1=1)(\omega_0=0)}$ is a two category model that simplifies $\text{NY}^{3(\omega_1=1)(\omega_0=0)}$ by assuming that no site evolves under a selective regime that is distinct from strict neutrality ($\omega_1 = 1.0$) or negative selection ($\omega_0 = 0.0$). This model is called M1 by Yang *et al.* [96] and has two free parameters less than $\text{NY}^{3(\omega_1=1)(\omega_0=0)}$.

Except NY^1 *vs.* $\text{NY}^{2(\omega_1=1)(\omega_0=0)}$, which have the same number of free parameters but model evolution in different ways (1 category with non-fixed ω versus 2 categories with fixed ω), these 5 NY-based models are nested (Fig. 3.2). Many variants have been proposed (see Yang *et al.* [96]). While the most popular and computationally tractable versions are those presented above, models that use a parametric distribution to describe the variation of ω across sites (e.g. models M7 and M8 in [96]) are also widely used.

3.2.6 *Accounting for among-site and time variability using Markov-modulated Markov (MMM) models*

Structural and functional constraints vary with time. Even though a given site evolves under positive selection in some clade, the very same site may become neutral or even positively selected in other clades. We have seen in the previous section how mixture models provide a unified framework to account for among-site variation. We shall see in this section how Markov-modulated Markov models [86] extend mixture models in a natural way, to incorporate time variability. These models are closely related to hidden Markov models (see [18] for an application in phylogenetics) and have been used for a long time in queueing theory [86]. They were introduced in phylogenetics by Tuffley and Steel [87], Lockhart *et al.* [53], Penny *et al.* [65], Galtier [19], and Huelsenbeck [37]. We show here that they provide a general framework, which deserves further exploration.

We use the same evolutionary categories that we had with mixtures and the same notation as in the previous section: Θ is the set of categories, θ is an element of Θ with probability π_θ, M_θ is the evolutionary model with the generator $\mathbf{Q}_\theta$ corresponding to category θ, and $\mathcal{M}_\Theta$ is the set of M_θ models. We assume that every model M_θ is homogeneous, stationary, and time-reversible, and satisfies equation (3.2); but $\mathbf{Q}_\theta$ generators are not normalized (equation (3.3)). Moreover, we assume that the stationary distribution of characters ($\mathbf{\Pi}_X = (\pi_x)$) is the same for all M_θ models. This latter assumption is not required for mixtures, but we shall see that it greatly simplifies MMM models.

The substitution process that governs the evolution of an individual site can now change with time. These category changes follow a homogeneous, stationary, and time-reversible Markovian process, as in the standard character evolution model, but the states are the evolutionary categories instead of the sequence characters. The stationary distribution of the categories is equal to $\mathbf{\Pi}_\Theta = (\pi_\theta)$, and the category generator, denoted as $\mathbf{C}$, satisfies equation (3.2). The general time reversible model for categories is analogous to the GTR model applied to DNA sequences and is defined by:

$$\mathbf{C}_{GTR} = \delta \begin{pmatrix} - & \pi_{\theta_2} R_{\theta_1 \leftrightarrow \theta_2} & \cdots & \pi_{\theta_{|\Theta|}} R_{\theta_1 \leftrightarrow \theta_{|\Theta|}} \\ \pi_{\theta_1} R_{\theta_1 \leftrightarrow \theta_2} & - & \cdots & \pi_{\theta_{|\Theta|}} R_{\theta_2 \leftrightarrow \theta_{|\Theta|}} \\ \cdots & \cdots & \ddots & \cdots \\ \pi_{\theta_1} R_{\theta_1 \leftrightarrow \theta_{|\Theta|}} & \cdots & \cdots & - \end{pmatrix}, \quad (3.11)$$

where each row sums to 0, and δ is an additional parameter that expresses the global rate of changes between categories. The R coefficients are normalized using equation (3.3) such that δ is the expected number of category changes during one time unit.

The whole process is a compound process, also called a Markov-modulated Markov (MMM) process. The evolutionary category of a given site evolves along the tree according to the category model. Thus the site evolves in the space of character states according to M_θ, where θ depends on the outcome of the category process. This MMM process can be seen as a single Markov process taking values in the Cartesian product of the two state spaces: $\Theta \times X = \{(\theta, x)\}$, with size $|\Theta| \times |X|$. We assume that the category states are ranked from θ_1 to $\theta_{|\Theta|}$, and that the compound states (θ_k, x) are ranked in lexicographic order. Let $\mathbf{I}_X$ be the identity matrix on the character space, and $\otimes$ the Kronecker product. The generator of the MMM process is denoted $\mathbf{Q}_{\mathbf{C}\odot\mathcal{M}_\Theta}$ in order to show that changes within the set of character models $\mathcal{M}_\Theta$ are driven by the category generator $\mathbf{C}$. We have:

$$\mathbf{Q}_{\mathbf{C}\odot\mathcal{M}_\Theta} = \mathbf{Diag}(\mathbf{Q}_{\theta_k}) + \mathbf{C} \otimes \mathbf{I}_X \tag{3.12}$$

$$= \begin{pmatrix} \mathbf{Q}_{\theta_1} & 0 & \dots & 0 \\ 0 & \mathbf{Q}_{\theta_2} & \dots & 0 \\ \dots & \dots & \ddots & \dots \\ 0 & 0 & \dots & \mathbf{Q}_{\theta_{|\Theta|}} \end{pmatrix} + \begin{pmatrix} C_{\theta_1\theta_1}\mathbf{I}_X & C_{\theta_1\theta_2}\mathbf{I}_X & \dots & C_{\theta_1\theta_{|\Theta|}}\mathbf{I}_X \\ C_{\theta_2\theta_1}\mathbf{I}_X & C_{\theta_2\theta_2}\mathbf{I}_X & \dots & C_{\theta_2\theta_{|\Theta|}}\mathbf{I}_X \\ \dots & \dots & \ddots & \dots \\ C_{\theta_{|\Theta|}\theta_1}\mathbf{I}_X & C_{\theta_{|\Theta|}\theta_2}\mathbf{I}_X & \dots & C_{\theta_{|\Theta|}\theta_{|\Theta|}}\mathbf{I}_X \end{pmatrix}.$$

Every compound state (θ_k, x) thus may:

(1) stay in category θ_k and be changed into (θ_k, y) with rate defined by $\mathbf{Q}_{\theta_k}$ (on the diagonal of the first matrix in sum (3.12)), or
(2) change of category and become (θ_j, x) at rate $C_{\theta_k\theta_j}$ (second matrix in sum (3.12)).

All rows in $\mathbf{Q}_{\mathbf{C}\odot\mathcal{M}_\Theta}$ sum to zero (this property holds for both matrices in (3.12)). Moreover, consider the probability distribution on the compound states: $\mathbf{\Pi}_{\Theta X} = (\pi_{(\theta_k,x)}) = (\pi_{\theta_k}\pi_x)$; it is easily seen that $\mathbf{\Pi}_{\Theta X}\mathbf{Q}_{\mathbf{C}\odot\mathcal{M}_\theta} = 0$ (this property holds again for both matrices in sum (3.12)), due to the equalities $\mathbf{\Pi}_\Theta\mathbf{C} = 0$ and $\mathbf{\Pi}_X\mathbf{Q}_{\theta_k} = 0$). Thus, $\mathbf{\Pi}_{\Theta X}$ is the unique stationary distribution of $\mathbf{Q}_{\mathbf{C}\odot\mathcal{M}_\Theta}$ provided that this Markov matrix is irreducible (i.e. every state can be reached from any starting state with non-zero probability). This last property holds in most cases, as it is a simple consequence of the irreducibility of $\mathbf{C}$ and of the $\mathbf{Q}_{\theta_k}$s. Note, however, that when category changes are not allowed (i.e. when $\delta = 0$ in equation (3.11)), the stationary distribution of $\mathbf{Q}_{\mathbf{C}\odot\mathcal{M}_\Theta}$ is no longer unique (even if the stationary distribution of each M_{θ_k} is still unique and equal to $\mathbf{\Pi}_X$). This special case actually reduces to a mixture model, as shall be discussed at the end of this section.

The normalization of $\mathbf{Q}_{\mathbf{C}\odot\mathcal{M}_\Theta}$ slightly differs from the normalization in equation (3.3). $\mathbf{Q}_{\mathbf{C}\odot\mathcal{M}_\Theta}$ is normalized such that the expected number of character changes per time unit is 1.0. As branch lengths are measured in expected number of character changes, category changes should not be accounted for. The normalization term is then equal to:

$$\mu_{\mathbf{C}\odot\mathcal{M}_\Theta} = -\sum_{k,x} \pi_{\theta_k}\pi_x\mathbf{Q}_{\theta_k,xx} = \sum_k \pi_{\theta_k}\mu_k, \tag{3.13}$$

where μ_k is the normalization term of $\mathbf{Q}_{\theta_k}$ that is obtained from equation (3.3).

Thus, MMM models do not differ in their structure from the standard models we described in Sections 3.2.1 to 3.2.3. Tree likelihood computations are performed using equations (3.1), (3.7), and (3.8), the characters (x, with probability π_x) being replaced by compound states ((θ_k, x), with probability $\pi_{\theta_k}\pi_x$). The only (slight) difference lies in equation (3.8): the partial likelihood $L_i^v((\theta_k, x))$ is 1.0 when v is a leaf and $\nu_i = x$, instead of $\nu_i = (\theta_k, x)$, as site categories are unobservable, even at the tips of the tree. Galtier and Jean-Marie [20] showed that the diagonalization of the (large) compound matrix $\mathbf{Q}_{\mathcal{C}\odot\mathcal{M}_\Theta}$ can be achieved in a fast way in some settings. However, because the state space is usually large (see applications below), MMM models can be computationally demanding.

We already mentioned the extreme case where the rate at which changes between categories occur is equal to zero, i.e. when the second matrix in sum (3.12) is null. In this situation, it is easily seen that the MMM model becomes equivalent to the mixture model defined by (M_{θ_k}) and (π_{θ_k}), as soon as the *a priori* probability of every compound state (θ_k, x) is equal to $\pi_{\theta_k}\pi_x$ in equation (3.7). When the rate of changes between categories is large, the distribution of categories along the tree becomes independent of the initial value at the tree root and is identical for all sites. This is equivalent to having a unique evolutionary category, i.e. there is no among-site nor across-lineage variability, and we fall back into the standard character evolution model (at least from a biological perspective; the behaviour of MMM models when $\delta \to \infty$ needs to be better characterized at the mathematical level).

3.2.7 *On/Off (two-state, DNA), covarion-like (DNA) and compound codon models*

The MMM approach can be used to define a broad variety of sequence evolution models. We outline a few examples to illustrate this. The 'simplest' MMM model is obtained by combining the Purine/Pyrimidine model presented in section 3.2.2 with 'On/Off' site categories, along the lines of Tuffley and Steel [87] and Huelsenbeck [37]. As explained earlier, in the 'On' category, sites are free to mutate, while in the 'Off' class, sites are invariant. Let r_θ be the substitution rate of category θ; we have $r_{\text{Off}} = 0.0$ and $r_{\text{On}} = \pi_{\text{On}}^{-1}$ due to the constraint $\sum \pi_\theta r_\theta = 1.0$ (the r_θs are relative substitution rates; therefore, they must be centred around 1.0, just as with Yang's rate across sites model, section 3.2.5). The 'On/Off' process is modelled by a Neyman matrix (equation (3.4)), which is not normalized but multiplied by δ to express the rate at which the category changes. Changes between characters within the 'On' category are also modelled with a Neyman matrix, which is multiplied by r_{On}. All together (including normalizations), the combination of these two models gives a MMM model with stationary distribution $\{\pi_{\text{On}}\pi_{\text{R}}, \pi_{\text{On}}\pi_{\text{Y}}, \pi_{\text{Off}}\pi_{\text{R}}, \pi_{\text{Off}}\pi_{\text{Y}}\}$ and generator:

$$\mathbf{Q}_{\text{OnOff}\odot\text{RY}} = \frac{1}{2\pi_{\text{On}}r_{\text{On}}}\begin{pmatrix} -r_{\text{On}}\pi_{\text{R}}^{-1} & r_{\text{On}}\pi_{\text{R}}^{-1} & 0 & 0 \\ r_{\text{On}}\pi_{\text{Y}}^{-1} & -r_{\text{On}}\pi_{\text{Y}}^{-1} & 0 & 0 \\ 0 & 0 & 0 & 0 \\ 0 & 0 & 0 & 0 \end{pmatrix} +$$

$$\frac{\delta}{2\pi_{\text{On}}r_{\text{On}}}\begin{pmatrix} -\pi_{\text{On}}^{-1} & 0 & \pi_{\text{On}}^{-1} & 0 \\ 0 & -\pi_{\text{On}}^{-1} & 0 & \pi_{\text{On}}^{-1} \\ \pi_{\text{Off}}^{-1} & 0 & -\pi_{\text{Off}}^{-1} & 0 \\ 0 & \pi_{\text{Off}}^{-1} & 0 & -\pi_{\text{Off}}^{-1} \end{pmatrix},$$

$$= \frac{1}{2}\begin{pmatrix} - & \pi_{\text{On}}^{-1}\pi_{\text{R}}^{-1} & \delta\pi_{\text{On}}^{-1} & 0 \\ \pi_{\text{On}}^{-1}\pi_{\text{Y}}^{-1} & - & 0 & \delta\pi_{\text{On}}^{-1} \\ \delta\pi_{\text{Off}}^{-1} & 0 & - & 0 \\ 0 & \delta\pi_{\text{Off}}^{-1} & 0 & - \end{pmatrix}. \tag{3.14}$$

This model requires estimating 3 free parameters (δ and 2 *a priori* probabilities, π_{On} and π_{R}).

Galtier's [19] model (see also Galtier and Jean-Marie [20]) extends Yang's [92] gamma-based mixture model which accounts for rate variability among sites (Section 3.2.5). Under this model, evolutionary categories are equally likely and evolve according to a Jukes and Cantor-like model, i.e. for all i, j, k, and l ($i \neq j$ and $k \neq l$), $\pi_{\theta_i} = \pi_{\theta_j}$ and $R_{\theta_i \leftrightarrow \theta_j} = R_{\theta_k \leftrightarrow \theta_l}$ in equation (3.11). The rates of character changes (r_θ) are defined by a gamma distribution with parameter γ, and we have $\sum r_\theta/|\Theta| = 1$. The generator of the substitution model within each category has the shape $\mathbf{Q}_\theta = r_\theta\mathbf{Q}$, where $\mathbf{Q}$ is the (normalized) substitution model that is shared by all categories. All together the (normalized) generator of this model is defined by:

$$\mathbf{Q}_G = \mathbf{Diag}(r_\theta \times \mathbf{Q}) + \delta\mathbf{C}_{\text{JC}} \otimes \mathbf{I}_X \tag{3.15}$$

$$= \begin{pmatrix} r_{\theta_1}\mathbf{Q} & 0 & \dots \\ 0 & r_{\theta_2}\mathbf{Q} & \dots \\ \dots & \dots & \ddots \end{pmatrix} + \delta\begin{pmatrix} -\mathbf{I}_X & (|\Theta|-1)^{-1}\mathbf{I}_X & \dots \\ (|\Theta|-1)^{-1}\mathbf{I}_X & -\mathbf{I}_X & \dots \\ \dots & \dots & \ddots \end{pmatrix}.$$

This model requires just one additional parameter (δ) compared to Yang's [92] mixture model, and was applied [19] to ribosomal RNA sequences (Fig. 3.1).

Finally, Guindon *et al.* [28] proposed a MMM model to account for selection regime changes among lineages. They combined the NY^3 model of codon substitution (Section 3.2.5) with the GTR-like model of equation (3.11) (Fig. 3.2).

The generator of this model is defined by:

$$\mathbf{Q}_{\mathbf{C}_{\mathrm{GTR}} \odot \mathrm{NY}^3} = \mathbf{Diag}(\mathbf{Q}_{\omega_\theta}) + \delta \mathbf{C}_{\mathrm{GTR}} \otimes \mathbf{I}_X \tag{3.16}$$

$$= \begin{pmatrix} \mathbf{Q}_{\omega_0} & 0 & 0 \\ 0 & \mathbf{Q}_{\omega_1} & 0 \\ 0 & 0 & \mathbf{Q}_{\omega_2} \end{pmatrix}$$

$$+ \delta \begin{pmatrix} - & \pi_{\theta_1} R_{\theta_0 \leftrightarrow \theta_1} \mathbf{I}_X & \pi_{\theta_2} R_{\theta_0 \leftrightarrow \theta_2} \mathbf{I}_X \\ \pi_{\theta_0} R_{\theta_0 \leftrightarrow \theta_1} \mathbf{I}_X & - & \pi_{\theta_2} R_{\theta_1 \leftrightarrow \theta_2} \mathbf{I}_X \\ \pi_{\theta_0} R_{\theta_0 \leftrightarrow \theta_2} \mathbf{I}_X & \pi_{\theta_1} R_{\theta_1 \leftrightarrow \theta_2} \mathbf{I}_X & - \end{pmatrix},$$

where $\mathbf{Q}_{\omega_0}$, $\mathbf{Q}_{\omega_1}$ and $\mathbf{Q}_{\omega_2}$ describe substitutions between codons under the three selection regimes defined by ω_0, ω_1 and ω_2. $\mathbf{Q}_{\mathbf{C}_{\mathrm{GTR}} \odot \mathrm{NY}^3}$ is normalized using equation (3.13). Guindon *et al.* [28] also tested a simplification of this combination using a F81-like model for the category changes ($R_{\theta_0 \leftrightarrow \theta_1} = R_{\theta_0 \leftrightarrow \theta_2} = R_{\theta_1 \leftrightarrow \theta_2}$) (Fig. 3.2). The GTR-like version of this model has five additional parameters (compared to NY^3): δ plus two equilibrium frequencies of selection regimes and two non-normalized R rates. The F81-like version has only three additional parameters: δ plus two equilibrium frequencies of selection regimes. We shall see in the following section how useful this compound model is for detecting biologically relevant site-specific changes of selection patterns during evolution.

3.3 Biological data sets

We use two data sets to illustrate the various substitution models described in the previous section. The first one is an alignment of both orthologous and paralogous coding sequences collected among plant genomes. The corresponding genes are involved in flower development. Their phylogeny is well established and displays two duplication events with unambiguous positions in the tree. The second data set is an alignment of homologous sequences coding for the envelope protein located at the surface of the HIV-1 virus. These viral sequences are especially interesting because they have been collected at various stages of the infection of an individual. Further details about the two data sets are given below.

3.3.1 *The role of Deficiens and Globosa genes in flower development*

A typical flower displays four whorls at the tip of a floral shoot. The first (outermost) whorl usually consists of leaf-like sepals. The second is composed of petals. The third and fourth whorls contain the male (stamen) and the female (carpel) reproductive organs, respectively. Knock-out experiments have been conducted in order to identify the genes responsible for such structures. These studies defined the 'ABC model' of floral organ identity. To simplify, A, B, and C-class genes are 'on, off, off' in the sepals, 'on, on, off' in the petals, 'off, on, on' in the stamens, and 'off, off, on' in the carpels. A, B, and C-class genes encode transcription factors and belong to the MADS-box gene family. Indeed, these sequences share a highly conserved DNA stretch of ~180 base pairs, the so-called MADS-box. This large family of genes has been studied extensively in order to shed light on the evolutionary origin of flowering plants, Darwin's famous 'abominable mystery'.

Deficiens (*DEF*) and Globosa (*GLO*) are B-class genes. They play a central role in specifying the petal, and may have been involved in the differentiation between non-flowering (gymnosperms) and flowering seed plants (angiosperms) [98]. The *DEF* and *GLO* clades are well defined from a phylogenetic viewpoint. They result from a duplication event that occurred within the lineage that led to the angiosperms [90]. Other duplication events occurred in various angiosperm lineages, most notably in the *DEF* clade [98].

In this chapter, we analyse a data set made of 89 *DEF* and *GLO* sequences. Each of these sequences is 627 base pairs long. An alignment of these sequences was kindly provided by Prof. Jim Leebens-Mack (University of Georgia, USA). This data set is well suited to tackle an important open question in molecular evolution: the fate of duplicated genes. Two hypotheses compete here [56]. The 'neofunctionalization' hypothesis states that one copy acquires a novel function while the other copy retains its original function. According to the 'subfunctionalization' hypothesis, both copies accumulate slightly deleterious mutations to the point at which the sum of the two copies have the same capacity as the ancestral gene. These two hypotheses imply very distinct patterns in terms of variation of selection regimes after the duplication event occurred. Most notably, under the subfunctionalization hypothesis the selection regimes that affect both gene copies are expected to be similar, while a strong contrast is expected under the neofunctionalization hypothesis. We will see that models that allow variations of the ω ratio across sites and lineages are specially well suited to bring insight to this problem.

3.3.2 *The singular dynamics of the envelope gene evolution during HIV-1 infection*

One of the most remarkable features of HIV-1 envelope (*env*) gene evolution is the speed at which it evolves. Indeed, its evolution rate is about five million times faster than the average rate in mammalian genes [14, 48]. A few years after the infection, orthologous *env* sequences display high levels of dissimilarity and share little resemblance to the ancestral sequence at the origin of the infection. Hence, when sampled at different timepoints, these sequences provide valuable information about the rates at which substitution events occur and their variations across different stages of the infection. HIV-1 *env* sequences thus meet all the criteria that define measurably evolving population ([14], see also Chapter 2 of this book).

In a pioneering work, Kaslow *et al.* [43] performed a longitudinal study involving more than 5,000 men infected by HIV-1. About ten years later, Shankarappa *et al.* [75] analysed the evolution of *env* sequences in nine patients. These sequences were collected at different time points, covering a period of 12 years.

This study clarified the links between the evolution of sequence diversity during the infection and important phenotypic changes of HIV-1. Ross and Rodrigo [69] later used standard codon-based models [62, 96] in a phylogenetic framework to decipher natural selection processes acting on these sequences. By applying models that allow the selection classes to vary among codon positions (see Section 3.2.5), they showed a significant positive correlation between the frequency of sites evolving under positive selection and disease duration, indicating that long term progressors have a strong immune response that forces the virus to evolve. In this chapter, our analysis focuses on a single patient (Patient 1). This patient was chosen randomly from the nine for whom data are available. The data set comprises 87 sequences. Each of these is 561 base pairs long.

An accurate description of the variations of selection regimes acting on the *env* protein during the infection is essential to understand the sources of the huge diversity of viral sequences. It has been shown [75] that, when the average of all sites is taken, the amino acid diversity increases during early stages of infection and decreases afterwards, when the selective pressure exerted by the immune system is weaker. Codon-based models give a much more precise picture of the variations of evolutionary patterns than the one given by the simple analysis of sequence diversity. Indeed, we will see that these models provide an adequate framework to classify sites into selection regimes. They also allow the identification of lineages that evolve under specific selection classes at individual sites.

3.4 The models in action: analysis of protein coding sequences

This section illustrates advances in modelling variations of substitution processes during the evolution of coding sequences. It focuses on the two reference data sets described in section 3.3. The exploration of these data using both well-known and quite recent statistical models focuses on among-site and time-dependent variations of substitution rates and selection regimes. We show how a thorough analysis of these sources of heterogeneity reveals important evolutionary features.

Such analysis usually relies on comparing how alternative models fit the data. Each of these pairwise comparisons tests for a 'biological hypothesis'. Several methods can be used to this end. If the two models are nested (i.e. the first model is equivalent to the second under some constraints on its parameters), twice the difference of log likelihood of these two models is asymptotically distributed as a χ^2 distribution or a mixture of χ^2 distributions (see [74] for more details). A difference of log likelihood of $\sim$2 is significant at the 5% level, if the two models differ by one parameter, $\sim$3 if the models differ by two parameters, $\sim$4 if the models differ by three parameters, etc. Note that two nested models which are compared must share the same tree topology. Indeed, if the two topologies differ, the more complex model cannot reduce to the simple one. In this chapter, the models that are compared do not systematically share the same topology. Fortunately, other methods, such as the Akaike criterion [1] (AIC) can be used for model comparison in such situation. AIC estimates the Kullback–Leibler information number which is a measure of the similarity between the model that generated the data (i.e. the true model) and the model that is used for the inference. This criterion penalizes models with additional parameters. A first model has a higher AIC than a second one if the difference of log likelihood exceeds the difference of the number of parameters in these two models.

Our experience is that the addition of a (set of) parameter(s) that capture(s) an important, and previously ignored feature of molecular evolution, systematically leads to an increase of the log likelihood much larger than the thresholds given above. As we shall see in the following section, most model comparisons appear to be highly significant, with log likelihood differences larger than 10 in most cases. Thus, in the following we do not discuss the testing approaches as any test would give the same conclusion. Increases of likelihood that are close to the threshold level generally correspond to biologically irrelevant features and/or to insufficient data.

3.4.1 *Among-site heterogeneity*

This section first focuses on the variability of substitution rates across amino acid positions. Both *DEF/GLO* and HIV-1 *env* data sets were analysed under four popular amino acid substitution models: PAM1 [13], JTT [41], Blosum62 [31], and WAG [89]. These four models were also coupled to a discrete gamma distribution (suffix: $+\Gamma$) with eight categories and fitted to the data. Maximum-likelihood topologies, branch lengths (and gamma shape parameters when needed) were estimated in the maximum-likelihood framework using PhyML [27].

The log likelihood obtained under the eight substitution models are displayed in Table 3.1. A first glance at these numbers confirms that taking variable rates

TABLE 3.1. **Log likelihood of amino acid substitution models.** $\widehat{\gamma}$ is the estimated value of the gamma shape parameter. Values around 1.0 suggest a moderate variability of rates across sites. Values around 0.5 suggest a strong heterogeneity. df is the number of free parameters of the model that are estimated from the data. Values of df presented here do not include the number of branch lengths, i.e. 175 for *DEF/GLO* and 171 for HIV-1 *env*.

	DEF/GLO				HIV-1 *env*		
model	lnL	$\widehat{\gamma}$	df	model	lnL	$\widehat{\gamma}$	df
WAG+Γ	**-17725.44**	**1.07**	1	**JTT+Γ**	**-2330.18**	**0.50**	**1**
JTT+Γ	**-17809.99**	**1.00**	1	**WAG+Γ**	**-2349.62**	**0.54**	**1**
Blosum62+Γ	**-17847.38**	**1.08**	1	**Blos62+Γ**	**-2395.16**	**0.49**	**1**
PAM1+Γ	**-17864.33**	**0.92**	1	JTT	-2417.24		0
WAG	-18448.49		0	WAG	-2424.76		0
Blos62	-18534.39		0	**PAM1+Γ**	**-2446.90**	**0.42**	**1**
JTT	-18578.87		0	Blos62	-2461.12		0
PAM1	-18691.30		0	PAM1	-2553.92		0

across sites into account largely improves the fit of the models to the data. When fitted to the *DEF/GLO* and the HIV-1 *env* sequences, the average increases of log likelihood are ~751 and ~90 units, respectively. Pairwise comparisons also confirm that models that include a gamma distribution are significantly more likely than models that do not.

Let $\mathcal{M}+\Gamma$ denote the set of models estimated using the gamma distribution. Class $\mathcal{M}$ denotes the models estimated without gamma distribution. For the *DEF/GLO* data set, the mean difference between log likelihood of models in $\mathcal{M}+\Gamma$ (i.e. the 'within difference') is ~50. The same statistic measured from models in $\mathcal{M}$ is equal to ~85. By contrast, the average of the differences of log likelihood between models that belong to different sets ('between' difference) is ~751. The differences of log likelihood related to variations of rates across sites are less contrasted with the HIV-1 *env* data set (Table 3.1). Some rate matrices alone (JTT and WAG) provide better fit to the data than a model that includes a gamma distribution (PAM1+Γ). Nonetheless, the 'within' differences of log likelihood among $\mathcal{M}+\Gamma$ and $\mathcal{M}$ are ~43 and ~49 respectively, to be compared to ~90, the 'between' difference. Therefore, the increase of fit due to the gamma distribution is much more important than the increase provided by some substitution rate matrices as compared to others.

Table 3.2 shows the log likelihood of phylogenetic models estimated under four popular nucleotide substitution models: JC [42], K2P [44], HKY [29], and GTR [83, 49] (Fig. 3.1). The 'within' differences of log likelihood computed from the *DEF/GLO* data set are ~212 and ~219 respectively. The 'between' difference is ~1746, which represents a very significant shift with respect to the 'within' differences. The same tendency is observed with the HIV-1 *env* data set: ~83 and ~87 ('within' differences) *vs.* ~173 ('between' difference). Hence, the increase of fit to the nucleotide data when including the gamma distribution is even more conspicuous than what is obtained with the corresponding protein alignment. The distinction between transitions and transversions also improves

TABLE 3.2. **Log likelihood of nucleotide substitution models.** See the caption of Table 3.1.

	DEF/GLO				HIV-1 *env*		
model	lnL	$\widehat{\gamma}$	df	model	lnL	$\widehat{\gamma}$	df
GTR+Γ	**−28939.75**	**0.92**	**9**	**GTR+Γ**	**−2961.39**	**0.31**	**9**
HKY+Γ	**−29071.53**	**0.92**	**5**	**HKY+Γ**	**−2978.92**	**0.31**	**5**
K2P+Γ	**−29082.95**	**0.92**	**2**	**K2P+Γ**	**−3010.68**	**0.30**	**2**
JC+Γ	**−29572.65**	**0.96**	**1**	GTR	−3100.94		8
GTR	−30641.43		8	HKY	−3119.72		4
HKY	−30839.53		4	K2P	−3162.49		1
K2P	−30884.70		1	**JC+Γ**	**−3202.44**	**0.31**	**1**
JC	−31284.01		0	JC	−3350.35		0

the fit of the models to the data in a very significant manner. This tendency is actually observed with most data sets. From a historical perspective, the use of the K2P instead of JC model has been the first, very significant, improvement of nucleotide substitution models. The next big step was undoubtedly the use of a distribution of rates across sites. Finally, note that the gamma shape parameter estimates are, on average, smaller when models are fitted to the nucleotide sequences. Hence, as expected (see Section 3.1.1), substitution rates are more heterogeneous among nucleotide sites than among amino acid positions.

We next analysed both data sets under the codon-based models described in Sections 3.2.2 and 3.2.5 (Fig. 3.2, Tables 3.3 and 3.4). Each codon model was fitted to the tree topology inferred using the GTR model of nucleotide substitution (including a gamma distribution of rates across sites). The comparison NY^1 *vs.* NY^3 tests for the variability of the ω ratio across sites. The likelihood ratio statistic for this model comparison asymptotically follows a χ^2_2 distribution (NY^3 tends to NY^1 if $\omega_0 \rightsquigarrow \omega_1 \rightsquigarrow \omega_2$). The large observed differences of log likelihood clearly reject the null hypothesis of homogeneity of the ω ratio across sites. This conclusion is valid for both data sets.

Comparing $NY^{2(\omega_1=1)(\omega_0=0)}$ and $NY^{3(\omega_1=1)(\omega_0=0)}$ tests for the presence of a selective regime that is distinct from strict neutrality ($\omega_1 = 1.0$) or strong negative selection ($\omega_0 = 0.0$). This model comparison tests for positive selection only if ω_2 in $NY^{3(\omega_1=1)(\omega_0=0)}$ is greater than 1.0. These two models are nested and the observed difference of log likelihood rejects the null hypothesis ('H_0: sequences evolve under $NY^{2(\omega_1=1)(\omega_0=0)}$'). The value of ω_2 is much larger than 1.0 for the HIV-1 *env* data set ($\omega_2 = 8.30$), suggesting the presence of strongly

TABLE 3.3. **Log likelihood of codon-based models (*DEF/GLO* data).** df is the number of free parameters of the model that are estimated from the data (Fig. 3.2). Values of df presented here do not include the number of branch lengths, i.e. 175 for *DEF/GLO* and 171 for HIV-1 *env*.

Model	df	Log likelihood	Estimated parameters
NY^3	15	−28631.47	$p_0 = 0.21$, $p_1 = 0.34$, $p_2 = 0.45$ $\omega_0 = 0.01$, $\omega_1 = 0.11$, $\omega_2 = 0.32$
$NY^{3(\omega_1=1)}$	14	−28743.97	$p_0 = 0.40$, $p_1 = 0.06$, $p_2 = 0.53$ $\omega_0 = 0.05$, $\omega_2 = 0.27$
$NY^{3(\omega_1=1)(\omega_0=0)}$	13	−29134.76	$p_0 = 0.07$, $p_1 = 0.19$, $p_2 = 0.74$ $\omega_2 = 0.18$
$NY^{2(\omega_1=1)(\omega_0=0)}$	11	−30919.02	$p_0 = 0.07$, $p_1 = 0.93$
NY^1	11	−29626.33	$\omega = 0.16$

TABLE 3.4. **Log likelihood of codon-based models (HIV-1 *env* data).** See the caption of Table 3.3.

Model	df	lnL	Estimated parameters
NY^3	15	−3036.86	$p_0 = 0.71$, $p_1 = 0.26$, $p_2 = 0.03$ $\omega_0 = 0.15$, $\omega_1 = 1.23$, $\omega_2 = 7.61$
$NY^{3(\omega_1=1)}$	14	−3037.26	$p_0 = 0.66$, $p_1 = 0.30$, $p_2 = 0.04$ $\omega_0 = 0.13$, $\omega_2 = 6.58$
$NY^{3(\omega_1=1)(\omega_0=0)}$	13	−3050.45	$p_0 = 0.39$, $p_1 = 0.56$, $p_2 = 0.04$ $\omega_2 = 8.30$
$NY^{2(\omega_1=1)(\omega_0=0)}$	11	−3095.77	$p_0 = 0.41$, $p_1 = 0.59$
NY^1	11	−3148.70	$\omega = 0.50$

positively selected sites. However, no sign of positive selection is found among the *DEF/GLO* data set as $\omega_2 = 0.18$. It is important to note that $NY^{2(\omega_1=1)(\omega_0=0)}$ *vs.* $NY^{3(\omega_1=1)(\omega_0=0)}$ is not the only model comparison that tests for traces of positive selection. Indeed, Yang *et al.* [96], Anisimova *et al.* [4], and others have shown that the comparison of slightly more realistic models (e.g. $NY^{2(\omega_1=1)}$ *vs.* $NY^{3(\omega_1=1)}$) provides more powerful tests of positive selection. Another potential pitfall with this approach is related to the confounding effect of recombination. For instance, recombination is widespread among HIV-1 sequences (e.g. [76]) and in the presence of high levels of recombination, the identification of sites experiencing positive selection may suffer from high false-positive rates [5]. Hence, the results of such likelihood analysis need to be interpreted with caution.

The increase of log likelihood from model $NY^{3(\omega_1=1)(\omega_0=0)}$ to $NY^{3(\omega_1=1)}$ is significant for both data sets. To understand this result, consider a site at which dozens of synonymous substitutions and only one non-synonymous change occurred. Models that constrain ω_0 to be 0 provide a poor description of such a site because, according to this model, non-synonymous substitutions never occur. Models with a small but positive ω_0 value give a much better description of such data. Hence, it is likely that both HIV-1 *env* and *DEF/GLO* alignments display very few sites where only synonymous changes occurred. The analysis of other HIV-1 *env* data sets has shown similar increases of likelihood when comparing $NY^{3(\omega_1=1)(\omega_0=0)}$ to $NY^{3(\omega_1=1)}$ [28]. Therefore, it is likely that imposing the constraint $\omega_0 = 0$ at certain sites and in every lineage is not biologically realistic in most cases.

Thanks to its flexibility, NY^3 is very useful to estimate the distribution of the ω ratio. Fitting this model to the *DEF/GLO* data set clearly shows that most ω ratios are centred around 0.1–0.3. Therefore, it is not surprising that models that force values of this ratio to be greater or equal to 1.0 provide a significantly worse description of this data set. Moreover, fitting a parametric distribution of the ω ratio to this data set would probably be more appropriate than NY^3 non-parametric approximation. Yang *et al.* [96] proposed a β density to approximate the distribution of ω in the $[0, 1]$ range (model 'M7'). They showed that, with fewer parameters than non-parametric models, such an approach provides an equally good fit to most data sets and an even better fit for a data set that mostly evolves under negative selection (see Table 6 in Yang *et al.*, [96]). The picture is quite distinct for the HIV-1 *env* data set. The non-parametric approximation of the ω distribution seems relevant here as the ratios estimated under NY^3 are very similar to those given by $NY^{3(\omega_1=1)}$ or $NY^{3(\omega_0=0)(\omega_1=1)}$. Hence, it is not surprising that these three models provide almost equally good explanations of the data.

3.4.2 *Application: classification of sites into selection regimes*

Codon-based models (Fig. 3.2) have been used extensively to identify specific regions in proteins that evolve under positive selection. For example, in the major histocompatibility complex, positive selection appears to be responsible for the excess of replacement substitutions in the antigen recognition site [38]. Positive selection has also been detected in abalone sperm lysins [51], primate lysozymes [57], regions involved in species-specific sperm-egg interaction [82], and in various viral proteins subject to immune surveillance [30, 62, 69].

The identification of positively selected sites usually relies on an empirical Bayesian approach. This method is based on the posterior probability for a site i to evolve under positive selection:

$$P(\omega > 1.0|i, D, \mathcal{M}_\Theta) = \frac{\sum_{\omega_\theta>1} \pi_\theta L_i(\omega_\theta; D)}{\sum_{\omega_\theta} \pi_\theta L_i(\omega_\theta; D)}, \tag{3.17}$$

where ω_θ is the ω ratio that corresponds to the M_θ component of the $\mathcal{M}_\Theta$ mixture model. $\mathcal{M}_\Theta$ is usually one of $NY^{3(\omega_1=1)(\omega_0=0)}$, $NY^{3(\omega_1=1)}$ or NY^3, and π_θ is an estimate of the equilibrium frequency of M_θ. The term $L_i(\omega_\theta; D)$ is generally a marginal likelihood with respect to the phylogenetic tree (T), including topology and branch lengths, as well as the free parameters of M_θ other than ω_θ (e.g. the transition/transversion ratio). This probability can also be calculated using Markov chain Monte Carlo methods [33]. This approach not only allows the posterior probability to be computed, it provides estimates of the distribution of the substitution model parameters too. Analysing the shape of these distributions gives useful hints about the quantity and the quality of information carried by the data. The computational burden involved here is often a limitation though. Hence, the most popular method to date [62] does not integrate over nuisance parameters such as the tree topology or branch lengths. The values of these parameters are usually maximum-likelihood estimates.

Yang, Wong, and Nielsen [97] recently proposed estimating a close form of (3.17) using a Bayes empirical Bayes approach. This method takes into account uncertainty in the estimation of the equilibrium frequencies of the selection classes (i.e. π_θ). It is also more tractable from a computational perspective than the fully Bayesian approach and generates less false positives when searching for positively selected sites, than methods that solely rely on the posterior probability (3.17). This approach is likely to become commonplace as it is implemented in the widely used 'codeml' programme from the PAML [94] package.

Nielsen and Yang [62] originally proposed a maximum *a posteriori* decision rule to identify positively selected sites. A site is said to be positively selected if the corresponding posterior probability is larger than the posterior probability of any other selection regime (defined by $\omega \leq 1.0$) at that site. In practice, however, a site is said to be positively selected if the corresponding posterior probability of positive selection is larger than a given threshold, typically 0.95. To test the stringency of this 0.95 threshold, Yang *et al.* [97] randomly generated sites that did not evolve under positive selection (i.e. $H_{0,i}$ is true for every i). They showed that a threshold of 0.95 on the posterior probability of the positive selection regime leads to a proportion of falsely rejected null hypotheses (type-I error) very close to 0 (i.e. $\alpha \simeq 0$, while $\alpha = 5\%$ is the value one would expect in a statistical test framework). This threshold approach then appeared to be very conservative.

During the last few years, lots of statistical methods have been developed for the analysis of microarray data. One typically asks the question 'given its expression profile, is this gene differentially expressed in the various experimental conditions tested here ?' for every gene included in the microarray experiment. In this context, it is specially important to control the frequency of type-I errors, more specifically the proportion of cases where one decides that the gene is differentially expressed while it is not in reality. Benjamini and Hochberg [8, 9] proved that the expected proportion of type-I errors among the significant results (or false detection rate, FDR) can actually be controlled. Controlling the FDR at a given α level is less stringent than a 1-α fixed threshold approach. Hence, more significant results are expected to be found while the reliability of the conclusions is still controlled by a sound statistical reasoning.

Newton *et al.* [60] later proposed a method to control the FDR from the posterior probabilities of the different classes of a mixture model. This approach can be easily adapted to the identification of positively selected sites [26]. Let $\beta_i = P(\omega \leq 1.0 | i, D, \mathcal{M}_\Theta)$ be the posterior probability that site i evolves under a regime that is distinct from positive selection. The goal here is to determine the value of the threshold ρ such that the expected proportion of false positives among the sites at which $\beta_i \leq \rho$ is less than some value α, the desired FDR. The expected rate of false detections among such a list of sites and given the data is:

$$FDR(\rho) = \frac{\sum_i \beta_i 1\{\beta_i \leq \rho\}}{\sum_i 1\{\beta_i \leq \rho\}},$$

where $1\{.\}$ is an indicator function and the sums run over all sites of the alignment. We therefore have to select $\rho \leq 1$ as large as possible so that $FDR(\rho) \leq \alpha$. Extensive simulations have shown that this method provides a substantial gain of power (i.e. more positively selected sites are detected) while being robust to model misspecification [26].

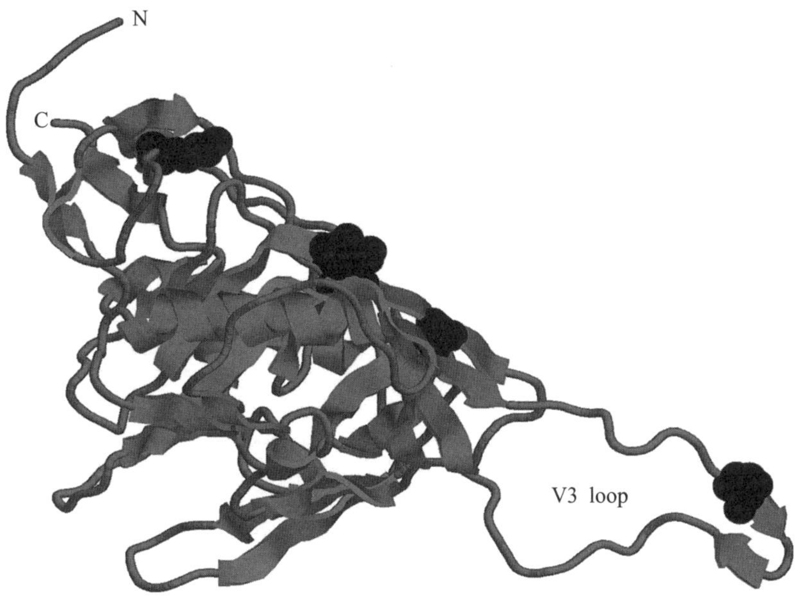

FIG. 3.3. **3D structure of the HIV-1** *env* **protein.** The black dots correspond to sites that are identified as positively selected. (Drawn with RasMol [73]).

Controlling the FDR at the $\alpha = 5\%$ level is standard. Both the FDR and the 0.95 fixed threshold methods converged to the same set of three sites under models $NY^{3(\omega_1=1)}$ and $NY^{3(\omega_1=1)(\omega_0=0)}$. However, under model NY^3, which is the most likely, five sites of the HIV-1 *env* data set are identified as positively selected according to the FDR approach, while only the same three sites are detected with the 0.95 fixed threshold method. Figure 3.3 shows the location of these five sites on the 3D structure of the HIV-1 *env* protein. One of the sites is located within the V3 variable loop region which is targeted by immunoglobulins [69]. The other sites are located in different areas but still on the surface of the molecule. Therefore, they are potential targets for the immune system, which would explain the evidence for positive selection. No *DEF/GLO* site evolves under positive selection according to the models tested here ($\omega_2 < 1$ under NY^3 and sub-models).

The approach described above is not only limited to the detection of positively selected sites. It can also be used to classify sites in any class of ω. It is also worth mentioning that if site i really belongs to class θ then the posterior probability of θ at that site is expected to be larger than the prior probability of the same class π_θ (see equation (3.17)). Hence, any attempt to classify a site i in a selection (or a substitution rate) class θ should be scrutinized with respect to the difference between prior and posterior probabilities of θ at site i.

To sum up, mixture models that allow ω to vary across sites are useful to decipher the natural selection processes involved at the molecular level. Most notably, these models are used to characterize the selection regimes that act at the individual-site level. However, such models use the same distribution of the ω ratio at each site to describe the heterogeneity across positions. In other words, these models assume that the variability of selection classes is the same across different regions of a protein. Huelsenbeck *et al.* [34] recently proposed an elegant solution that removes this constraint. They modelled the variation of selective processes among sites using a Dirichlet process in a Bayesian framework. Using Markov chain Monte Carlo, they were able to estimate the distributions of ω at individual codon sites. The analysis of several data sets suggests that these distributions vary extensively across sites. Hence, this model provides a more realistic picture of the selective regimes and their heterogeneity across positions of a sequence. This approach is also much more computationally demanding than fitting the models described in this section, which is usually done under the maximum likelihood framework. Hence, it is warranted to test if the new model discovers biologically relevant features that the standard approach fails to detect.

3.4.3 *Among-site and lineage heterogeneity in a unified framework*

Variability across sites is not the only source of heterogeneity of substitution processes. Substitution rates and selection regimes also vary across lineages. In sections 3.2.6 and 3.2.7, we have seen that Markov-modulated Markov models (MMM) account for both variability across sites and across lineages. This section focuses on variability of the selection classes and applies the MMM approach, combined with standard codon-based models (Fig. 3.2), to our two illustrative data sets.

Standard codon-based models are nested within the corresponding MMM versions (see Fig. 3.2). For instance, $\mathbf{C}_{\text{F81}}\odot\text{NY}^3$ tends to NY^3 when the rate of switches between selection regimes (i.e. δ in equation (3.11)) tends to 0.0. The distribution of the likelihood ratio statistic when testing $\delta = 0$ asymptotically follows a 50:50 mixture of χ^2_0 and χ^2_1 under the null hypothesis. The $\mathbf{C}_{\text{F81}}$ versions of the MMM models are also nested within the corresponding $\mathbf{C}_{\text{GTR}}$ models. $\mathbf{C}_{\text{GTR}}\odot\text{X}$ and $\mathbf{C}_{\text{F81}}\odot\text{X}$ (where X corresponds to $\text{NY}^{3(\omega_1=1)(\omega_0=0)}$, $\text{NY}^{3(\omega_1=1)}$ or NY^3) are the same model when the three R rates in the $\mathbf{C}_{\text{GTR}}$ matrix (see equation (3.11)) are equal. Hence, $\Delta = 2[\ln\text{L}(\mathbf{C}_{\text{GTR}}\odot\text{X}) - \ln\text{L}(\mathbf{C}_{\text{F81}}\odot\text{X})]$ follows a χ^2_2 distribution under the null hypothesis that sequences evolve under $\mathbf{C}_{\text{F81}}\odot\text{X}$.

MMM versions of the standard codon-based models were fitted to the data. Log-likelihood and values of the substitution parameters are given in Tables 3.5 and 3.6. The comparison between log likelihoods of the NY^3 model (Tables 3.3 and 3.4) and the corresponding MMM models is systematically significant and is impressive with the *DEF/GLO* data set. Hence, the selection patterns vary extensively across lineages and sites. Differences of log likelihood between $\mathbf{C}_{\text{X}}\odot\text{NY}^{3(\omega_1=1)(\omega_0=0)}$, $\mathbf{C}_{\text{X}}\odot\text{NY}^{3(\omega_1=1)}$ and $\mathbf{C}_{\text{X}}\odot\text{NY}^3$ (where X is either F81 or GTR,

TABLE 3.5. **Log likelihood of Markov-modulated Markov models (*DEF/GLO* data set).** Values of $R_{0\leftrightarrow 1}$, $R_{0\leftrightarrow 2}$ and $R_{1\leftrightarrow 2}$ are normalized such that δ is the expected number of changes in selection class during one time unit. Values of likelihood and model parameters estimated under $\mathbf{C}_{\text{GTR}}\odot\text{NY}^{3(\omega_1=1)}$ and $\mathbf{C}_{\text{GTR}}\odot\text{NY}^{3(\omega_1=1)(\omega_0=0)}$ are very similar to those given by $\mathbf{C}_{\text{GTR}}\odot\text{NY}^3$. The same holds for models $\mathbf{C}_{\text{F81}}\odot\text{NY}^{3(\omega_1=1)}$ and $\mathbf{C}_{\text{F81}}\odot\text{NY}^{3(\omega_1=1)(\omega_0=0)}$ when compared to $\mathbf{C}_{\text{F81}}\odot\text{NY}^3$.

Model	df	lnL	Estimated parameters
$\mathbf{C}_{\text{GTR}}\odot\text{NY}^3$	18	−28130.85	$\delta = 0.38$
			$R_{0\leftrightarrow 1} = 0.66$ $R_{0\leftrightarrow 2} = 3.10^{-3}$, $R_{1\leftrightarrow 2} = 5.22$
			$p_0 = 0.38$, $p_1 = 0.46$, $p_2 = 0.16$
			$\omega_0 = 0.01$, $\omega_1 = 0.12$, $\omega_2 = 1.24$
$\mathbf{C}_{\text{F81}}\odot\text{NY}^3$	16	−28200.33	$\delta = 0.22$
			$R_{0\leftrightarrow 1} = R_{0\leftrightarrow 2} = R_{1\leftrightarrow 2} = 1.59$
			$p_0 = 0.46$, $p_1 = 0.33$, $p_2 = 0.20$
			$\omega_0 = 2.10^{-3}$, $\omega_1 = 0.20$, $\omega_2 = 0.73$

TABLE 3.6. **Log likelihood of Markov-modulated Markov models (HIV-1 *env* data set).** See the caption of Table 3.5.

Model	df	lnL	Estimated parameters
$\mathbf{C}_{\text{GTR}}\odot\text{NY}^3$	18	−3018.98	$\delta = 2.62$
			$R_{0\leftrightarrow 1} = 1.94$, $R_{0\leftrightarrow 2} = 2.10^{-4}$, $R_{1\leftrightarrow 2} = 7.78$
			$p_0 = 0.66$, $p_1 = 0.30$, $p_2 = 0.05$
			$\omega_0 = 0.06$, $\omega_1 = 0.80$, $\omega_2 = 9.84$
$\mathbf{C}_{\text{F81}}\odot\text{NY}^3$	16	−3021.14	$\delta = 2.17$
			$R_{0\leftrightarrow 1} = R_{0\leftrightarrow 2} = R_{1\leftrightarrow 2} = 2.25$
			$p_0 = 0.70$, $p_1 = 0.25$, $p_2 = 0.05$
			$\omega_0 = 0.05$, $\omega_1 = 0.94$, $\omega_2 = 8.70$

results not shown) are much smaller than the differences between these three models implemented in a mixture model framework (Table 3.3). This result is not surprising as allowing for site-specific switches of selection regimes adds more flexibility to fit a codon substitution model to the data. Indeed, we have seen above (see Section 3.4.1) that a site at which dozens of synonymous substitutions and only one non-synonymous change occurred is not properly described by a mixture model that constrains ω_0 to be 0. However, such site-pattern is well explained if the same codon-based model is combined with a process that accounts for site-specific changes between negative and positive selection classes.

Free values of ω are more extreme when estimated under MMM models than under mixture models. For instance, traces of (weak) positive selection are detected among the *DEF/GLO* data set under $\mathbf{C}_{\text{GTR}} \odot \text{NY}^3$ while these sequences mostly evolve under a strong negative selection according to NY^3. The reason is that mixture models interpret similar amounts of non-synonymous and synonymous substitutions as the consequence of an underlying neutral process. If non-synonymous and synonymous substitutions are clustered on distinct lineages in the tree, MMM models will rather interpret this pattern as the succession of positive and negative selection episodes.

Differences of log likelihood between $\mathbf{C}_{\text{F81}} \odot \text{NY}^3$ and $\mathbf{C}_{\text{GTR}} \odot \text{NY}^3$ are less important than those observed between mixture and MMM models. Nevertheless, they are highly significant when considering the *DEF/GLO* data set. Therefore, the three different changes between selection regimes do not occur at the same rate for this data set. The rate of switches between the smallest and the largest values of ω is much lower than the two other rates. Hence, it is likely that the site-specific evolution of the ω ratio does not involve drastic changes of selection regimes. Moderate variations of this parameter during the course of evolution seem to be the most common.

The $\mathbf{C}_{\text{F81}}$ and $\mathbf{C}_{\text{GTR}}$ matrices are normalized such that δ is the expected number of selection class changes during one time unit. The $\mathbf{Diag}(\mathbf{Q}_{\omega_\theta})$ matrix being normalized such that one expected codon substitution occur in one time unit (see equation (3.13)), δ also corresponds to the ratio between the rate of changes between selection classes and the rate of substitutions between codons. Hence, the rate of switches between selection regimes is $\sim$3 times slower than the substitution rate among the *DEF/GLO* data set. Surprisingly, the switching rate is $\sim$2 to $\sim$4 times larger than the substitution rate in the HIV-1 *env* data set. This result is somewhat surprising as the expected number of switches between selection regimes that can be inferred from sequence comparison should not exceed the expected number of codon substitutions. Further investigations would be needed to understand this finding. A plausible explanation could be that the value of δ is poorly estimated.

3.4.4 *Application: visualization of time-dependent variations at individual sites*

MMM models can be used to evaluate the posterior probability of a given selection regime at any node of the phylogeny, at a given position of the alignment. It is also possible to compute this posterior probability anywhere on a given edge. Hence, measuring these probabilities at multiple positions in the tree allows us to follow the site-specific variations of selection regimes. It is worth mentioning that these site-specific patterns of variations are inferred from the data and not specified *a priori*, in contrast to Yang and Nielsen [95] branch-site models.

Let e be an edge of length l_e. $\nu(\lambda l_e)$ is a (non-existing) node located on edge e, at a fraction $\lambda \in [0,1]$ of l_e. $M(\nu(\lambda l_e))$ is the model observed at node $\nu(\lambda l_e)$. The posterior probability of model M_θ at $\nu(\lambda l_e)$ and site i is defined by:

$$P(M(\nu(\lambda l_e)) = M_\theta | i, e, T, D, \mathcal{M}_\Theta) = \frac{\pi_\theta L_i(M(\nu(\lambda l_e)) = M_\theta, T; D)}{\sum_\theta \pi_\theta L_i(M(\nu(\lambda l_e)) = M_\theta, T; D)}.$$

For each edge e in the tree and each site i, we compute:

$$P(M_\theta | i, e, T, D, \mathcal{M}_\Theta) = \frac{1}{N} \sum_{k=0}^{k=N-1} P\left(M\left(\nu\left(\frac{k+\frac{1}{2}}{N} l_e\right)\right) = M_\theta | i, T, D, \mathcal{M}_\Theta\right),$$

with N usually set to 10. This equation summarizes the posterior probability of model $\mathcal{M}_\Theta$ on edge e, at site i.

The posterior probabilities of the third selection class (which corresponds to strong positive selection with HIV-1 *env* and a nearly neutral process of evolution with *DEF/GLO*) were computed under model $\mathbf{C}_{\text{F81}} \odot \text{NY}^3$ (Fig. 3.2) for both data sets. These probabilities are then displayed on the corresponding phylogenies at each site of the alignment. Figures 3.4 and 3.5 show the patterns obtained

FIG. 3.4. **Patterns of variations of the selection regimes along five distinct sites of the HIV-1 *env* protein.** The thickness of each edge is proportional to the posterior probability of the third selection class. The $\mathbf{C}_{\text{F81}} \odot \text{NY}^3$ model was fitted to the data and $\omega_2 = 8.70$, indicating a strong positive selection in the third class.

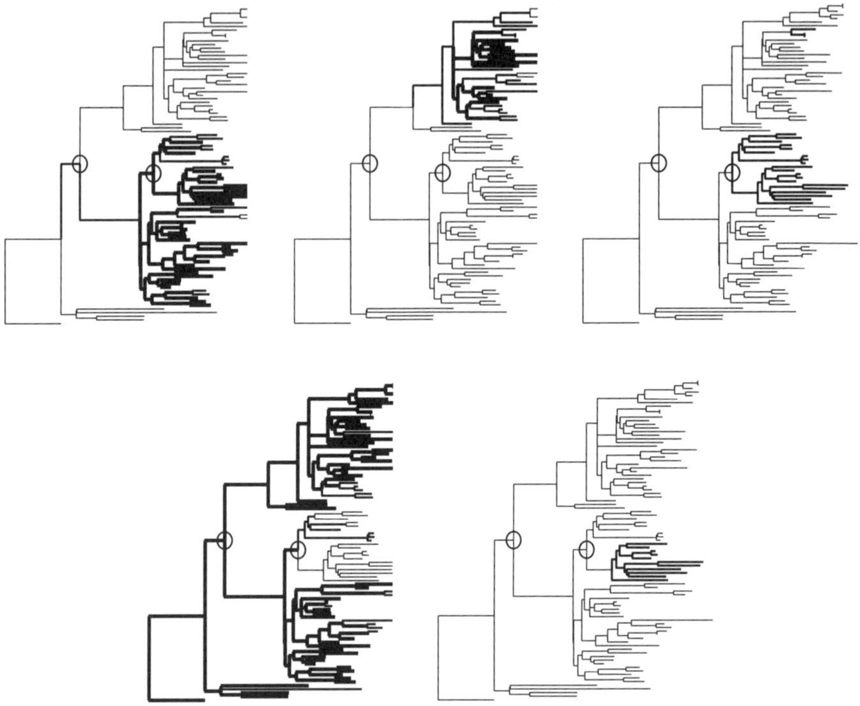

FIG. 3.5. **Patterns of variations of the selection regimes along five distinct sites of the *DEF/GLO* protein.** The circles correspond to duplication events. The duplication near the root of the tree separates the *DEF* and *GLO* clades. The shallow duplication is the most important duplication event that occurred in the *DEF* lineage. The edge width is proportional to the posterior probability of the third selection class. The $\mathbf{C}_{\text{F81}} \odot \text{NY}^3$ model was fitted to the data and $\omega_2 = 0.73$, indicating a nearly neutral process of evolution.

from five sites for each data set. These sites display typical patterns of site-specific variation of selection regimes in each data set. The analysis of the HIV-1 *env* data set shows clear traces of positive selection among a limited number of lineages in the tree. According to models that do not allow the selection regimes to vary across lineages, these sites are not positively selected. However, a closer analysis of these positions shows that non-synonymous substitutions are generally clumped on a few branches of the phylogeny instead of being scattered on the whole tree [28]. It is therefore very likely that these sites were positively selected at some stage of their evolution. Many sites of the *DEF/GLO* data set also display switches between selection patterns (Fig. 3.5).

From a biological perspective, it is interesting to note that, in some cases, positive selection occurs at early stages of the HIV-1 infection and vanishes afterwards. Other sites show very distinct patterns with positive selection occurring in intermediate or late stages of the infection. Such observations raise several questions about the complex interactions between HIV-1 genome evolution, virus reproductive fitness, and immune response. Are these episodes of positive selection the consequences of a transient immune response? Or do they facilitate the entry of the virus in the host cells? The residues that display these peculiar evolutionary patterns are located on peripheral regions of the tree-dimensional structure of the *env* protein. This observation suggests that the transient immune response hypothesis is more likely than the replicative fitness one.

Patterns of changes between selection classes displayed by the *DEF/GLO* data set (Fig. 3.5) also shed some light on important evolutionary mechanisms. The positions of these changes seem strongly correlated to those of duplication events, even though this hypothesis remains to be statistically tested. It is interesting to note that the changes close to duplication events do not systematically occur in the same direction. Indeed, most changes are from a strong negative to a weak selection process, but a few others are from weak to strong negative selection. These changes also generate asymmetrical patterns: the two lineages generated by the duplication event most often evolve under distinct selection regimes. These results suggest that the question of the neofunctionalization or subfunctionalization to explain the fate of duplicated genes should not be tackled at the gene level. Indeed, while the asymmetrical nature of the changes of selection processes supports the neofunctionalization hypothesis, different sites display distinct patterns which are not compatible with a single biological hypothesis to describe the evolution of the whole gene.

3.5 Discussion

We discussed and applied mixture and Markov-modulated Markov approaches to account for rate and selection regime heterogeneity. These mathematical tools have been used to deal with a number of other biological questions. At the DNA level, Huelsenbeck and Nielsen [36] used mixture models to represent differences in the transition/transversion ratio, while Pagel and Mead [64] analysed a large 22-gene data set and showed that a 4-component mixture of GTR+Γ models greatly increases the fit to the data and improves phylogeny reconstruction. Several authors also used mixtures to represent the heterogeneity of site evolution in proteins, depending either on the secondary structure and exposition [23] or on the biochemical context [46, 50].

Markov-modulated Markov models were not the first to be used to describe among-site and lineage heterogeneity of substitution processes. Indeed, efforts have been made to describe variations of selection patterns using new types of mixture models [95]. Under such models, namely, the branch-site models, it is first necessary to determine which lineages are likely to evolve under positive selection using *a priori* knowledge. These mixture models then assume that such lineages evolve under a negative, neutral, or positive selection process while the other parts of the tree are only allowed to evolve under negative selection or a neutral process. The branch-site models have been successfully used to study molecular mechanisms involved in genetic co-option [11] or gene duplication [10]. A number of studies [3, 7, 52, 53, 55, 58, 81] were dedicated to the related problem of building statistical tests to detect sites showing evidence for heterotachy. The proposed tests, basically, select sites that do not evolve in a standard homotachous way but do not focus on modelling heterotachy.

We already mentioned several applications of the Markov-modulated Markov approach: Huelsenbeck [37] tested the 'On/Off' model on a number of DNA data sets and showed a significant likelihood improvement, in comparison with a standard gamma plus invariant model; Pupko and Galtier [68] used the Galtier [19] model to detect sites showing rapid adaptation after a speciation event; Guindon *et al.* [28] developed the combination of codon-based NY^3 and category GTR-like models to analyse viral sequences, as described in Sections 3.2.7, 3.4.3 and 3.4.4. However, the Markov-modulated Markov approach, despite its elegance and conceptual simplicity, has still not been much used, probably because of its computational cost.

A recent paper by Kolaczkowski and Thornton [45] attracted attention to heterotachy. The authors simulated a mixture model where sites were equally distributed between two components, each corresponding to a four-taxon tree belonging to the Felsenstein zone [15]. Let a, b, c, and d be the taxa; in both trees, the $\{a, b\}$ pair was separated from the $\{c, d\}$ pair, but in one tree a and c corresponded to long branches and b and d to short branches, while in the second tree, the a and c branches were short and the b and d ones were long. Kolaczkowski and Thornton showed that with data simulated this way, parsimony and maximum-likelihood methods performed poorly, but maximum-likelihood was the worst, which was somewhat surprising as maximum-likelihood outperforms parsimony in the Felsenstein zone. A number of responses to this article were published, showing that these data are quite special, both from a mathematical and biological standpoint [66, 77, 78]. It was also surprising that Markov-modulated Markov models in the line of Galtier [19] did not perform any better than standard models with these data. Spencer *et al.* [77] explained this fact, which is due to the reduced number (i.e. 2) of branch length configurations being used for their simulations, while Galtier or Tuffley and Steel models implicitly assume that all configurations are possible and are equally likely. Kolaczkowski and Thorton's findings outline the limits of the current heterotachy models. They give a certain level of flexibility but do not allow for individual and distinct evolution of the sites. These models still include a unique tree with unique branch length assignment, which shows the common history of the sites. Sites evolve under different rates and these rates may change during the course of evolution, but these events are rare and penalized in likelihood calculations. Further work should be done to determine whether these models are flexible enough with real data. We showed the usefulness of these models for studying evolution at the molecular level; it is still unclear whether and how they should be used for reconstructing phylogenies [53, 65, 79], which is another important direction for further research.

Acknowledgements

Many thanks to Maria Anisimova, Avner Bar-Hen, Samuel Blanquart, Nicolas Galtier, Allen Rodrigo, Mike Steel, and an anonymous reviewer for their help and comments. This work was supported by ACI-NIM and ACI-IMPBIO.

References

[1] Akaike, H. (1974). A new look at the statistical model identification. *IEEE Transactions on Automatic Control*, **19**, 716–723.

[2] Altschul, S., Gish, W., Miller, W., Myers, E., and Lipman, D. (1990). Basic local alignment search tool. *Journal of Molecular Biology*, **215**, 403–410.

[3] Ané, C., Burleigh, J., McMahon, M., and Sanderson, M. (2005). Covarion structure in plastid genome evolution: a new statistical test. *Molecular Biology and Evolution*, **22**, 914–924.

[4] Anisimova, M., Bielawski, J., and Yang, Z. (2001). The accuracy and power of likelihood ratio tests to detect positive selection at amino acid sites. *Molecular Biology and Evolution*, **18**, 1585–1592.

[5] Anisimova, M., Nielsen, R., and Yang, Z. (2003). Effect of recombination on the accuracy of the likelihood method for detecting positive selection at amino acid sites. *Genetics*, **164**, 1229–1236.

[6] Aris-Brosou, S. and Yang, Z. (2002). Effects of models of rate evolution on estimation of divergence dates with special reference to the metazoan 18S ribosomal RNA phylogeny. *Systematic Biology*, **51**, 703–714.

[7] Baele, G., Raes, J., de Peer, Y. Van, and Vansteelandt, S. (2006). An improved statistical method for detecting heterotachy in nucleotide sequences. *Molecular Biology and Evolution*, **23**, 1397–1405.

[8] Benjamini, Y. and Hochberg, Y. (1995). Controlling the false discovery rate – a practical and powerful approach to multiple testing. *Journal of the Royal Statistics Society: Series B (Statistical Methodology)*, **57**, 289–300.

[9] Benjamini, Y. and Hochberg, Y. (2000). The adaptive control of the false discovery rate in multiple hypothesis testing with independent statistics. *Journal of Educational and Behavioral Statistics*, **25**, 60–83.

[10] Bielawski, J. and Yang, Z. (2003). Maximum likelihood methods for detecting adaptive evolution after gene duplication. *Journal of Structural and Functional Genomics*, **3**, 201–212.

[11] Bielawski, J. and Yang, Z. (2004). A maximum likelihood method for detecting functional divergence at individual codon sites, with application to gene family evolution. *Journal of Molecular Evolution*, **59**, 121–132.

[12] Bryant, D., Galtier, N., and Poursat, M.-A. (2005). Likelihood calculations in phylogenetics. In *Mathematics of Evolution & Phylogenetics* (ed. O. Gascuel), pp. 33–62. Oxford University Press, Oxford.

[13] Dayhoff, M., Schwartz, R., and Orcutt, B. (1978). A model of evolutionary change in proteins. In *Atlas of Protein Sequence and Structure* (ed. M. Dayhoff), Volume 5, pp. 345–352. National Biomedical Research Foundation, Washington, D. C.

[14] Drummond, A., Pybus, O., Rambaut, A., Forsberg, R., and Rodrigo, A. (2003). Measurably evolving populations. *Trends in Ecology and Evolution*, **18**, 481–488.

[15] Felsenstein, J. (1978). Cases in which parsimony and compatibility methods will be positively misleading. *Systematic Zoology*, **27**, 401–410.

[16] Felsenstein, J. (1981). Evolutionary trees from DNA sequences: a maximum likelihood approach. *Journal of Molecular Evolution*, **17**, 368–376.

[17] Felsenstein, J. (2003). *Inferring Phylogenies.* Sinauer Associates, Inc., Sunderland.

[18] Felsenstein, J. and Churchill, G.A. (1996). A hidden Markov model approach to variation among sites in rate of evolution. *Molecular Biology and Evolution*, **13**, 93–104.

[19] Galtier, N. (2001). Maximum-likelihood phylogenetic analysis under a covarion-like model. *Molecular Biology and Evolution*, **18**, 866–873.

[20] Galtier, N. and Jean-Marie, A. (2004). Markov-modulated Markov chains and the covarion process of molecular evolution. *Journal of Computational Biology*, **11**, 727–733.

[21] Gaucher, E., Miyamoto, M., and Benner, S. (2001). Function-structure analysis of proteins using covarion-based evolutionary approaches: Elongation factors. *Proceedings of the National Academy of Sciences of the United States of America*, **98**, 548–552.

[22] Golding, G. B. (1983). Estimates of DNA and protein sequence divergence: an examination of some assumptions. *Molecular Biology and Evolution*, **1**, 125–142.

[23] Goldman, N., Thorne, J., and Jones, D. (1998). Assessing the impact of secondary structure and solvent accessibility on protein evolution. *Genetics*, **149**, 445–458.

[24] Goldman, N. and Yang, Z. (1994). A codon-based model of nucleotide substitution for protein-coding DNA sequences. *Molecular Biology and Evolution*, **11**, 725–736.

[25] Gu, X., Fu, Y.X., and Li, W.H. (1995). Maximum likelihood estimation of the heterogeneity of substitution rate among nucleotide sites. *Molecular Biology and Evolution*, **12**, 546–557.

[26] Guindon, S., Black, M., and Rodrigo, A. (2006). Control of the false discovery rate applied to the detection of positively selected amino acid sites. *Molecular Biology and Evolution*, **23**, 919–926.

[27] Guindon, S. and Gascuel, O. (2003). A simple, fast and accurate algorithm to estimate large phylogenies by maximum likelihood. *Systematic Biology*, **52**, 696–704.

[28] Guindon, S., Rodrigo, A., Dyer, K., and Huelsenbeck, J. (2004). Modeling the site-specific variation of selection patterns along lineages. *Proceedings of the National Academy of Sciences of the United States of America*, **101**, 12957–12962.

[29] Hasegawa, M., Kishino, H., and Yano, T. (1985). Dating of the Human-Ape splitting by a molecular clock of mitochondrial-DNA. *Journal of Molecular Evolution*, **22**, 160–174.

[30] Haydon, D., Bastos, A., Knowles, N., and Samuel, A. (2001). Evidence for positive selection in foot-and-mouth disease virus capsid genes from field isolates. *Genetics*, **157**, 7–15.

[31] Henikoff, S. and Henikoff, J. (1992). Amino acid substitution matrices from protein blocks. *Proceedings of the National Academy of Sciences of the United States of America*, **89**, 10915–10919.

[32] Ho, S., Phillips, M., Drummond, A., and Cooper, A. (2005). Accuracy of rate estimation using relaxed-clock models with a critical focus on the early Metazoan radiation. *Molecular Biology and Evolution*, **22**, 1355–1363.

[33] Huelsenbeck, J. and Dyer, K. (2004). Bayesian estimation of positively selected sites. *Journal of Molecular Evolution*, **58**, 661–672.

[34] Huelsenbeck, J., Jain, S., Frost, S., and Pond, S. (2006). A Dirichlet process model for detecting positive selection in protein-coding DNA sequences. *Proceedings of the National Academy of Sciences of the United States of America*, **103**, 6263–6268.

[35] Huelsenbeck, J., Larget, B., and Swofford, D. (2000). A compound Poisson process for relaxing the molecular clock. *Genetics*, **154**, 1879–1892.

[36] Huelsenbeck, J. and Nielsen, R. (1999). Variation in the pattern of nucleotide substitution across sites. *Journal of Molecular Evolution*, **48**, 86–93.

[37] Huelsenbeck, J. P. (2002). Testing a covariotide model of DNA substitution. *Molecular Biology and Evolution*, **19**, 698–707.

[38] Hughes, A. and Nei, M. (1988). Pattern of nucleotide substitution at major histocompatibility complex class I loci reveals overdominant selection. *Nature*, **335**, 167–170.

[39] Hughes, A., Ota, T., and Nei, M. (1990). Positive darwinian selection promotes charge profile diversity in the antigen-binding cleft of class I major-histocompatibility-complex molecules. *Molecular Biology and Evolution*, **7**, 515–524.

[40] Jin, L. and Nei, M. (1990). Limitations of the evolutionary parsimony method of phylogenetic analysis. *Molecular Biology and Evolution*, **7**, 82–102.

[41] Jones, D., Taylor, W., and Thornton, J. (1992). The rapid generation of mutation data matrices from protein sequences. *Computer Applications in the Biosciences*, **8**, 275–282.

[42] Jukes, T. and Cantor, C. (1969). Evolution of protein molecules. In *Mammalian Protein Metabolism* (ed. H. Munro), Volume III, Chapter 24, pp. 21–132. Academic Press, New York.

[43] Kaslow, R., Ostrow, D., Detel, R., Phair, J., Polk, B., and Rinaldo, C. (1987). The Multicenter AIDS Cohort Study: rationale, organization, and selected characteristics of the participants. *American Journal of Epidemiology*, **126**, 310–318.

[44] Kimura, M. (1980). A simple method for estimating evolutionary rates of base substitutions through comparative studies of nucleotide sequences. *Journal of Molecular Evolution*, **16**, 111–120.

[45] Kolaczkowski, B. and Thornton, J. (2004). Performance of maximum parsimony and likelihood phylogenetics when evolution is heterogeneous. *Nature*, **431**, 980–984.

[46] Koshi, J. and Goldstein, R. (1998). Models of natural mutations including site heterogeneity. *Proteins*, **32**, 289–295.

[47] Kosiol, C. and Goldman, N. (2004). Different versions of the Dayhoff rate matrix. *Molecular Biology and Evolution*, **22**, 193–199.

[48] Kumar, S. and Subramanian, S. (2002). Mutation rates in mammalian genomes. *Proceedings of the National Academy of Sciences of the United States of America*, **99**, 803–808.

[49] Lanave, C., Preparata, G., Saccone, C., and Serio, G. (1984). A new method for calculating evolutionary substitution rates. *Journal of Molecular Evolution*, **20**, 86–93.

[50] Lartillot, N. and Philippe, H. (2004). A Bayesian mixture model for across-site heterogeneities in the amino-acid replacement process. *Molecular Biology and Evolution*, **21**, 1095–1109.

[51] Lee, Y., Ota, T., and Vaquier, V. (1995). Positive selection is a general phenomenon in the evolution of abalone sperm lysin. *Molecular Biology and Evolution*, **12**, 231–238.

[52] Lockhart, P., Huson, D., Maier, U., Fraunholz, M., de Peer, Y. Van, Barbrook, A., Howe, C., and Steel., M. (2000). How molecules evolve in eubacteria. *Molecular Biology and Evolution*, **17**, 835–838.

[53] Lockhart, P., Steel, M., Barbrook, A., Huson, D., Charleston, M., and Howe, C. (1998). A covariotide model explains apparent phylogenetic structure of oxygenic photosynthetic lineages. *Molecular Biology and Evolution*, **15**, 1183–1188.

[54] Lopez, P., Casane, D., and Philippe, H. (2002). Heterotachy, an important process of protein evolution. *Molecular Biology and Evolution*, **19**, 1–7.

[55] Lopez, P., Forterre, P., and Philippe, H. (1999). The root of the tree of life in the light of the covarion model. *Journal of Molecular Evolution*, **49**, 496–508.

[56] Lynch, M. and Conery, J. (2000). The evolutionary fate and consequences of duplicated genes. *Science*, **290**, 1151–1155.

[57] Messier, W. and Stewart, C.-B. (1997). Episodic adaptative evolution of primate lysozymes. *Nature*, **385**, 151–154.

[58] Misof, B., Anderson, C., Buckley, T., Erpenbeck, D., Rickert, A., and Misof, K. (2002). An empirical analysis of mt 16s rRNA covarion-like evolution in insects: site-specific rate variation is clustered and frequently detected. *Journal of Molecular Evolution*, **56**, 330–340.

[59] Nei, M. and Gojobori, T. (1986). Simple methods for estimating the number of synonymous and nonsynonymous nucleotide substitutions. *Molecular Biology and Evolution*, **3**, 418–426.

[60] Newton, M., Noueiry, A., Sarkar, D., and Ahlquist, P. (2004). Detecting differential expression with a semiparametric hierarchical mixture method. *Biostatistics*, **5**, 155–176.

[61] Neyman, J. (1971). Molecular studies of evolution: a source of novel statistical problems. In *Statistical decision theory and related topics* (ed. S. Gupta and J. Yackel), pp. 1–27. Academic Press, New York.

[62] Nielsen, R. and Yang, Z. (1998). Likelihood models for detecting positively selected amino acid sites and application to the HIV-1 envelope gene. *Genetics*, **148**, 929–936.

[63] Ohta, T. (1993). Pattern of nucleotide substitutions in growth hormone-prolactin gene family: a paradigm for evolution by gene duplication. *Genetics*, **134**, 1271–1276.

[64] Pagel, M. and Meade, A. (2004). A phylogenetic mixture model for detecting pattern-heterogeneity in gene sequence or character-state data. *Systematic Biology*, **53**, 571–581.

[65] Penny, D., McComish, B., Charleston, M., and Hendy, M. (2001). Mathematical elegance with biochemical realism: the covarion model of molecular evolution. *Journal of Molecular Evolution*, **53**, 711–723.

[66] Philippe, H., Zhou, Y., Brinkmann, H., Rodrigue, N., and Delsuc, F. (2005). Heterotachy and long-branch attraction in phylogenetics. *BMC Evolutionary Biology*, http://www.biomedcentral.com/1471–2148/5/50.

[67] Pollock, D., Taylor, W., and Goldman, N. (1999). Co-evolving protein residues: maximum likelihood analysis and relationship to structure. *Journal of Molecular Biology*, **287**, 187–198.

[68] Pupko, T. and Galtier, N. (2002). A covarion-based method for detecting molecular adaptation: application to the evolution of primate mitochondrial genomes. *Proceedings of The Royal Society B: Biological Sciences*, **269**, 1313–1316.

[69] Ross, H. and Rodrigo, A. (2002). Immune-mediated positive selection drives human immunodeficiency virus type 1 molecular variation and predicts disease duration. *Journal of Virology*, **76**, 11715–11720.

[70] Sanderson, M. (1997). A nonparametric approach to estimating divergence times in the absence of rate constancy. *Molecular Biology and Evolution*, **14**, 1218–1231.

[71] Sanderson, M. (2002). Estimating absolute rates of molecular evolution and divergence times: a penalized likelihood approach. *Molecular Biology and Evolution*, **19**, 101–109.

[72] Sarich, V. and Wilson, A. (1967). Immunological time scale for hominid evolution. *Science*, **158**, 1200–1203.

[73] Sayle, R. and Milner-White, J. (1995). RasMol: Biomolecular graphics for all. *Trends in Biochemical Sciences*, **20**, 374.

[74] Self, S. and Liang, K. (1987). Asymptotic properties of maximum likelihood estimators and likelihood ratio tests under nonstandard conditions. *Journal of the American Statistical Association*, **82**, 605–610.

[75] Shankarappa, R., Margolick, J., Gange, S., Rodrigo, A., Upchurch, D., Farzadegan, H., Gupta, P., Rinaldo, C., Learn, G., He, X., Huang, X.-L, and Mullins, J. (1999). Consistent viral evolutionary changes associated with the progression of human immunodeficiency virus type 1 infection. *Journal of Virology*, **73**, 10489–10502.

[76] Shrinner, D., Rodrigo, A., Nickle, D., and Mullins, J. (2004). Pervasive genomic recombination of HIV-1 *in vivo*. *Genetics*, **167**, 1573–1583.

[77] Spencer, M., Susko, E., and Roger, A. (2005). Likelihood, parsimony, and heterogeneous evolution. *Molecular Biology and Evolution*, **22**, 1161–1164.

[78] Steel, M. (2005). Should phylogenetic models be trying to 'fit an elephant'? *Trends in Genetics*, **21**, 307–309.

[79] Steel, M., Huson, D., and Lockhart, P. (2000). Invariable site models and their use in phylogeny reconstruction. *Systematic Biology*, **49**, 225–232.

[80] Susko, E., Field, C., Blouin, C., and Roger, A. (2003). Estimation of rates-across-sites distributions in phylogenetic substitution models. *Systematic Biology*, **52**, 594–603.

[81] Susko, E., Inagaki, Y., Field, C., Holder, M., and Roger, A. (2002). Testing for differences in rates-across-sites distributions in phylogenetic subtrees. *Molecular Biology and Evolution*, **19**, 1514–1523.

[82] Swanson, W., Yang, Z., Wolfner, M., and Aquadro, C. (2001). Positive darwinian selection drives the evolution of several female reproductive proteins in mammals. *Proceedings of the National Academy of Sciences of the United States of America*, **98**, 2509–2514.

[83] Tavaré, S. (1986). Some probabilistic and statistical problems on the analysis of DNA sequences. *Lectures on Mathematics in the Life Sciences*, **17**, 57–86.

[84] Thompson, J., Higgins, D., and Gibson, T. (1994). CLUSTAL W: improving the sensitivity of progressive multiple sequence alignment through sequence weighting, position-specific gap penalties and weight matrix choice. *Nucleic Acids Research*, **22**, 4673–4680.

[85] Thorne, J., Kishino, H., and Painter, I. (1998). Estimating the rate of evolution of the rate of molecular evolution. *Molecular Biology and Evolution*, **15**, 1647–1657.

[86] Trivedi, K. (2001). *Probability and Statistics with Reliability, Queuing, and Computer Science Applications.* Wiley, Chichester.

[87] Tuffley, C. and Steel, M. (1998). Modelling the covarion hypothesis of nucleotide substitution. *Mathematical Biosciences*, **147**, 63–91.

[88] Uzzell, T. and Corbin, K. (1971). Fitting discrete probability distributions to evolutionary events. *Science*, **172**, 1089–1096.

[89] Whelan, S. and Goldman, N. (2001). A general empirical model of protein evolution derived from multiple protein families using a maximum-likelihood approach. *Molecular Biology and Evolution*, **18**, 691–699.

[90] Winter, K.-U., Saedler, H., and Theissen, G. (2002). On the origin of class B floral homeotic genes: functional substitution and dominant inhibition in *Arabidopsis* by expression of an orthologue from the gymnosperm *Gnetum. The Plant Journal*, **31**, 457–475.

[91] Yang, Z (1993). Maximum-likelihood estimation of phylogeny from DNA sequences when substitution rates differ over sites. *Molecular Biology and Evolution*, **10**, 1396–1401.

[92] Yang, Z. (1994). Maximum likelihood phylogenetic estimation from DNA sequences with variable rates over sites: approximate methods. *Journal of Molecular Evolution*, **39**, 306–314.

[93] Yang, Z. (1995). A space-time process model for the evolution of DNA sequences. *Genetics*, **193**, 993–1005.

[94] Yang, Z. (1997). PAML: a program package for phylogenetic analysis by maximum likelihood. *Computer Applications in the Biosciences*, **13**, 555–556.

[95] Yang, Z. and Nielsen, R. (2002). Codon-substitution models for detecting molecular adaptation at individual sites along specific lineages. *Molecular Biology and Evolution*, **19**, 908–917.

[96] Yang, Z., Nielsen, R., Goldman, N., and Krabbe Pedersen, A.-M. (2000). Codon-substitution models for heterogeneous selection pressure at amino acid sites. *Genetics*, **155**, 431–449.

[97] Yang, Z., Wong, W., and Nielsen, R. (2005). Bayes empirical Bayes inference of amino acid sites under positive selection. *Molecular Biology and Evolution*, **22**, 1107–1118.

[98] Zahn, L., Leebens-Mack, J., DePamphilis, C., Ma, H., and Theissen, G. (2005). To B or Not to B a flower: the role of DEFICIENS and GLOBOSA orthologs in the evolution of the angiosperms. *Journal of Heredity*, **96**, 225–240.

[99] Zuckerkandl, E. and Pauling, L. (1962). *Horizons in Biochemistry*, Chapter Molecular disease, evolution, and genic heterogeneity, pp. 189–225. Elsevier, Amsterdam.

4

PHYLOGENETIC INVARIANTS

Elizabeth S. Allman and John A. Rhodes

Abstract

Under many common models of sequence evolution along trees, frequencies of base patterns in extant taxa satisfy certain polynomial relationships known as 'phylogenetic invariants'. Though introduced in 1987 for phylogenetic inference, invariants remained difficult to construct, and the inefficiency of simple inference schemes based on known linear ones was discouraging. Recently there has been much progress in producing phylogenetic invariants, and in understanding their structure. Potentially useful connections between specific topological features in a tree (vertices and nodes) and specific invariants have emerged. We introduce some of the mathematical ideas underlying current understanding of invariants, with an emphasis on a geometric viewpoint and rank computations. We also highlight new insights arising from invariants, including better understanding of maximum-likelihood estimation and proofs of the identifiability of certain substitution models, such as the covarion and mixture models.

4.1 Introduction

Probabilistic models for the evolution of biological sequences are used throughout phylogenetics, both for theoretical analysis and for practical inference. Basic assumptions in these models lead naturally to expressing their predictions through *polynomial* expressions. This simple observation leads to the insight that polynomial algebra can provide alternative perspectives in phylogenetics.

Phylogenetic invariants were introduced in 1987 in two independent works, by Cavender and Felsenstein [13], and by Lake [48]. For DNA sequences, phylogenetic invariants are polynomial relationships that must hold between the frequencies of various base patterns in idealized data, which is perfectly in accord with a particular model and tree. By testing whether such polynomials for various trees were 'nearly zero' when evaluated on the observed frequencies of patterns in real data sequences, it was hoped that one could infer which tree best explained the data.

A number of difficulties, which will be surveyed later in this chapter, prevented invariants from being quickly developed into useful inference tools. In particular, while Lake's linear invariants had some desirable statistical properties, practical inference based on them performed poorly on sequences of a length typical of real data. This perhaps led some to question the value of invariants in general, even though few serious attempts at using higher degree invariants for inference were made. Indeed, thorough knowledge of non-linear invariants was largely lacking for DNA models, with the notable exception of the results on group-based models that began with Evans and Speed [24]. As the need for more general models to adequately describe data had become clearer, invariants that incorporated the added complexity were simply not known.

Recently, however, much progress has occurred in understanding phylogenetic models algebraically. Our knowledge of phylogenetic invariants has grown to include models of sufficient generality to encompass some of those currently used for inference. Most importantly, for those models which are well understood, a close relationship holds between specific invariants and particular local topological features of trees, such as edges or nodes. While more remains to be done in determining the structure of invariants for additional models of interest, there is now enough understanding to consider again how we might use invariants, either for inference or for theoretical analysis.

This chapter is divided into two parts. In the first, we discuss constructions of invariants, explain how they can be interpreted, and survey results on the extent to which all invariants for various models are known. We begin with a careful development of some invariants for the general Markov model, in order both to be concrete and to emphasize that invariants make interpretable statements about statistical models. In the second part, we turn to applications of invariants. These include recent investigations focused on understanding when maximum-likelihood inference may face multiple local optima, and on establishing the identifiability of tree topologies for certain mixture models. We end with more speculative uses for practical inference. We hope to convey that the perspectives invariants offer on phylogenetic models can be valuable in many settings, and that more applications remain to be discovered.

The mathematical field most appropriate to studying phylogenetic invariants is *algebraic geometry*, which is rich and well-developed, but far from the typical background of most phylogenetics researchers. In this chapter we provide only a gentle introduction to its terminology when necessary, and our presentation of some results omits more technical details. We hope this creates an overview that will be especially useful for those who might be more interested in thinking about how to use invariants than in how to find them.

A first example. For a concrete introduction to viewing a probabilistic model in phylogenetics algebraically, consider the following illustrative example:

An ancestral sequence at the root r of a tree gives rise to two descendant sequences, at leaves a and b of the tree shown in Fig. 4.1. We model evolution at a single site in a sequence, with the idea that each site evolves according to the same model, but independently (the i.i.d. assumption).

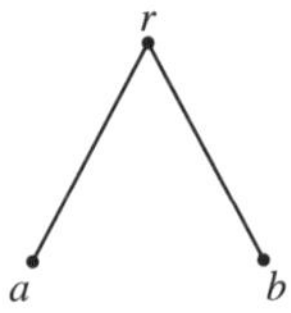

FIG. 4.1. Two taxa a and b descend from a common ancestor r.

For the ancestral sequence at r we specify the probabilities $\boldsymbol{\pi} = (\pi_1, \pi_2, \pi_3, \pi_4)$ with which the four bases ($A = 1$, $G = 2$, $C = 3$, $T = 4$) might appear at a particular site, or equivalently by the i.i.d. assumption, the relative frequencies at which bases appear across all sites. For each edge of the tree, we model the evolutionary process by specifying probabilities of various substitutions occurring. Thus for edge e_1, leading from r to a, we specify a 4×4 matrix M_1 whose (i, j)-entry is the conditional probability of observing base j in the sequence at a given that the ancestral base at r was i. Similarly, a matrix M_2 describes the mutation process on edge e_2, leading from r to b. The parameters of the model, the 4-*state general Markov model*, are the tree of Fig. 4.1 along which we model evolution, and the entries of $\boldsymbol{\pi}$, M_1, and M_2.

From the model parameters we compute the probability of each possible observation. The probability of seeing base j in a site at a and base k in the same site at b is

$$\text{Prob}(a = j,\ b = k) = p_{jk} = \sum_{i=1}^{4} \pi_i M_1(i, j) M_2(i, k). \tag{4.1}$$

The *joint distribution* of bases $P = (p_{jk})$, then, can be thought of as a 4×4 matrix, each of whose entries is a 4-term degree-3 polynomial in the parameters of the model. These 16 polynomials parameterizing the model reflect all the modeling assumptions, including the substitution probabilities, and the tree topology of Fig. 4.1.

In order to produce a clear and instructive example, we simplify the model further (at the expense of biological plausibility) by restricting it to the situation where the ancestral sequence is composed of only the base A, so that $\boldsymbol{\pi} = (1, 0, 0, 0)$. This *ancestral-A model* leads to a simplification of equation (4.1) so that the joint distribution of bases is given by the 16 quadratic polynomials

$$p_{jk} = M_1(1, j) M_2(1, k). \tag{4.2}$$

Now from inspecting equation (4.2), we observe that

$$p_{jk} p_{mn} - p_{jn} p_{mk} = 0, \tag{4.3}$$

since each term in this difference can be expressed in terms of the parameters as

$$M_1(1, j) M_2(1, k) M_1(1, m) M_2(1, n).$$

Thus for every choice of j, k and m, n we have found a polynomial,

$$f_{jk,mn}(P) = p_{jk}p_{mn} - p_{jn}p_{mk},$$

that will evaluate to 0 when $P = (p_{jk})$ is any true distribution of bases arising from the ancestral A-model, without regard to the particular numerical values appearing in the Markov matrix parameters. These polynomials are called *invariants* for the ancestral-A model on a 2-taxon tree.[1]

More generally, an invariant for a model is a polynomial that gives zero when evaluated on any distribution arising from that model, regardless of the parameter values leading to that distribution. On a distribution that does not arise from the model, an invariant typically evaluates to give a non-zero result. Since the invariants found here will, in fact, vanish on distributions arising from ancestral-G, ancestral-C, and ancestral-T models also, they are better termed as invariants for an ancestral-1-base model. Even so, by allowing two or more ancestral states it is easy to construct numerical examples of distributions on which these invariants will not be zero.

To see why model invariants might be useful, imagine aligned DNA sequences from taxa a and b. We wish to test whether this data might have been produced from the ancestral-A model on the tree above. We record the observed distribution $\widehat{P}$, a 4×4 array giving frequencies of aligned bases in the two sequences. If we believe the model provides a good description of the data, then we suspect $\widehat{P} \approx P$, where P is a true distribution arising from the model for some unknown choice of parameters. Thus for any model invariant, f, since $f(P) = 0$, we should find that $f(\widehat{P}) \approx 0$.

Thus we might simply evaluate the model's invariants on the observed distribution $\widehat{P}$ and, if we get values close to zero, take that as evidence that the ancestral-A model might describe the data well. If we get values far from zero, we could take that as evidence against the ancestral-A model providing a good fit to the data.

This is schematically indicated in Fig. 4.2, where we imagine two alternative models leading to different sets of invariants. In order to choose which model may best describe a data point $\widehat{P}$, we wish to determine if $\widehat{P}$ is closer to the zero set of one collection of invariants or the other.

In this way polynomial invariants for more elaborate phylogenetic models might provide a method of inference that *circumvents determination of numerical parameters*. In particular, the tree topology may be of more intrinsic interest than the numerical parameters in a phylogenetic model. If invariants can be found that test for each possible tree topology for a set of taxa, evaluating them on an observed distribution to see if they nearly vanish might enable us to infer the topology.

---

[1]This model is actually a familiar one in statistics, outside of phylogenetics; it is the independence model for a 2-way table P. The invariants above are commonly expressed in a slightly different form, using an *odds ratio*: $\frac{p_{jk}p_{mn}}{p_{jn}p_{mk}} = 1$.

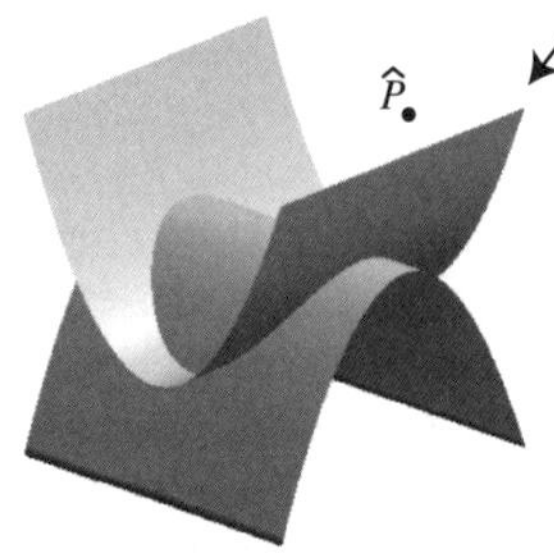

FIG. 4.2. The f_i and h_i are invariants for two alternative models. All joint distributions arising from the first model lie in the 'surface' defined by $f_i(P) = 0$, and similarly for the second. To decide which model better explains a data point $\widehat{P}$, we attempt to judge whether $f_i(\widehat{P}) \approx 0$ or $h_i(\widehat{P}) \approx 0$.

This idea, focused on determining the tree topology from larger sets of sequences, was the one introduced in both [13] and [48]. There are difficulties in applying this idea as naively as described here; nonetheless, it is a good one to keep in mind for motivation. In a nutshell, invariants have the potential to tell us something about whether an observed distribution might have arisen from a particular model, without having any need to infer numerical parameters.

Notice that in this example there are two sets of polynomials. The first, appearing in equation (4.2), are the *parameterization polynomials*, expressing the true distribution our model predicts in terms of the model parameters. The second, the invariants of the model, appearing in equation (4.3), describe the relationships that must hold within a distribution resulting from the parameterization. The parameterization polynomials are straightforward to produce, since they express the model as we have designed it. The invariants are consequences of the parameterization polynomials, but how to produce them or interpret their meaning is much less obvious for most models.

Finally, note that the idea of invariants need not be limited to phylogenetic models. Indeed, they can be studied in other statistical settings where polynomial parameterizations arise. The complexity and structure of phylogenetic models, however, makes the subject particularly rich in this setting.

Part 1. Finding Invariants

Discussing constructions of invariants first requires a more detailed specification of some phylogenetic models. Before proceeding, however, we note there is one invariant whose existence is easy to explain.

Consider any probabilistic model which allows only finitely many outcomes. The distribution will take the form of an array, where each entry is the probability of one possible outcome. For instance, for DNA substitution models for n-taxa, the joint distribution can be given by an n-dimensional $4 \times \cdots \times 4$ array. The vanishing of the *stochastic invariant*,

$$\sum_{i_1, i_2, \ldots, i_n} p_{i_1 i_2 \ldots i_n} - 1,$$

where the summation is over all entries of the distribution, states that the probabilities of all possible outcomes must add to 1. It is therefore an invariant for every such model.

4.2 Phylogenetic models on a tree

For convenience, we will assume all trees are binary (i.e. trivalent at all internal nodes, except possibly bivalent at a root).

Let T be an n-leaf unrooted binary tree, with its leaves labeled by a collection of taxa $X = \{a_1, a_2, \ldots, a_n\}$. We may introduce a root r by either choosing some existing node of T, or subdividing some edge of T and choosing the new node as the root, obtaining the rooted tree T^r. In a rooted tree T^r we view all edges as directed away from r.

The *κ-state general Markov (GM) model* on T^r is a model of character evolution parameterized by:

1. A root distribution vector $\boldsymbol{\pi}_r = (\pi_1, \pi_2, \ldots, \pi_\kappa)$. We interpret π_i as the probability that the character is in state i in the ancestral taxon r. Thus $\pi_i \geq 0$ and $\sum_{i=1}^{\kappa} \pi_i = 1$. For simple DNA models, $\kappa = 4$.
2. For each directed edge e of the rooted tree, a $\kappa \times \kappa$ Markov matrix M_e. We interpret the (i,j)-entry of M_e as giving the conditional probability that the character is in state j at the descendant end of e given that it was in state i at the ancestral end. Thus $M_e(i,j) \geq 0$ and $\sum_{j=1}^{\kappa} M_e(i,j) = 1$.

A key feature of the model is that we may observe states only at the leaves of the tree; states at all internal nodes are *hidden*.

Rather than give a general formula for the joint distribution arising from this model, we indicate its form through an example, with a specific tree. Considering the tree of Fig. 4.3, with M_i denoting the Markov matrix for edge e_i, we find the entries of the joint distribution P are

$$P(i,j,k,l,m) = p_{ijklm} = \sum_{s=1}^{\kappa} \sum_{t=1}^{\kappa} \sum_{u=1}^{\kappa} \sum_{v=1}^{\kappa} [\pi_s M_1(s,i) M_2(s,t) M_3(t,u) \times$$
$$M_4(u,j) M_5(u,k) M_6(t,v) M_7(v,l) M_8(v,m)]. \quad (4.4)$$

Note that these κ^5 parameterization polynomials in $8\kappa(\kappa-1)+\kappa-1$ variables reflect not only the assumptions of the general Markov model, but also the form of the tree in Fig. 4.3. Indeed, from the parameterization one can even reconstruct the tree, as it algebraically encodes the topology.

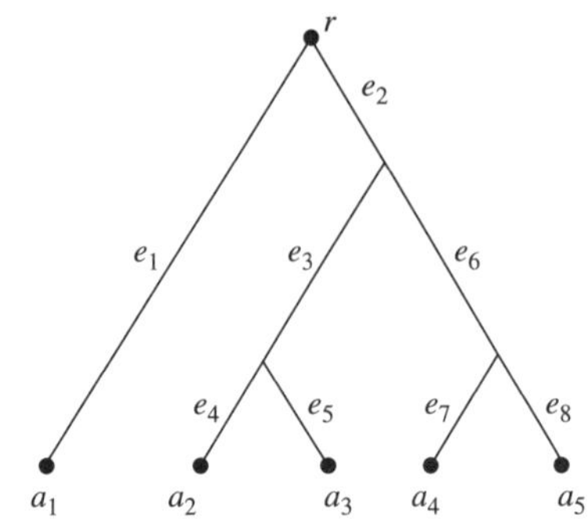

FIG. 4.3. A 5-taxon tree.

Most of the other models we consider are submodels of GM, in that they merely place additional restrictions on the form of the numerical parameters. The 2-*state symmetric model*, or Cavender–Farris–Neyman model, assumes $\kappa = 2$, $\boldsymbol{\pi} = (.5, .5)$, and that every Markov matrix has the form

$$M_e = \begin{pmatrix} 1 - a_e & a_e \\ a_e & 1 - a_e \end{pmatrix},$$

where a_e is a scalar parameter. This is an example of a *group-based* model (see [60] for an explanation of this terminology). Note that with this assumption, the overall polynomial form of equation (4.4) is retained, but the degree of each term drops by one, and the polynomial involves only the variables a_e, for each edge e.

Other group-based models of particular interest include the Kimura 3-parameter model, a 4-state model that assumes $\boldsymbol{\pi} = (.25, .25, 25, .25)$ and

$$M_e = \begin{pmatrix} d_e & a_e & b_e & c_e \\ a_e & d_e & c_e & b_e \\ b_e & c_e & d_e & a_e \\ c_e & b_e & a_e & d_e \end{pmatrix},$$

where $d_e = 1 - a_e - b_e - c_e$. Specializing by requiring $b_e = c_e$ yields the Kimura 2-parameter model, and requiring $a_e = b_e = c_e$ yields the Jukes–Cantor (or 4-state symmetric) model.

It is common in other contexts to use phylogenetic models which have a continuous time formulation, where one specifies a rate matrix Q and edge lengths t_e to describe the substitution process, with the Markov matrix on an edge being $M_e = \exp(Qt_e)$. Usually the rate matrix for all edges is taken to be the same, which is a strong assumption of commonality about the substitution process over the entire tree. Indeed, typical implementations in software of the general time-reversible model (GTR) are of this sort. Note that such models do not have polynomial parameterizations, but rather ones involving exponentials.

In studying invariants, in order that the parameterization maps be polynomial, we do *not* assume a continuous-time model of base substitution, nor commonality of rates across the tree. Rather we use a discrete notion of time, in which the full evolutionary process on an edge of the tree is lumped together to be described by a single matrix. As a result, the models dealt with when studying or utilizing invariants are, in this respect, more general than those used in most software.

One might view the generality of GM as either a strength (if one doubts that the assumptions of a model such as the GTR are justified for a data set) or a weakness (if one believes those assumptions are justified, and extra generality in the model leads to the possibility of overfitting data). Regardless, note that a model such as the GTR is a submodel of GM, in that it merely places additional (non-algebraic) restrictions on the form of allowable parameters. Thus, whatever invariants allow us to say about the GM model will imply statements about its submodels such as GTR. In Section 4.9, for example, we describe an application of invariants to some continuous-time models.

We also note that the GM model does not allow any 'rate variation' across sites, so a model such as GTR+I+Γ is not a submodel. Later, in Section 4.9, we return to a discussion of rate variation, explaining in more detail how invariants can be used to understand both rate-matrix models with variation in rates across sites, and also the covarion model.

4.3 Edge invariants and matrix rank

An invariant f for the GM model on the particular tree T^r of Fig. 4.3 is a polynomial in κ^5 variables, the indeterminate entries p_{ijklm} of a $\kappa \times \kappa \times \kappa \times \kappa \times \kappa$ array P. Furthermore, when P is given numerical values P_0 produced by some choice of parameter values in equations (4.4), we have $f(P_0) = 0$. Even a glance at equations (4.4), however, indicates we have little chance of finding any invariants by the 'inspection' approach we used for the ancestral-A model of Section 4.1.

To construct a first class of invariants for this model, we proceed by building on the example of the Introduction. We again consider the much simpler situation of Fig. 4.1 and equation (4.1), in order to rederive its invariants in a more sophisticated way. Notice first that the 16 versions of equation (4.1) can be combined into a single matrix equation

$$P = M_1^T \operatorname{diag}(\boldsymbol{\pi}) M_2, \tag{4.5}$$

where $\operatorname{diag}(\boldsymbol{\pi})$ denotes a matrix with the vector $\boldsymbol{\pi}$ placed along the diagonal and with 0 in all off-diagonal entries.

For the ancestral-A model, we make the additional assumption that $\boldsymbol{\pi} = (1, 0, 0, 0)$, so that $\operatorname{diag}(\boldsymbol{\pi})$ has only one non-zero entry. With this assumption, then, equation (4.5) implies that the matrix P must have rank at most 1, for $\operatorname{diag}(\boldsymbol{\pi})$ is a matrix of rank 1, and the rank of a product of matrices is at most the minimal rank of the factors. But from linear algebra there is a well-known algebraic condition on the entries of a matrix of rank 1: A matrix has rank 1 if, and only if, its 2×2 minors (determinants of submatrices chosen by picking 2 rows and 2 columns) are all zero. Since these minors are precisely the polynomials of equation (4.3), we have recovered our previous invariants for the ancestral-A model on a 2-taxon tree.

To develop this viewpoint further, we consider an ancestral-AG model on the same tree; that is, we assume the GM model with $\boldsymbol{\pi} = (\pi_A, 1 - \pi_A, 0, 0)$. Now since diag($\boldsymbol{\pi}$) has rank 2, again using that the rank of a product is at most the minimal rank of its factors, equation (4.5) establishes that the rank of P is at most 2. Thus all 3×3 minors of P give invariants. For an ancestral-AGC model, similar reasoning shows P has rank at most 3, and so $\det(P) = 0$ is the sole invariant we obtain.

For the 4-state GM model on the 2-leaf tree, where we place no restrictions on $\boldsymbol{\pi}$, we similarly conclude that P must have rank at most 4. However, since P is 4×4, there is no real content in this observation, since the rank of any matrix is bounded by its dimensions. Thus we obtain no invariants from this viewpoint (and indeed none exist for this model on the 2-taxon tree, except the stochastic one.)

To summarize the viewpoint so far, and ultimately to obtain invariants for more taxa, it will be helpful to consider a slight broadening of the model. We step beyond the phylogenetic setting, but still base our model on the graphical depiction of Fig. 4.1. We imagine 3 discrete random variables, associated to the nodes r, a, b. The variable at r may take on any of κ states, while those at a and b may take on any of λ and μ states, respectively. A κ element root distribution vector $\boldsymbol{\pi}$ specifies probabilities of states at r, while $\kappa \times \lambda$ and $\kappa \times \mu$ Markov matrices give transition probabilities to the various states at a and b. Finally, we observe states only at a and b, with those at r being hidden.

Under this model, we see that equation (4.5) still applies to give the joint distribution. We also see that since the diagonal matrix has rank at most κ, P will also have rank at most κ, and thus all $(\kappa + 1) \times (\kappa + 1)$ minors of P must vanish. Provided $\lambda, \mu > \kappa$, so that P is big enough for such minors to exist, we have found some invariants of the model.

These invariants, which test for matrix rank, have a direct statistical interpretation: They express the basic assumption of this model, that the stochastic processes occurring along the two edges leading from r are *independent, conditioned on the state at* r.

To use this observation to find invariants for the κ-state GM model, we must consider a tree with more taxa, such as that to the left in Fig. 4.4. Suppose the root r is located at the left of the internal edge. Then the GM model has as parameters a root distribution vector, and five $\kappa \times \kappa$ Markov matrices.

We can ignore some of the structure in the model by grouping together taxa, letting $a = \{a_1, a_2\}$ and $b = \{a_3, a_4\}$. The random variable associated to a now has κ^2 states, the pairs of states for a_1 and a_2, and similarly for b. The graphical depiction of the model is now that of the right side of Fig. 4.4, which is identical to Fig. 4.1. For this 'coarsened' model we can express the $\kappa \times \kappa^2$ matrix

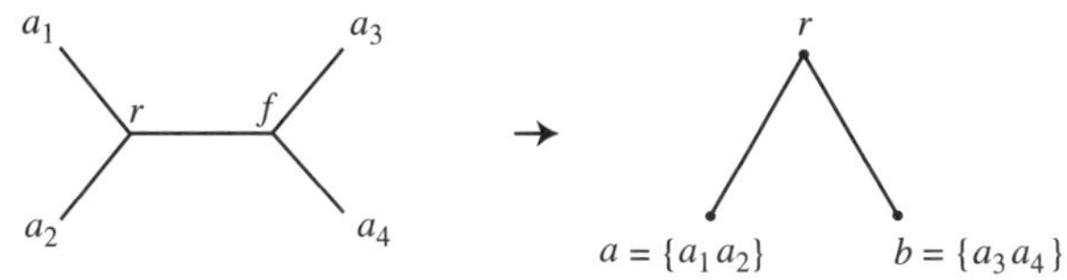

FIG. 4.4. A 4-taxon tree, with taxa a_1, a_2, a_3, a_4, rooted at r, and its coarsening to a simpler model.

parameters M_1 and M_2 in terms of the GM parameters:

$$M_1(i,(j,k)) = M_{ra_1}(i,j)M_{ra_2}(i,k),$$

$$M_2(i,(j,k)) = \sum_{l=1}^{\kappa} M_{rf}(i,l)M_{fa_3}(l,j)M_{fa_4}(l,k).$$

Coarsening the GM model in this way corresponds to changing the way we view the joint distribution array P. Though initially we viewed P as a $\kappa \times \kappa \times \kappa \times \kappa$ array, we now 'flatten' it to a $\kappa^2 \times \kappa^2$ matrix

$$\text{Flat}(P)((i,j),(k,l)) = P(i,j,k,l).$$

Note that we have merely rearranged the way we view entries of P; the entries themselves are unchanged.

This coarsened GM is now an instance of a model for which we have already found invariants. We can therefore immediately see that all $(\kappa+1) \times (\kappa+1)$ minors of Flat(P) are invariants of the GM model on this tree, since the flattening of P must have rank at most κ. These invariants should be interpreted as expressing a conditional independence statement that the state-change process on the branches leading from r to a_1 and a_2 is independent of that on the edges leading from r to a_3 and a_4, conditioned on the state at r.

Despite appearances, these invariants do not actually depend on the location of r at one end of the internal edge of the tree. It can be shown that for a dense subset of all parameters, the GM model with one specified root location on a tree T produces the same joint distributions as the GM model with a different root location on T. This means we can freely move the root to a location convenient for our construction.

Note that the arrangement of entries in Flat(P), and thus the invariants we have found, depend only on the split of taxa $\{a_1, a_2\}, \{a_3, a_4\}$ induced by the internal edge of the tree. We thus refer to these as *edge invariants* associated to the single internal edge of the tree.

This construction easily generalizes to larger trees. We can pick any internal edge of T and flatten P according to the resulting split. For a concrete example, consider the 2-state GM model on the 5-taxon tree of Fig. 4.5. Denoting states by 0 and 1, from the $2 \times 2 \times 2 \times 2 \times 2$ joint-distribution array P, we obtain two

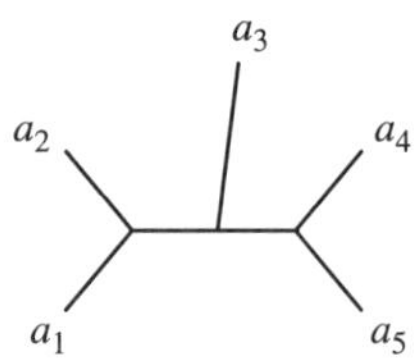

FIG. 4.5. A 5-taxon tree.

edge flattenings. The $\{a_1, a_2\}, \{a_3, a_4, a_5\}$ split gives

$$\begin{pmatrix} p_{00000} & p_{00001} & p_{00010} & p_{00011} & p_{00100} & p_{00101} & p_{00110} & p_{00111} \\ p_{01000} & p_{01001} & p_{01010} & p_{01011} & p_{01100} & p_{01101} & p_{01110} & p_{01111} \\ p_{10000} & p_{10001} & p_{10010} & p_{10011} & p_{10100} & p_{10101} & p_{10110} & p_{10111} \\ p_{11000} & p_{11001} & p_{11010} & p_{11011} & p_{11100} & p_{11101} & p_{11110} & p_{11111} \end{pmatrix},$$

and the $\{a_1, a_2, a_3\}, \{a_4, a_5\}$ split gives

$$\begin{pmatrix} p_{00000} & p_{00001} & p_{00010} & p_{00011} \\ p_{00100} & p_{00101} & p_{00110} & p_{00111} \\ p_{01000} & p_{01001} & p_{01010} & p_{01011} \\ p_{01100} & p_{01101} & p_{01110} & p_{01111} \\ p_{10000} & p_{10001} & p_{10010} & p_{10011} \\ p_{10100} & p_{10101} & p_{10110} & p_{10111} \\ p_{11000} & p_{11001} & p_{11010} & p_{11011} \\ p_{11100} & p_{11101} & p_{11110} & p_{11111} \end{pmatrix}.$$

By what we have seen, all 3×3 minors of each of these matrices are invariants of the GM model on this particular tree.

4.4 Vertex invariants and tensor rank

The edge invariants of the GM model that are described in the last section express a conditional independence statement: character state-changes in the two parts of a tree separated by an edge are independent of one another, conditioned on the state of the character at some point along the edge.

Other invariants for the GM model express a similar sort of conditional independence statement, but focus on an internal node of the tree rather than an edge. To explain them, we first focus on the simplest tree for which they can arise, the 3-taxon tree with only one internal node, as in Fig. 4.6.

Here we imagine the central node is the root. Numerical parameters for the model then are the root distribution $\boldsymbol{\pi}_r$ and three $\kappa \times \kappa$ Markov matrices M_1, M_2, and M_3 giving probabilities of changes in state along the three edges leading from the root. The joint distribution for this model is a $\kappa \times \kappa \times \kappa$ array $P = (p_{ijk})$,

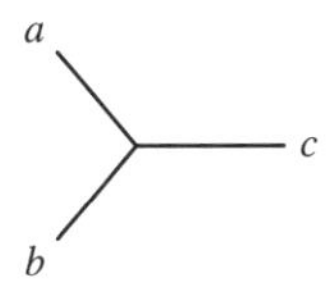

FIG. 4.6. The 3-taxon tree.

where

$$p_{ijk} = \sum_{l=1}^{\kappa} \pi_l M_1(l,i) M_2(l,j) M_3(l,k). \tag{4.6}$$

Since the matrix notation used in equation (4.5) is insufficient for describing a 3-dimensional array, we take an alternate approach. We first introduce arrays representing intermediate steps in equation (4.6): for each state l at the internal node, let P_l be the $\kappa \times \kappa \times \kappa$ array with ijk-entry $M_1(l,i)M_2(l,j)M_3(l,k)$. Notice that P_l is simply a joint distribution for an 'ancestral-base-l' model, similar to that of the introduction, but now for a 3-taxon tree.

The arrays P_l have a particularly simple structure, though. All entries are found by taking the various products of entries from the lth rows of M_1, M_2, and M_3. In other words, P_l is the *tensor product* of three rows. This parallels the situation for the 2-taxon tree in the last section, where the ancestral-A model had joint distribution $P = (p_{ij})$, with

$$p_{ij} = M_1(1,i)M_2(1,j)$$

so

$$P = \mathbf{r}_1^T \mathbf{r}_2,$$

where $\mathbf{r}_1$ was the first row of M_1 and $\mathbf{r}_2$ the first row of M_2. Just as this P was a rank 1 matrix, we call the 3-dimensional array P_l a *rank 1 tensor*. More formally, a 3-dimensional array is said to have rank 1 if it is the tensor product of 3 non-zero vectors.

When a 3-dimensional joint distribution is a rank 1 tensor, the fact that its entries are simple products of the form given here is just a manifestation of independence of the states for the 3 indices. Indeed, a rank 1 joint distribution occurs exactly when a model assumes a single state at the internal node of the graphical model of Fig. 4.6, with independent state changes on each edge leading away.

Now for the full model on the 3-taxon tree, we have that P is the weighted sum of κ rank 1 tensors,

$$P = \sum_{l=1}^{\kappa} \pi_l P_l,$$

with one summand for each of the κ possible states at the internal node. As the *tensor rank* of an array is the smallest number of rank 1 tensors needed to express it as a sum, P is thus a tensor of rank at most κ since, just as before, it is a sum of rank 1 tensors. Emphasizing the statistical viewpoint, the joint distribution P is a tensor of rank at most κ precisely because of independence of the state changes on the edges of the tree, conditioned on the κ possible states at the root. This parallels the 2-taxon, matrix situation of the last section.

This gives a good way of thinking of invariants for the GM model on the 3-taxon tree: they should be interpreted as making a conditional independence statement about state changes on the 3 edges emerging from the internal node. But how do we explicitly find invariants for this model?

For edge invariants, we could use the classical results on the relationship between matrix rank and vanishing of minors. Although by general principles of algebraic geometry, analogs of matrix minors must exist for testing tensor rank, they are only explicitly known for tensors of a few special sizes.[2]

To find invariants for the 3-taxon tree, we need a direct construction, as supplied in [1]. While that paper gives a variety of invariants of different forms, the most important are the ones arising from *commutation relations* that are derived from an observation that certain expressions built from the joint distribution give commuting matrices. Even for the 4-state model, these invariants are rather complicated when expressed in ordinary polynomial notation; though each term is only of degree 5, there are hundreds of terms.

However, they can be given a concise expression using matrices. To illustrate a typical form, for any choice of state k let $P_{abk} = (p_{ijk})$ be the kth 'slice' of P, a matrix obtained by only considering those entries in the 3-dimensional array P with a fixed index of k in the 3rd position (corresponding to state k at taxon c). Then for any choice of i, j, k it can be shown that the matrix equations

$$P_{abk}\,\mathrm{Cof}(P_{abj})^T P_{abi} = P_{abi}\,\mathrm{Cof}(P_{abj})^T P_{abk} \tag{4.7}$$

must hold if P arises from the GM model. Here $\mathrm{Cof}(M)^T$ refers to the transpose of the co-factor matrix of M, which is a standard construction from linear algebra. As this equation expresses the equality of two $\kappa \times \kappa$ matrices, it gives κ^2 individual invariants from equating entries. Since each entry of the co-factor matrix is a polynomial of degree $\kappa - 1$, these invariants are of degree $\kappa + 1$.

When $\kappa = 2$, a calculation shows that all of these polynomials simply give 0. In fact, for the 2-state GM model on a 3-taxon tree, one can show the only invariant is the stochastic one, so this is as it should be.

For the 4-state model, however, one can verify that these polynomials are not zero. In fact, minor variations on the construction can produced 1728 linearly-independent degree 5-invariants. Other means [36, 49] can show this

[2]Tensor rank is a more subtle notion than one might expect from familiarity with the matrix concept. In particular, analogues of matrix minors will test for *border rank* rather than rank, since the closure of tensors of a certain rank may contain ones of higher rank. This phenomenon does not occur for matrices.

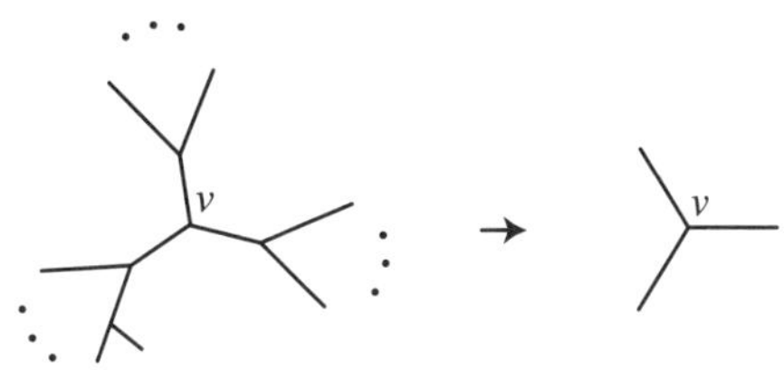

FIG. 4.7. Flattening a model at a vertex ν.

is the dimension of the full space of degree 5-invariants, and that except for the stochastic invariant there are essentially no others of lower degree.[3]

With some invariants in hand for the 3-taxon tree, a 'flattening' approach can be used again to give invariants for n-taxon binary trees. Picking any internal vertex v of the tree, we combine the taxa into three groups, as indicated in Fig. 4.7.

Coarsening our model in this way corresponds to rearranging the entries of the n-dimensional joint distribution array into a 3-dimensional array with size $\kappa^{n_1} \times \kappa^{n_2} \times \kappa^{n_3}$, where $n = n_1 + n_2 + n_3$. For a κ-state model, this flattened array must be a tensor of rank at most κ, since just as before it is a sum of rank 1 tensors, with one summand for each possible state at the internal node. Invariants for this coarsened model, which must also be invariants for the original model, are referred to as *vertex invariants*. With a bit of additional work [6], one can obtain explicit formulas for all vertex invariants provided one has them for the 3-taxon tree.

4.5 Algebraic geometry and computational algebra

Once invariants, such as the edge and vertex invariants for the GM model discussed above, have been found for a particular model, a natural question is whether there are others. To be able to discuss this properly, we informally introduce a little of the viewpoint and language of algebraic geometry.

Suppose we are given a collection of M polynomial functions, $g_1, g_2, \ldots, g_M$ depending on N variables $x_1, x_2, \ldots, x_N$. Allowing the variables to range over the complex numbers, we have a function

$$\begin{aligned} \phi : \mathbb{C}^N &\longrightarrow \mathbb{C}^M, \\ (x_1, x_2, \ldots, x_N) &\longmapsto (g_1(x_1, \ldots, x_N), \ldots, g_M(x_1, \ldots, x_N)). \end{aligned}$$

We have in mind here that the g_i are the parameterization polynomials for the joint distribution of a phylogenetic model, the x_i are the parameters, and ϕ gives us the full joint distribution array for any parameter choice.

[3]While it is known that some additional invariants of degree 9 are also needed to obtain all invariants for the 3-taxon model, the full situation is not yet completely understood [6].

The image of ϕ, the set $\phi(\mathbb{C}^N)$, is a parameterized subset of $\mathbb{C}^M$, which we view as some sort of high-dimensional 'surface', which is smooth at most points, but perhaps has some singularities. For example, the cartoon depiction of Fig. 4.2 represents two such 'surfaces', though in practice dimensions are usually much higher.

We now try to describe $\phi(\mathbb{C}^N)$ *implicitly*, as the zero set of polynomials. Introducing variables $P = (p_1, p_2, \dots, p_M)$, we look for polynomials in the p_i that vanish when $P = \phi(x_1, \dots, x_N)$. Optimally, we determine the entire set $I = \{f\}$ of all polynomials f in the variables p_i, such that

$$P = (p_1, \dots, p_M) \in \phi(\mathbb{C}^N) \text{ implies } f(p_1, \dots, p_M) = 0.$$

Thus the vanishing of such an f on a point P would provide some evidence that it is in the image of ϕ.[4]

Indeed, this is precisely what we have been trying to do for phylogenetic models. In that context the set I of polynomials implicitly defining the set of joint distributions $\phi(\mathbb{C}^N)$ are the phylogenetic invariants.[5]

For phylogenetic models, or statistical models in general, the numerical parameters usually represent probabilities. It thus might seem more reasonable to require that parameters be in some subset of the interval $[0, 1]$, or at least in $\mathbb{R}$. However, in algebraic geometry it is well understood that many technical issues are easier dealt with when we allow variables to range over $\mathbb{C}$. For finding all possible invariants this has little consequence due to the following:

Fact 1. For polynomial maps ϕ and f as above, $f(\phi(x_1, \dots, x_N)) = 0$ for all choices of $(x_1, \dots, x_N)$ in an open subset of $\mathbb{R}^N$ if, and only if, $f(\phi(x_1, \dots, x_N)) = 0$ for all choices of $(x_1, \dots, x_N)$ in $\mathbb{C}^N$.

Thus while phylogenetic invariants express polynomial relationships that must hold for joint distributions arising from stochastically-meaningful parameter values, they are exactly the same relationships that hold for all complex parameter values.

Suppose we knew several polynomials in the set I, that evaluate to zero on $\phi(\mathbb{C}^N)$. From these, say $f_1(P), f_2(P), \dots, f_k(P) \in I$, we can find many more, since for any choice of polynomials $h_i(P)$, the polynomial $\sum_{i=1}^k h_i(P) f_i(P)$ will then vanish wherever all the f_i do. Thus any such combination of invariants will also be an invariant. In the language of algebra, this means the set I of polynomials vanishing on $\phi(\mathbb{C}^M)$ forms an *ideal*. For a phylogenetic model, we call the collection of all invariants the *phylogenetic ideal*.

---

[4]A more optimistic hope would be that

$$P = (p_1, \dots, p_M) \in \phi(\mathbb{C}^N) \text{ if, and only if, } f(p_1, \dots, p_M) = 0 \text{ for all } f \in I,$$

but this is usually not possible. The common zero set of all $f \in I$ is closed, while $\phi(\mathbb{C}^M)$ may not be, and thus the zero set may contain additional points.

[5]Some writers refer to these merely as 'invariants,' reserving 'phylogenetic' for those invariants we refer to as *topologically informative*. We use 'phylogenetic invariant' to mean any invariant for a phylogenetic model.

But we must be more explicit about the role of the tree parameter, T, in a phylogenetic model. Even if we have fixed a model to consider, such as GM, the form of the parameterization map depends intimately on T. We signify this by denoting the parameterization map by ϕ_T, and its image by $\phi_T(\mathbb{C}^M)$. The phylogenetic ideal is the set of polynomials vanishing on this image, and so also depends on T. We typically denote the phylogenetic ideal by I_T, as we consider different trees. We omit from our notation a reference to the model, such as GM, since this is usually fixed throughout a discussion.

Since an ideal I is generally an infinite set of polynomials, to specify its elements we can ask for a list of *generators*, that is, a set of polynomials $\{f_1, f_2, \dots\}$ such that if $f \in I$ then $f = \sum_i h_i f_i$ for some choices of polynomials h_i. Fortunately, only finitely many generators are needed:

Fact 2. Any ideal of complex polynomials in M variables has a finite set of generators.

Thus, to find all invariants for a phylogenetic model and tree T, it is enough to determine a finite set of generators of the ideal I_T. For most ideals there is no canonical choice of a set of generators; different sets might generate the same ideal. In the phylogenetic setting we will of course prefer that our generators can be given a statistical explanation, such as the conditional independence interpretations of the edge and vertex invariants introduced earlier.

Given a collection S of polynomials in variables $P = (p_1, \dots, p_M)$, define the *algebraic variety* associated to S as

$$V(S) = \{P \in \mathbb{C}^M \mid f(P) = 0 \text{ for all } f \in S\}.$$

Thus the variety is simply the set of common zeros of the polynomials in S.

In particular, for phylogenetic models, we refer to $V_T = V(I_T)$, the common zero set of all phylogenetic invariants, as the *phylogenetic variety.* The phylogenetic variety will typically be larger than $\phi_T(\mathbb{C}^M)$, including points in the topological closure of the image of the parameterization. Thus the phylogenetic variety is made up of all 'joint-distributions' arising from complex parameter values, together with some additional points nearby.

When studying a model in the framework of algebraic geometry, finding generators for the phylogenetic ideal is certainly the most desirable goal. However, proving that one has found generators is often technically quite difficult, and a weaker result may be the best we can achieve.

Let V be an algebraic variety and I the ideal of all polynomials vanishing on V. Suppose S is some other set of polynomials having the same zero set as I, so that $V(S) = V$. Then we say S *defines* V *set-theoretically.* In such a circumstance $S \subset I$, but we may have $S \subsetneq I$, and even that S fails to generate I. While having a collection of set-theoretic defining polynomials for a variety does give us a way to test whether a point lies on a variety, we do not necessarily know all such tests unless we have generators of I.

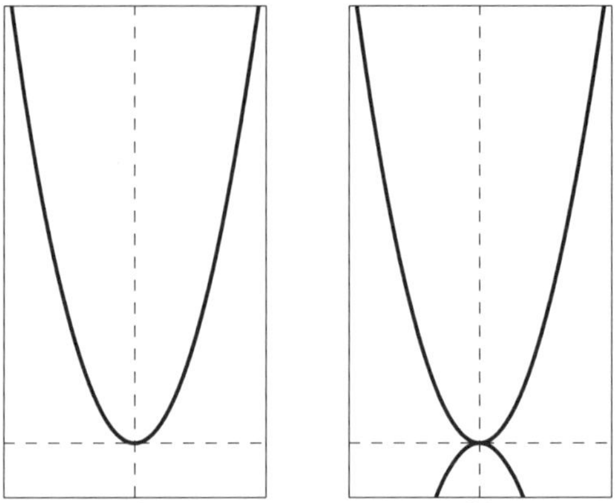

FIG. 4.8. The real points in varieties (a) defined by $p_1^2 - p_2 = 0$, or by $(p_1^2 - p_2)^2 = 0$, and (b) defined by $(p_1^2 - p_2)(p_1^2 + p_2) = 0$.

In order to clarify this terminology, we give a simple example, outside a phylogenetic setting. Consider the parameterization

$$\phi : \mathbb{C} \to \mathbb{C}^2$$
$$\phi(x) = (x, x^2).$$

The real points in the image of ϕ are shown in Fig. 4.8(a). It is easy to guess, correctly, that the ideal I of all polynomials vanishing on $\phi(\mathbb{C})$ is generated by the single polynomial $p_1^2 - p_2$. But notice that $(p_1^2 - p_2)^2$ also defines the variety set-theoretically, even though it does not generate the ideal. A third invariant is $p_1^4 - p_2^2 = (p_1^2 - p_2)(p_1^2 + p_2)$, which defines too large a variety, the union of the one of interest and its reflection below the p_1-axis. Although phylogenetic models with their many variables are necessarily more complicated, this simple example illustrates the main points: one might characterize ideal generators as the 'least complicated description' of a variety, and set-theoretic defining polynomials as a 'good description'. Other sets of polynomials have extraneous common zeros.

In principle, for a particular model on a particular tree, passing from a parameterization to an implicit description of a variety can be done by a computation involving variable elimination. This *implicitization problem* is described more fully in the excellent and accessible introduction to algebraic geometry [20], or more specifically in the phylogenetic setting by [37]. Computational algebra software implementing Gröbner basis algorithms, such as Maple, or the more specialized and powerful packages such as Macaulay2 [34] or Singular [35], can thus sometimes be used to explore invariants, form conjectures, and prove results.

However, several caveats on using computational algebra with phylogenetic models are in order. First, despite the impressive abilities of these packages, the large number of variables involved in phylogenetic problems can make the computations intractable except for small trees and some of the less-complicated models. Second, the form of the invariants one finds this way can depend on computational choices that are made along the way, such as the term order necessary for any Gröbner basis computation. Therefore one will usually still want to find an interpretation, or natural construction, of the invariants produced computationally. Despite this, such computational explorations have played important roles in quite a few recent works focused on both finding and using invariants. Such software is an extremely valuable tool.

In many early papers on invariants, dimension counting was applied to determine how many invariants might be 'needed' for a particular model. If a model depended on N numerical parameters (with no redundancy), and gave a joint distribution with M entries, then the phylogenetic variety should be an N-dimensional object in M-dimensional space, i.e. of *codimension* $L = M - N$. Thus one might look for L phylogenetic invariants to define the variety set-theoretically.

Unfortunately, an algebraic variety of codimension L may require more than L set-theoretic defining polynomials. Although for some neighbourhood of any point there will be L polynomials defining the part of the variety in that neighbourhood, those polynomials may have additional common zeros outside of the neighbourhood that are not part of the variety. There may not be *any* set of L polynomials defining the variety globally.

This issue was first clearly brought up rather recently, in [37]. (See also the expository papers [38, 47].) In [66], as a consequence of the determination of all invariants for some group-based models, Sturmfels and Sullivant established that this issue did in fact arise for some standard phylogenetic models; previously given sets of invariants had many extraneous zeros. The authors argued strongly for the determining of the full ideal of invariants, or at least set-theoretic defining polynomials.

As a result of this history, one must be careful in interpreting literature that refers to 'complete sets of invariants which are algebraic generators' of the ideal. The concept of algebraic generators is a weaker one than set-theoretic defining polynomials, allowing extraneous zeros such as those in Fig. 4.8(b) when the variety of interest is (a). While such local defining polynomials might still be useful for future applications, it is likely that one needs some understanding of the locations of their extraneous zeros.

There are many open mathematical questions concerning phylogenetic ideals and varieties, some of which have been surveyed for algebraic geometers in [23]. Here we mention only one issue whose relevance will immediately be clear.

As mentioned, the vanishing of the invariants for a particular model and tree does not just distinguish joint distributions arising from parameter values that are probabilistically meaningful, but also those arising from complex parameters. This is not because of any lack of understanding of all invariants on our part, but rather due to the fundamental features of defining sets by the vanishing of polynomials. The field of *real algebraic geometry*, in which polynomial inequalities as well as equalities play a role, would be a more appropriate setting in which to work if we hope to understand points coming from real parameter values. Although polynomial inequalities were used in both of the papers [13, 48] inaugurating the study of invariants, more recent works have not dealt with them. Real algebraic geometry presents greater technical difficulties than complex algebraic geometry, but it may provide greater understanding as well.

4.6 Invariants for specific models

Invariants have been found for phylogenetic models by many means, ranging from insightful observations, to exact algebraic computations, to more brute-force numerical computations.

Many papers focused on determining linear invariants for various models [12, 30, 31, 32, 42, 51, 65], partly because of the behaviour of linear invariants for rate-variation models that had been noted in [48] and will be discussed in Section 4.7. Other investigations, including [26, 27, 28, 29], found higher degree invariants. Already in [13] it was pointed out that some of these invariants encode statements of independence of substitutions in different parts of the tree, a theme that was further elaborated on in [21, 55].

Rather than survey these works in detail, we instead focus on some results obtained more recently. We hope this will provide a clearer overview of what invariants are and how they might be useful.

4.6.1 *Group-based models*

Group-based models, such as the Kimura 3-parameter model, and their submodels, such as the Jukes–Cantor and Kimura 2-parameter models, have a particularly nice mathematical structure which aids us in determining invariants. Since a full explanation could require a chapter in itself, we provide only an overview.

The key to analysing group-based models is the powerful tool of Fourier analysis. This was first recognized in Hendy's discovery of the Hadamard conjugation in [33, 41] for the 2-state symmetric model, where the underlying group is $\mathbb{Z}_2$. (See [40] for a more recent overview.) The relationship between the Kimura 3-parameter model and the group $\mathbb{Z}_2 \times \mathbb{Z}_2$, and the utilization of the associated Fourier transform, formed the basis of Evans and Speed's [24] insightful construction of invariants for the model. Fourier ideas were further explored for arbitrary group-based models in the work of Székely, Steel, and Erdős [67], which was then exploited for constructing invariants in [63]. See also [25]. Phylogenetic invariants for group-based models, then, appeared to be well-understood.

Recently, however, the question was considered of whether these constructions gave essentially all invariants: could one produce an explicit list of generators for the phylogenetic ideal for a group-based model? This was addressed by Sturmfels and Sullivant in [66].

The Fourier transform developed in the earlier works cited above amounts to a linear change of variables for the parameterization map, in both inputs and outputs. The result of this transformation is that the complicated polynomial formulas for the parameterization map become quite simple: they can be given by monomials (one-term polynomials) in the transformed variables. Varieties parameterized by monomial functions are called *toric varieties* in algebraic geometry, and form a class that is particularly amenable to detailed analysis.

Using this, Sturmfels and Sullivant were able to show that all invariants for a particular tree could be constructed from invariants from the two smaller trees obtained by breaking an edge, together with some invariants associated to the edge itself. This 'breaking' or 'gluing' process reduced the problem of explicitly finding all invariants for an arbitrary tree to that for star trees, with only one internal node. Thus, after an analysis for the 3-leaf tree was completed, generators of the ideal for any binary tree could be explicitly given. We quote only a summary form of their result [66].

Theorem 4.1 *For a binary tree T, the ideal of phylogenetic invariants for the models $\mathcal{M}$ below is generated by the stochastic invariant, together with an explicit set of polynomials of the given degrees:*
$\mathcal{M}$ = *2-state symmetric, degree 2;*
$\mathcal{M}$ = *4-state Jukes–Cantor, degree* $1, 2, 3$;
$\mathcal{M}$ = *4-state Kimura 2-parameter, degree* $1, 2, 3, 4$;
$\mathcal{M}$ = *4-state Kimura 3-parameter, degree* $2, 3, 4$.

In addition to the explicit nature of the theorem, and the insight of the underlying analysis, there are two larger lessons to be drawn from these results.

First, the work shows that all invariants for group-based models arise from local features in the tree—from edges and nodes. As one considers trees with additional taxa, there will be larger sets of invariants, but their construction remains straightforward. Because the number of invariants needed to generate the phylogenetic ideal grows at least exponentially with the number of taxa, if invariants are to be useful for large trees, some local understanding of their meaning is valuable. Being able to tie generating invariants to specific topological features within a tree is likely to be essential for any application they may have.

Second, as mentioned in Section 4.5, it could be seen that for the 2-state symmetric model on a 4-leaf tree the 'complete sets of algebraic generators' of the invariants found in earlier works had many extraneous zeros. Indeed, the natural set of generators of the ideal of invariants for this model had more than the codimension number of polynomials in it, and any subset had extraneous zeros. This clearly showed that finding generators of the phylogenetic ideal, or at least set-theoretic defining polynomials, is necessary for adequate understanding of a phylogenetic variety.

Although we omit a detailed exposition of the precise form and construction of the invariants for group-based models, the 'Small Trees' web site [9] provides a valuable entryway for those interested in seeing or using them. It gives a compilation of invariants, Fourier transforms, and other information for trees of up to 5 taxa, with and without a molecular clock assumption. Input files for both Maple and Singular are helpfully provided.

4.6.2 *The general Markov model*

A separate thread of work on invariants was also undertaken recently, for the general Markov model, some of whose invariants were introduced in Sections 4.3 and 4.4. This model has many more parameters than the group-based models, and in studying it we lack the tool of Fourier analysis on a group. Nonetheless, fairly complete results have been obtained.

For the GM model, a single invariant for a 4-taxon tree was first given in [61]. The underlying idea was a suitable encoding of the 4-point condition for metric trees of [8], using log-det distances, building on an approach taken in [13]. Remarks in [59] point out that many additional invariants can be produced from the entries of certain matrix equations built from the joint distribution array. All these invariants depend only on two-dimensional marginalizations of the joint distribution (i.e. comparisons of sequences two at a time), as the underlying reasoning takes a generalized distance viewpoint.

The edge invariants for the GM model, which have been described in Section 4.3, are not inspired by any distance reasoning. Recall that they can be interpreted as statements of the independence of the substitution process on parts of the tree separated by an edge, conditioned on the state at some point along that edge.

For the 2-state GM model on a binary tree, the edge invariants in fact provide generators of the phylogenetic ideal, as was conjectured in [52] and proved in [6].

Theorem 4.2 *For the* 2*-state GM model on any n-leaf binary tree T, let P denote an n-dimensional array of indeterminants representing the joint distribution array. Then the ideal of phylogenetic invariants is generated by the stochastic invariant, together with all 3×3 minors of the matrix edge flattenings $\mathrm{Flat}_e(P)$ for all interior edges e of T.*

For instance in the 5-taxon tree example discussed at the end of Section 4.3, the 448 minors of size 3×3 of the two matrices shown, the edge invariants, provide a set of generators of the ideal.

Although the proof of this theorem requires mathematical techniques we will not discuss here, the result has a concrete, accessible interpretation: to each internal edge e of a tree we can associate both an explicit collection of cubic polynomials (the edge invariants for e) and a split of the taxa (into the two sets separated by e). These polynomials will be zero for any joint distribution arising from an n-taxon model on a tree inducing the same split of taxa. Moreover, these polynomials are essentially the only polynomial relationships that hold for all joint distributions arising from the fixed tree. Thus the structure of invariants for the GM model is determined by local features of the tree.

For the κ-state GM model, with $\kappa > 2$, our understanding is not quite as complete, but partial results again indicate a prominent role for invariants associated to local tree topology. The best current result is the following from [6].

Theorem 4.3 *Suppose a set of polynomials set-theoretically defining the variety associated to the GM model on a* 3*-taxon tree were given. Then an explicit* *construction will produce a set of polynomials set-theoretically defining the phylogenetic variety for any n-taxon binary tree.*

Although this statement fails to highlight it, the construction of the explicit polynomials it refers to involves precisely the vertex invariants and edge invariants as discussed earlier. A large tree is viewed as many star trees joined together, and from invariants for each star tree, set-theoretic defining polynomials for the large tree are constructed.

We also note that while an understanding of set-theoretic defining polynomials for the 3-leaf tree is not complete, good partial results are available in [1] for the 4-state GM model.

Theorem 4.4 *Let S be the set of 1728 degree-5 polynomials, constructed as discussed in Section 4.4, which are invariants for the 4-state GM model on the 3-leaf tree. Then $V(S)$, the variety they define, is the union of the phylogenetic variety and possibly a set of extraneous zeros which lies in an explicitly describable set.*

The extraneous zeros mentioned in this last theorem can even be shown to be far from points on the phylogenetic variety arising from biologically-relevant parameter values.

We emphasize that the results for group-based and GM models parallel one another, in that all invariants ultimately arise from edges and nodes of the tree. Explicit polynomials tied to these features either generate the ideal or at least set-theoretically define the variety. However, the methods of proof are quite different. For group-based models, in addition to using the Fourier transform, the arguments are combinatorial in flavor and depend on an understanding of toric varieties. For the GM model, linear algebra and representations are the main ingredients.

4.6.3 *The strand symmetric model*

While the elegant mathematical structure of the group-based models facilitates an understanding of invariants, their restrictive assumptions are not always viewed as biologically realistic. While the GM model is also well-structured for understanding invariants, it might be considered to be too flexible, with too many parameters, for some phylogenetic applications. It is desirable, then, to look for biologically motivated models between these whose invariants can be successfully determined.

One potentially valuable one is the *strand symmetric* model introduced by Cassanellas and Sullivant in [10]. This 4-state model can be viewed as an amalgamation of a 2-state group-based model with a 2-state GM model, and thus its study can build on our understanding of each of those.

Specifically, with the fixed ordering of bases A,G,T,C, the model assumes a root distribution vector of the form

$$\boldsymbol{\pi} = (\pi_1, \pi_2, \pi_1, \pi_2),$$

so that frequency of any base matches its complement in the Watson–Crick pairing. This symmetry with respect to the pairing is also assumed for all Markov matrices on edges, so that they have the form

$$M_e = \begin{pmatrix} a & b & c & d \\ e & f & g & h \\ c & d & a & b \\ g & h & e & f \end{pmatrix}.$$

Since the rows of these matrices must sum to 1, there are 6 parameters introduced for each edge. Note that with this ordering of the bases the matrices have a block structure with 2×2 GM blocks arranged in a pattern reflecting the 2-state symmetric model.

As one might expect, the symmetry of this model leads to the existence of some linear invariants for any tree. Focusing next on the 3-taxon tree, a number of invariants of degree 3 and 4 can be constructed. However, it is not known whether these generate the phylogenetic ideal, or even set-theoretically define the phylogenetic variety, echoing the incompleteness of the corresponding result for the GM model. However, through the use of a computational algebra package, it can be seen that they generate all invariants of degree at most 4.

Finally, to handle trees relating more taxa, it is established that producing a set of invariants set-theoretically defining the variety for a 3-taxon tree would suffice to allow construction of invariants set-theoretically defining the variety for an arbitrary binary tree. This emphasizes once again that for those models for which we have made substantial progress in understanding invariants, we can tie particular invariants to particular local features of the tree.

4.6.4 *Stable base distribution models*

Another attempt to consider a biologically motivated model less general than the GM, but more general than group-based ones, appeared in [3]. The motivation was to understand what invariants might be valid for any model assuming a stable base distribution throughout the tree. In the course of this investigation, several nested models with this feature are formulated, including 1) an *algebraic time-reversible model* (ATR), which is similar to the GTR model but unlike the GTR has a polynomial parameterization map and, 2) a *stable base distribution model* (SBD) that assumes only that all Markov matrices fix the root distribution.

In the case of characters with 2 states, these models become the same, and generalize a model that had appeared earlier in [26]. In this simple situation the parameterization map is even explicitly invertible by a rational function; parameters can be recovered from a joint distribution by quite simple formulas. However, for a larger number of states our understanding is quite incomplete.

While some invariants are constructed for these models, little is understood about the full phylogenetic ideal or variety. Perhaps the most interesting result is a construction of a specific invariant that involves the hyperdeterminant of a 3-dimensional array, making a connection between phylogenetic invariants and what mathematicians refer to as invariant theory. Though only of degree 6 for the 2-state model, unfortunately this invariant is of degree 408 in the 4-state case.

Part 2. Using Invariants

In this section we turn from questions of determining invariants for various models, to questions of how they might be used. Although invariants have had key roles in several contributions to theoretical understanding, for data analysis it is still less clear how they can be exploited. While their potential is attractive, much more needs to be done to develop ways to use them with data.

4.7 Invariants and statistical tests

In the decade following the first appearance of phylogenetic invariants in [13] and [48], many papers appeared building upon the idea. In particular, a number of these works dealt primarily with linear invariants for various models.

A compelling reason for the emphasis on linear invariants was the hope that they might be particularly useful for certain types of rate variation models. Suppose an invariant $f(P)$ for a specific model on a specific tree T is found which is linear and homogeneous (without constant term). Then since $f(c_1P_1 + \cdots + c_kP_k) = c_1f(P_1) + \cdots + c_kf(P_k)$, this polynomial will also vanish on any linear combination of joint distributions arising from the model. But linear combinations such as $c_1P_1 + \cdots + c_kP_k$ arise naturally when we consider *mixture models*, where sites are distributed among classes, and each class has its own set of parameters for the same model and tree. Then P_i represents the joint distribution for the ith class, and c_i the class size parameter. Thus a linear invariant for a model on a tree will also be a linear invariant for any *rates-across-sites* extension of that model on the same tree. We need not even make any assumptions about the nature of the distribution of sites among rate classes. This observation on linear invariants for mixture models holds for both discrete and continuous distributions of rates.[6]

If an invariant for a model is *topologically informative*, in the sense that it vanishes on all joint distributions arising from the model for some tree topologies and not others, then it could be the basis of a statistical test to distinguish between the topologies. Topologically-informative linear invariants, then, could give tests for topologies that would be insensitive to across-site rate variation. Tests of various sorts based on linear invariants were proposed in [11, 48], and investigated more thoroughly in [50].

Although higher degree invariants for a model are typically not invariants for rates-across-sites extensions, some attention was also given to how they might be used in a statistical framework. In [13], one of the quadratic invariants

[6] A higher degree invariant for a specific 2-class mixture model was first constructed in [29]. While this demonstrated that higher-degree invariants might be sought for mixture models, until recently it remained an isolated result.

constructed encoded an independence statement, that substitutions in one part of a tree were independent of those in another part of the tree separated from it by an edge. Thus the possibility of a statistical analysis based on 2-way contingency tables, as is typically done to test for independence, was suggested. This idea was pursued further in [55]. Using general formulas for multinomial distributions to estimate variances of quadratic invariants was suggested in [21]. However, as far as we know, no firmly-grounded statistical test based on general non-linear invariants has been suggested.

Several comparison studies [44, 45, 46] of the effectiveness of various phylogenetic inference methods included Lake's linear invariants. Using simulated data, Lake's method was found to be less efficient than many other methods, in that it required much longer data sequences to perform well. Note that Lake's method had been shown to be statistically consistent, so that provided data was in accord with the underlying model, as the length of data sequences approaches infinity the probability of inferring the correct tree approaches 1. Despite this theoretical strength, on sequences of a length typical for real data sets, Lake's invariants failed to reliably infer the correct tree even when no underlying model assumptions were violated.

In retrospect, this is not so surprising. Linear invariants only can test if a data point is in the smallest linear subspace containing the phylogenetic variety. Though higher degree invariants could potentially yield much more information than linear ones, a statistical framework for using them was largely lacking.

Indeed, how to use higher degree invariants in a statistically meaningful way is still an open question, and one needing exploration. There is evidence [9] that naive approaches to identifying topologies using all invariants can be effective on simulated data even with relatively short sequence length. Thus the inefficiency of Lake's linear invariants should not be interpreted as a sign that invariants in general are necessarily inefficient.

4.8 Invariants and maximum-likelihood

In current software, when phylogenetic inference is performed using a maximum-likelihood approach, the maximization of the likelihood function is undertaken by numerical search for optimal model parameters. For a possible tree topology, an attempt is made to find optimal numerical parameters such as base distributions, mutation rates, and edge lengths, and then the tree is varied and a new search for optimal parameters is undertaken.

Various algorithms can be used for the two aspects of this search (for numerical parameters and for topology), but rarely can one be certain that the true maximum has been located. For a fixed tree, a good algorithm will ensure locally optimal numerical parameters will be found, but the possibility of missing a global optimum remains. In addition, because the number of possible tree topologies will be quite large when the number of taxa is big, it may be impossible to consider all topologies, and so heuristic searches of tree spaces may overlook the optimal tree.

While many packages incorporate methods to avoid being trapped at non-global optima as they search, they generally come with no guarantee. Comparing the performance of one algorithm against another may shed some light on the issue, but cannot really give us full understanding if we have no way to verify that any maximum we have found is the true one.

Beginning with the work of Yang [69], a number of papers have sought to better understand the maximum-likelihood (ML) problem through exact optimization in simple settings.[7] In particular Chor and his collaborators introduced the use of phylogenetic invariants as an aid in this optimization problem.

To see why invariants might be useful for exact ML optimization, recall the construction of the likelihood function for a fixed n-leaf tree whose leaves are labeled by taxa. We first express the joint distribution of bases by an n-dimensional array P, as in Section 4.2. With variables $\mathbf{u} = (u_1, u_2, \ldots, u_L)$ representing the numerical parameters of the model, each entry of $P = P(\mathbf{u}) = (p_{i_1 i_2 \ldots i_n}(\mathbf{u}))$ is thus expressed by a polynomial parameterization function.

Given aligned sequences for the taxa, we record the observed distribution of bases as an n-dimensional array $\widehat{P} = (\widehat{p}_{i_1 i_2 \ldots i_n})$. The log-likelihood function is then

$$\ln L(\mathbf{u}) = \sum (\widehat{p}_{i_1 i_2 \ldots i_n}) \ln(p_{i_1 i_2 \ldots i_n}(\mathbf{u})).$$

To find maxima of this function, we can first look for critical points, where all partial derivatives are zero. Thus differentiating with respect to each variable u_j we obtain the system of equations

$$0 = \sum \frac{\widehat{p}_{i_1 i_2 \ldots i_n}}{p_{i_1 i_2 \ldots i_n}(\mathbf{u})} \frac{\partial p_{i_1 i_2 \ldots i_n}(\mathbf{u})}{\partial u_j}, \quad j = 1, \ldots, L.$$

Now since each $p_{i_1 i_2 \ldots i_n}(\mathbf{u})$ is a polynomial, these are rational equations. Clearing denominators, they give rise to a system of polynomial equations in the unknown parameters $\mathbf{u}$. If they can be solved, then among the solutions lie all local maxima of the likelihood function. Note that the polynomials $p_{i_1 i_2 \ldots i_n}(\mathbf{u})$ are typically of high degree (e.g. of degree approximately the number of edges in the tree), and clearing denominators could therefore lead to equations of very high degree.

While solving such a system of equations by hand is not usually possible, one might hope that a computer algebra package could handle it. Unfortunately, the polynomial system one obtains, even for a simple model on a small tree, may be intractable for current software.

However, this optimization problem can be reformulated as a constrained optimization problem that may be tractable. Rather than seek optimal parameters $\mathbf{u}$, we instead seek optimal values for the entries $p_{i_1 i_2 \ldots i_n}$ of P. We'd like to constrain P so that it lies in the image of the parameterization map, so we impose the slightly weaker condition that it lie in the phylogenetic variety. Thus we require that all phylogenetic invariants vanish on P. The ML problem

[7] Though this is often referred to as seeking *analytic* solutions to ML, we avoid that terminology as the methods are in fact generally algebraic.

becomes one of maximizing

$$\ln L(P) = \sum (\widehat{p}_{i_1 i_2 \ldots i_n}) \ln(p_{i_1 i_2 \ldots i_n})$$

subject to the constraints

$$f(P) = 0 \quad \text{for } f \in I_T.$$

Note that the model parameters do not appear here; we view the entries of P as the variables. Moreover, since the phylogenetic ideal I_T is finitely generated, only finitely many constraint equations $f_i(P) = 0$, $i = 1, \ldots, K$, are actually needed here.

Formulating this problem using Lagrange multipliers, all critical points are found by solving the system given by the K constraint equations together with the κ^n equations from the entries of

$$\nabla \ln L(P) = \sum_{i=1}^{K} \lambda_i \nabla f_i(P).$$

Explicitly, these last equations are simply

$$\frac{\widehat{p}_{i_1 i_2 \ldots i_n}}{p_{i_1 i_2 \ldots i_n}} = \sum_{i=1}^{K} \lambda_i \frac{\partial f_i}{\partial p_{i_1 i_2 \ldots i_n}}.$$

Though we again need to clear denominators to obtain polynomial equations, note that the resulting equations may well be of much lower degree then the ones obtained from the original parameter formulation of the ML problem, especially if the degrees of phylogenetic invariants are not that large.

This last observation gives some hope that with judicious use of a computer algebra system we might able to solve this constrained optimization problem. Indeed this is the case, at least for some small trees and simple models.

In [17] this approach was used to show that maximum likelihood estimation of trees could be quite ill-behaved. For a 2-state symmetric model on a 4-leaf tree, a number of examples of observed distributions $\widehat{P}$ were constructed for which the ML problem on a particular tree topology had a continuum of global maxima. For some of these, the global maxima even tied with a continua of global maxima for the other possible tree topologies as well. Proving these results for the specific examples required algebraic methods of solution of the above constrained optimization problem. The symmetry of the model results in some linear invariants which first allow a reduction in the number of variables $p_{i_1 i_2 \ldots i_n}$. Because the model is group-based, higher degree (quadratic) invariants could be constructed using the Fourier transform in the form of the Hadamard conjugation.

The paper [15] gives a more positive result on maximum likelihood, focusing on the 2-state symmetric model on a 3-taxon tree with a molecular clock, as had Yang in [69]. For this model, a linear invariant resulting from the molecular clock hypothesis is found through Hadamard conjugation. Using the constrained optimization formulation of the ML problem, the authors were able not only to recover Yang's result on uniqueness of the ML optimum for this model on a fixed tree, but to extend it to allow variation in rates across sites, with mild restriction on the distribution of the rate parameter.

In [18, 19], the 2-state symmetric model with a molecular clock hypothesis is considered again, but now on 4-taxon trees. Hadamard conjugation again facilitates the derivation of invariants from the molecular clock hypothesis, though these must be derived separately for each of the possible rooted 4-taxon tree shapes, a 'fork' and a 'comb', and are quadratic rather than linear. The constrained optimization formulation of the ML problem is then solved, by a mix of insightful reductions and computer calculation. For the fork a unique maximum is found, whose coordinates can even be given as rational expressions in the entries of $\widehat{P}$. For the comb, the result is a bit more complicated, but the system is ultimately seen to have a finite number of solutions. However, all but one of these solutions is complex or outside the range $[0, 1]$, so again there is a unique maximum with statistical meaning.

In [16], this sort of analysis is pushed to a 4-state Jukes–Cantor model, on rooted 3-leaf trees. By working with transformed 'path-set' variables arising through Hadamard conjugation, rather than the variables $p_{i_1 i_2 \ldots i_n}$, the authors are able to avoid explicit use of constraint equations. Still, a symbolic algebra software package is needed to find critical points in the unconstrained formulation. They show that the ML problem has a finite number of optima, though some of the parameter values may not be meaningful in the context of the model.

Whenever a statistical model is parameterized through polynomial equations, one might take a similar algebraic approach to ML optimization. In [43], Hoşten, Khetan, and Sturmfels provide a general framework for using algebra to find exact solutions of ML problems. Computational approaches to both the constrained and unconstrained formulations are given. The authors further report that the constrained version generally performs better, though to take that approach requires one first finding model invariants, which of course may be quite difficult.

That paper also contains several phylogenetic calculations as examples. In one, for real data, the ML tree using a 4-state Jukes–Cantor model with 4 taxa is found, with the existence of a second local maximum established for that data also. This further indicates that multiple local maxima are a genuine issue in practical inference by maximum-likelihood. In another example, the result of [16] is reproved, this time in a constrained formulation.

The recent volume [53] provides a broader view of algebraic perspectives on statistics, with particular focus on applications to computational biology. Included in it is further background on the connections between algebra and general maximum-likelihood estimation.

4.9 Invariants and identifiability of complex models

While invariants were originally proposed for inferring trees from data, they can also be used to give theoretical results that such inference is possible. Separate from the question of what inference method performs best for data analysis, is the more fundamental question concerning the limits of what can be inferred under perfect conditions.

A statistical model is said to be *identifiable* if from any joint distribution arising from the model it is possible to recover all parameters or, in other words, if the parameterization map of the model is injective. Identifiability is important because it plays a key role in proofs that methods of inference such as maximum likelihood are statistically consistent. If, for instance, two different tree topologies could give rise to the same joint distribution under some model, it is intuitively clear that inferring the 'correct' tree from data cannot be done reliably.

In practice, for phylogenetic models one must modify the strict notion of identifiability. For instance, allowing no substitutions to occur on an internal edge would lead to non-identifiability of the tree topology for 4-taxon trees, since each of the 3 fully-resolved 4-taxon trees as well as a 4-leaf star tree could all lead to the same joint distribution. Allowing too much substitution along internal edges, so that states become completely 'randomized' and uncorrelated in different parts of the tree, can also lead to loss of phylogenetic signal and non-identifiability of topology. Even when the tree parameter is identifiable for a model, numerical parameters may not be. For instance, for the GM model one can permute the states at an internal node of the tree, adjusting parameters appropriately, without changing the joint distribution [1, 14], so that numerical parameters are not identifiable unless one places additional restrictions on them. But while understanding the issues of non-identifiability mentioned so far is important, these are rather mild problems that can be dealt with by imposing biologically plausible assumptions on parameter values.

Identifiability of the tree parameter is often of primary interest in phylogenetics. For many basic models, such as the Jukes–Cantor, Kimura, or even GM, tree identifiability can be shown by first defining an appropriate phylogenetic distance, and then using the 4-point condition [8]. However, for models without a known distance formula, such as the covarion model [68], this approach is not possible. General mixture models, in which different classes of sites undergo substitutions according to different numerical parameter values for a model, but with the same tree parameter, also lack a distance. In both these situations tree identifiability has been an open question.

Note that while identifiability of the GTR+I+Γ model was shown in [54], the approach makes use of the assumption that the rate-parameters are described by a known distribution in such a way that the 4-point condition can still be applied. If the rate-parameter distribution is unknown for GTR+rates-across-sites model, then [64] established the topology is not identifiable for certain (non-explicit) parameter choices.

How general non-identifiability of a tree might be is quite important, both for knowing whether a particular model might be usable for inference, and for understanding under what circumstances tree inference might simply be impossible.

Phylogenetic invariants were recently used to study the problem of identifiability of the tree parameter for a variety of models in [4]. General theorems are produced that guarantee tree identifiability for most parameter choices for both the covarion model and many mixture models, provided the number of classes is small.

In order to study a variety of models at once, a substitution model is introduced that is much like the general Markov, but which allows λ states for the characters at internal nodes of the tree, and κ states at the leaves, with $\lambda \geq \kappa$. For DNA models with several classes, the states at the internal nodes might be indexed by pairs (i, j), where i refers to the base A,G,C,T and j to a rate-class, while at the leaves the states are simply the bases. Thus if there are n rate classes, then we have $\lambda = 4n$ states for all ancestral taxa, but only $\kappa = 4$ states for the currently extant taxa. The idea behind this is simply that while each site is in some rate-class, we cannot observe that class when data is collected; only the base can be recorded. The generality of this framework encompasses not only rates-across sites models, in which no site can change class, but also covarion models, where rate-class switching can occur.

While most invariants for such a model, even on a 4-leaf tree, are beyond our current knowledge, some can be found through a generalization of the edge invariant construction for the GM model. It can then be shown that these invariants are sufficient to identify the tree topology for *generic* choices of parameters, provided $\lambda < \kappa^2$. 'Generic' is given a precise meaning of 'all except those in a proper subvariety'. Since such a subvariety is necessarily of lower dimension than the parameter space, this means that if parameters are chosen randomly, according to any reasonable notion of randomness, they will be generic and the tree topology can be identified from the resulting joint distribution.

This result is for a model much more general than typically of interest in phylogenetics. Further arguments are given to show that when more usual mixture models are viewed as submodels of this general model they inherit identifiability of trees for generic parameters of their own.

In particular for κ-state models, even a GM+GM+$\cdots$+GM model, with a mixture of $\kappa - 1$ classes each described by the GM model but with unrelated numerical parameters, has identifiable tree topology for generic parameters. For DNA models, then, trees are identifiable for generic parameters of models with 3 unrelated GM classes. The result further specializes to a model such as the GTR, where a common rate matrix is assumed for the substitutions on all edges, allowing up to 3 classes of sites with scaled rates.

While the framework of invariants seems best suited to studying models with a finite number of rate-classes, much research literature refers to continuous distributions of rates. Indeed, the commonly-used GTR+Γ model assumes a continuous distribution. In fact, though, software implementations usually use discretized versions of Γ with only a few classes (although more than 3). Thus models with a finite number of rate-classes are common in practice.

It should be emphasized that there is no reason to believe identifiability for generic parameters should not hold for rate-class models with more than $\kappa - 1$

classes, provided the number of classes is not too large. The current restriction to $\kappa - 1$ classes is an artifact of having incomplete knowledge of all invariants for the models. A better understanding of what limits must be placed on the number of classes to preserve generic identifiability is still needed.

In addition to giving results on mixture models, [4] leads to establishing generic identifiability of the tree topology for certain covarion models, such as that of Tuffley and Steel [68] and extensions. Covarion models are biologically quite attractive in that they describe sites passing between being invariable and being free to vary as they evolve over a tree. However, identifiability of trees had not previously been established for them, despite their implementation in software [33].

For some of the results described here, such as for the covarion model and the GTR+rate-classes models, the underlying model is *not* one with a polynomial parameterization. These are inherently continuous time models, involving matrix exponentials in their parameterization formulas. Nonetheless, because they are submodels of a more general polynomially-parameterized model, they can be effectively studied through invariants.

Another investigation [5] of invariants for mixture models has focused on the GM+I model, with 2 classes, one evolving according to GM and the other held invariable. Although identifiability of the tree for generic parameters in this model follows from [4], a focus on this more specific model allows additional invariants to be found, giving a refined analysis. Note that some questions of identifiability for this model had been studied previously in [7], in which it was shown the tree was not identifiable from marginalizations of the joint distribution to 2 taxa (i.e. from pairwise sequence comparisons).

An interesting consequence of studying invariants for GM+I is a set of explicit formulas that can recover the proportion of invariable sites with any given base from the joint distribution. For the more restrictive Kimura 3-parameter model with invariable sites, such a formula was found in [62] by a rather different argument using 'capture/recapture' reasoning. For the GM+I model an understanding of the invariants naturally leads to determinantal formulas to recover these parameters. For example, in the 2-state case on a 4 taxon tree, with states 0 and 1, the proportion of invariable characters of state 0 is given as a quotient:

$$\pi_0^I = \frac{\begin{vmatrix} p_{0000} & p_{0001} & p_{0010} \\ p_{0100} & p_{0101} & p_{0110} \\ p_{1000} & p_{1001} & p_{1010} \end{vmatrix}}{\begin{vmatrix} p_{0101} & p_{0110} \\ p_{1001} & p_{1010} \end{vmatrix}}.$$

Here subscripts indicating states corresponding to the taxa ordered as a, b, c, d, where the tree has split $ab|cd$. Similar formulas are valid for the 4-state characters, or even κ-state.

Note that such formulas are far from unique, since they can be modified by the addition of any invariant for the model without affecting the value the formula will yield when evaluated at a distribution. Nonetheless, there is a possibility that such formulas might be useful for quick estimation of parameters from data.

Identifiability by means of invariants also appeared in [2], which focused on the use of invariants only for quartets (subsets of 4 taxa) to determine a fit of n data sequences to a tree. Although the precise results require some technical conditions, they can be roughly summarized as indicating that while quartet invariants can indicate a unique n-taxon tree, additional invariants are needed to assure the n-dimensional joint distribution is fit well by the model. This clarifies the loss of information inherent in quartet methods of inference.

4.10 Other directions

4.10.1 *A tree construction algorithm*

A first step toward a novel invariant-based inference method was taken by Eriksson in [22], with a software implementation for DNA sequence data. The underlying idea uses only the edge invariants for the GM model. Following an algorithmic approach reminiscent of neighbour joining, the method iteratively builds a tree by finding good taxa, or clades, to join together, and thus has good running times.

In the initial step, all splits that separate two taxa from the rest are considered. If all the edge invariants for a hypothetical split come close to vanishing, then that is evidence that the two taxa should be joined. However, evaluating these invariants would simply be a test that the corresponding flattening of the observed distribution is close to a rank 4 matrix. Thus, rather than actually evaluate the many edge invariants for each flattening, the algorithm instead uses a numerical approach to determine how close each flattening is to a matrix with rank 4.

This problem of measuring how well a matrix can be approximated by one of fixed rank is well understood, provided closeness is measured by the Frobenius (i.e. L^2 on matrix entries) norm. The singular value decomposition of matrices provides a good numerical approach both to finding such approximations, and measuring error. Thus the algorithm avoids both the issues of how to use the large number of invariants associated to one edge to get a combined measure of support for that edge, and how one would interpret such a measure in a statistically meaningful way.

Although the performance of Eriksson's SVD method on simulated data was not as good as neighbour joining or maximum-likelihood, as a first attempt it gave several reasons to be hopeful. First, the simulation studies were in some sense biased against the new method: data was simulated according to a more restricted model than the GM model underlying the SVD algorithm, so that one might expect the generality of the GM model allowed too much flexibility in parameters for optimal tree recovery. It would be interesting to see how the algorithm's performance compares on simulated data that violates some of the common assumptions of the competing methods. For data arising without stable base frequencies throughout evolution, or with substitution rates on different edges of the tree varying substantially, the GM model may be valid where something like the GTR is not. Indeed, in such a situation the SVD method could be proved to be statistically consistent, unlike standard implementations of other methods, which do not allow such flexibility in models.

Second, the SVD method is based only on consideration of edge invariants, and not of vertex invariants. In a sense, it is dealing with a model even more general than GM, by placing no assumptions at all on how substitutions occur around the time of speciation events. While one might expect better performance if vertex invariants are somehow utilized, it is unfortunately not immediately clear how to do so. There is no simple analogue of the SVD for determining best approximations of 3-dimensional tensors of specified rank, so new ideas are likely to be needed.

Although more needs to be done to develop this approach, there is also much potential to do so. The focus on the relationship of invariants to local tree structure, as well as the introduction of the SVD to provide an alternative to naive evaluation for 'near-vanishing' of polynomials, can guide future work.

4.10.2 *Invariants for gene order models*

In [56, 57, 58], a new direction in the application of invariants was given by Sankoff and Blanchette, to inferring phylogenetic trees from gene order data. Not only are parsimony approaches to inference in this setting computationally slow even for quite small trees, but they can also produce incorrect results if there are large differences in branch lengths in the tree. Since invariants are based on a model, and are designed to 'ignore' specific parameter values such as branch lengths, they might provide a useful new approach.

First a simplified probabilistic model is given to describe gene order data with n genes. Focusing on any particular gene, the various states for the model are the possible genes that might be its successor in the ordering. Assuming equal probabilities of all such changes on a given edge of the tree, an $(n-1)$-state model generalizing the Jukes–Cantor one is produced. Thus linear invariants are well understood, and can be explicitly produced for a small tree and small n. From simulation using parameters inferred from the data, distributions for the values of these invariants can be produced, and significance levels assigned to the values they produce when evaluated on the data. For real data, the method produces plausible results, in line with a parsimony approach focusing only on adjacent genes. While there is possibly some improvement in inference, examples are too few to be conclusive.

As the authors noted, little had been done with probabilistic models for evolving gene order, and the simple model they used is only a very rough approximation that might be improved. They also investigated only linear invariants, the construction of which was already well known for this model, noting their insensitivity to rate variation. Producing a more sophisticated model and determining its invariants might well enable better inference, though how difficult that might be is unclear.

4.11 Concluding remarks

Much progress has been made in understanding invariants of various phylogenetic models. Only recently has it been possible to claim we know all invariants for some models, or even a large number of invariants for models general enough to include those commonly used in inference. For group-based models and the GM model our knowledge is now extensive, and a pleasing and potentially useful relationship between invariants and local tree features has emerged. Even for certain very general mixture models, we have learned of some non-linear invariants that are topologically informative.

Moreover invariants have proved their usefulness in addressing two fundamental theoretical issues. They played an important role in investigating the possibility of multiple maxima of the likelihood functions, making it possible to formulate the problem as one of constrained optimization so that exact solutions could be found. They also were the key tool in establishing the identifiability of trees for general mixture models, with a small number of classes, for generic parameters.

How invariants might be useful in practical inference is now a question ready for renewed exploration. Earlier disappointments in the performance of linear invariants should not be discouraging, since that small subclass of invariants offers little insight into how higher degree ones might perform. For naive approaches to using invariants for inference to be developed into useful and well-founded methods, we need to find both good ways of evaluating a large number of invariants, and good statistical approaches to judging whether the results are near to zero. But as the SVD algorithm has shown, we might let invariants guide our thinking yet use other computational ideas in developing an inference method.

Simply put, we do not yet know how to use invariants to address practical problems. Although their potential seems clear, the development of ways to use invariants, either heuristically or in well-founded statistical tests, needs the attention of a wider group of researchers.

References

[1] Allman, E. S., and Rhodes, J. A. (2003). Phylogenetic invariants for the general Markov model of sequence mutation. *Mathematical Biosciences*, **186**, 113–144.

[2] Allman, E. S. and Rhodes, J. A. (2004). Quartets and parameter recovery for the general Markov model of sequence mutation. *Applied Mathematics Research eXpress*, **2004**(4), 107–131.

[3] Allman, E. S. and Rhodes, J. A. (2006). Phylogenetic invariants for stationary base composition. *Journal of Symbolic Computation*, **41**(2), 138–150.

[4] Allman, E. S., and Rhodes, J. A. (2006). The identifiability of tree topology for phylogenetic models, including covarion and mixture models. *Journal of Computational Biology*, **13**(5), 1101–1113. `arXiv:q-bio.PE/0511009`.

[5] Allman, E. S. and Rhodes, J. A. (2007). Identifying evolutionary trees and substitution parameters for the general Markov model with invariable sites. arXiv:q-bio:PE/0702050.

[6] Allman, E. S. and Rhodes, J. A. (2007). Phylogenetic ideals and varieties for the general Markov model. To appear in, *Advances in Applied Mathematics*, arXiv:math.AG/0410604.

[7] Baake, E. (1998). What can and what cannot be inferred from pairwise sequence comparisons? *Mathematical Biosciences*, **154**(1), 1–21.

[8] Buneman, P. (1971). The recovery of trees from measures of dissimilarity. In *Mathematics in the Archeological and Historical Sciences*, pp. 387–395. Edinburgh University Press, Edinburgh.

[9] Casanellas, M., Garcia, L. D., and Sullivant, S. (2005). Catalog of small trees. In *Algebraic Statistics for Computational Biology* (ed. L. Pachter and B. Sturmfels), pp. 291–304. Cambridge University Press, Cambridge. http://www.math.tamu.edu/~lgp/small-trees/.

[10] Casanellas, M. and Sullivant, S. (2005). The strand symmetric model. In *Algebraic Statistics for Computational Biology* (ed. L. Pachter and B. Sturmfels), pp. 305–321. Cambridge University Press, Cambridge.

[11] Cavender, J. A. (1989). Mechanized derivation of linear invariants. *Molecular Biology and Evolution*, **6**, 301–316.

[12] Cavender, J. A. (1991). Necessary conditions for the method of inferring phylogeny by linear invariants. *Mathematical Biosciences*, **103**, 69–75.

[13] Cavender, J. A. and Felsenstein, J. (1987). Invariants of phylogenies in a simple case with discrete states. *Journal of Classification*, **4**, 57–71.

[14] Chang, J. T. (1996). Full reconstruction of Markov models on evolutionary trees: identifiability and consistency. *Mathematical Biosciences*, **137**(1), 51–73.

[15] Chor, B., Hendy, M., and Penny, D. (2001). Analytic solutions for three-taxon ML_{MC} trees with variable rates across sites. In *Algorithms in Bioinformatics (Århus, 2001)*, Volume 2149 of *Lecture Notes in Computer Science*, pp. 204–213. Springer, Berlin.

[16] Chor, B., Hendy, M., and Snir, S. (2006). Maximum likelihood Jukes-Cantor triplets: Analytic solutions. *Molecular Biology and Evolution*, **23**(3), 626–632. arXiv:q-bio.PE/0505054.

[17] Chor, B., Hendy, M. D., Holland, B. R., and Penny, D. (2000). Multiple maxima of likelihood in phylogenetic trees: an analytic approach. *Molecular Biology and Evolution*, **17**, 1529–1541.

[18] Chor, B., Khetan, A., and Snir, S. (2003). Maximum likelihood on four taxa phylogenetic trees: Analytic solutions. *RECOMB'03*, pp. 76–83. ACM Press, New York.

[19] Chor, B. and Snir, S. (2004). Molecular clock fork phylogenies: Closed form analytic maximum likelihood solutions. *Systematic Biology*, **53**(6), 963–967.

[20] Cox, D., Little, J., and O'Shea, D. (1997). *Ideals, Varieties, and Algorithms* (2nd edn.). Springer-Verlag, New York.

[21] Drolet, S. and Sankoff, D. (1990). Quadratic tree invariants for multivalued characters. *Journal of Theoretical Biology*, **144**, 117–129.

[22] Eriksson, N. (2005). Tree construction using singular value decomposition. In *Algebraic Statistics for Computational Biology* (ed. L. Pachter and B. Sturmfels), pp. 347–358. Cambridge University Press, Cambridge.

[23] Eriksson, N., Ranestad, K., Sturmfels, B., and Sullivant, S. (2004). Phylogenetic algebraic geometry. In *Projective Varieties with Unexpected Properties; Siena, Italy*, (Eds. Ciro Ciliberto, Antony V. Geramita, Brian Harbourne, Rosa Maria Miró–Roig, and Kristian Ranestrad) pp. 237–256. de Gruyter, Berlin. `arXiv:math.AG/0407033`.

[24] Evans, S. N. and Speed, T. P. (1993). Invariants of some probability models used in phylogenetic inference. *Annals of Statistics*, **21**(1), 355–377.

[25] Evans, S. N. and Zhou, X. (1998). Constructing and counting phylogenetic invariants. *Journal of Computational Biology*, **5**(4), 713–724.

[26] Ferretti, V., Lang, B. F., and Sankoff, D. (1994). Skewed base compositions, asymmetric transition matrices, and phylogenetic invariants. *Journal of Computational Biology*, **1**(1), 77–92.

[27] Ferretti, V. and Sankoff, D. (1993). The empirical discovery of phylogenetic invariants. *Advances in Applied Probability*, **25**(2), 290–302.

[28] Ferretti, V. and Sankoff, D. (1995). Phylogenetic invariants for more general evolutionary models. *Journal of Theoretical Biology*, **173**, 147–162.

[29] Ferretti, V. and Sankoff, D. (1996). A remarkable nonlinear invariant for evolution with heterogeneous rates. *Mathematical Biosciences*, **134**(1), 71–83.

[30] Fu, Y. (1995). Linear invariants under Jukes' and Cantor's one-parameter model. *Journal of Theoretical Biology*, **173**, 339–352.

[31] Fu, Y. and Li, W. (1992). Construction of linear invariants in phylogenetic inference. *Mathematical Biosciences*, **109**, 201–228.

[32] Fu, Y. and Li, W. (1992). Necessary and sufficient conditions for the existence of linear invariants in phylogenetic inference. *Mathematical Biosciences*, **108**, 203–218.

[33] Galtier, N. (2001). Maximum-likelihood phylogenetic analysis under a covarion-like model. *Molecular Biology and Evolution*, **18**(5), 866–873.

[34] Grayson, D. R. and Stillman, M. E. (2002). Macaulay2, a software system for research in algebraic geometry. Available at `http://www.math.uiuc.edu/Macaulay2/`.

[35] Greuel, G.-M., Pfister, G., and Schönemann, H. (2001). SINGULAR 2.0. A Computer Algebra System for Polynomial Computations, Centre for Computer Algebra, University of Kaiserslautern. `http://www.singular.uni-kl.de`.

[36] Hagedorn, T. R. (2000). A combinatorial approach to determining phylogenetic invariants for the general model. Technical report, Centre de recherches mathmatiques.

[37] Hagedorn, T. R. (2000). Determining the number and structure of phylogenetic invariants. *Advances in Applied Mathematics*, **24**(1), 1–21.

[38] Hagedorn, T. R. and Landweber, L. F. (2000). Phylogenetic invariants and geometry. *Journal of Theoretical Biology*, **205**, 365–376.

[39] Hendy, M. D. (1989). The relationship between simple evolutionary tree models and observable sequence data. *Systematic Zoology*, **38**, 310–321.

[40] Hendy, M. D. (2005). Hadamard conjugation: An analytic tool for phylogenetics. In *Mathematics of Evolution and Phylogeny* (ed. O. Gascuel), pp. 143–177. Oxford University Press, Oxford.

[41] Hendy, M. D. and Penny, D. (1989). A framework for the quantitative study of evolutionary trees. *Systematic Zoology*, **38**, 297–309.

[42] Hendy, M. D. and Penny, D. (1996). Complete families of linear invariants for some stochastic models of sequence evolution, with and without the molecular clock assumption. *Journal of Computational Biology*, **3**(1), 19–31.

[43] Hoşten, S., Khetan, A., and Sturmfels, B. (2005). Solving the Likelihood Equations. *Foundations of Computational Mathematics. The Journal of the Society for the Foundations of Computational Mathematics.* **5**(4), 389–407. `arXiv:math.ST/0408270`.

[44] Huelsenbeck, J. P. (1995). Performance of phylogenetic methods in simulation. *Systematic Biology*, **44**(1), 17–48.

[45] Huelsenbeck, J. P. and Hillis, D. M. (1993). Success of phylogenetic methods in the four-taxon case. *Systematic Biology*, **42**(3), 247–264.

[46] Jin, L. and Nei, M. (1990). Limitations of the evolutionary parsimony method of phylogenetic analysis. *Molecular Biology and Evolution*, **7**(1), 82–102.

[47] Kim, J. (2000). Slicing hyperdimensional oranges: The geometry of phylogenetic estimation. *Molecular Phylogenetics and Evolution*, **17**(1), 58–75.

[48] Lake, J. A. (1987). A rate independent technique for analysis of nucleic acid sequences: Evolutionary parsimony. *Molecular Biology and Evolution*, **4**(2), 167–191.

[49] Landsberg, J. M. and Manivel, L. (2004). On the ideals of secant varieties of Segre varieties. *Foundations of Computational Mathematics*, **4**(4), 397–422.

[50] Navidi, W. C., Churchill, G. A., and von Haeseler, A. (1993). Phylogenetic inference: Linear invariants and maximum likelihood. *Biometrics*, **49**(2), 543–555.

[51] Nguyen, T. and Speed, T. P. (1992). A derivation of all linear invariants for a nonbalanced transversion model. *Journal of Molecular Evolution*, **35**, 60–76.

[52] Pachter, L. and Sturmfels, B. (2004). Tropical geometry of statistical models. *Proceedings of the National Academy of Sciences, USA*, **101**(46), 16132–16137 (electronic).

[53] Pachter, L. and Sturmfels, B. (ed.) (2005). *Algebraic Statistics for Computational Biology*. Cambridge University Press, Cambridge.

[54] Rogers, J. S. (2001). Maximum likelihood estimation of phylogenetic trees is consistent when substitution rates vary according to the invariable sites plus gamma distribution. *Systematic Biology*, **50**(5), 713–722.

[55] Sankoff, D. (1990). Designer invariants for large phylogenies. *Molecular Biology and Evolution*, **7**(3), 255–269.

[56] Sankoff, D. and Blanchette, M. (1999). Phylogenetic invariants for genome rearrangements. *Journal of Computational Biology*, **6**(3/4), 431–445.

[57] Sankoff, D. and Blanchette, M. (1999). Probability models for genome rearrangements and linear invariants for phylogenetic inference. In *Proceeedings of the Third Annual International Conference on Computational Molecular Biology (RECOMB 99)*, pp. 302–309. ACM Press, New York.

[58] Sankoff, D. and Blanchette, M. (2000). Comparative genomics via phylogenetic invariants for Jukes-Cantor semigroups. In *Stochastic models (Ottawa, ON, 1998)*, pp. 399–418. American Mathematical Society, Providence.

[59] Semple, C. and Steel, M. (1999). Tree representations of non-symmetric group-valued proximities. *Advances in Applied Mathematics*, **23**(3), 300–321.

[60] Semple, C. and Steel, M. (2003). *Phylogenetics*, Volume 24 of *Oxford Lecture Series in Mathematics and its Applications*. Oxford University Press, Oxford.

[61] Steel, M. (1994). Recovering a tree from the leaf colourations it generates under a Markov model. *Applied Mathematics Letters*, **7**(2), 19–23.

[62] Steel, M., Huson, D., and Lockhart, P. J. (2000). Invariable sites models and their uses in phylogeny reconstruction. *Systematic Biology*, **49**(2), 225–232.

[63] Steel, M., Székely, L., Erdös, P. L., and Waddell, P. (1993). A complete family of phylogenetic invariants for any number of taxa under Kimura's 3ST model. *New Zealand Journal of Botany*, **31**(31), 289–296.

[64] Steel, M., Székely, L. and Hendy, M. D. (1994). Reconstructing trees from sequences whose sites evolve at variable rates. *Journal of Computational Biology*, **1**(2), 153–163.

[65] Steel, M. A. and Fu, Y. X. (1995). Classifying and counting linear phylogenetic invariants for the Jukes-Cantor model. *Journal of Computational Biology*, **2**(1), 39–47.

[66] Sturmfels, B. and Sullivant, S. (2005). Toric ideals of phylogenetic invariants. *Journal of Computational Biology*, **12**(2), 204–228. `arXiv:q-bio.PE/0402015`.

[67] Székely, L. A., Steel, M. A., and Erdős, P. L. (1993). Fourier calculus on evolutionary trees. *Advances in Applied Mathematics*, **14**(2), 200–210.

[68] Tuffley, C. and Steel, M. (1998). Modeling the covarion hypothesis of nucleotide substitution. *Mathematical Biosciences*, **147**(1), 63–91.

[69] Yang, Z. (2000). Complexity of the simplest phylogenetic estimation problem. *Proceedings of the Royal Society of London B: Biological Sciences*, **267**, 109–116.

III

TREE SHAPE, SPECIATION, AND EXTINCTION

5

SOME MODELS OF PHYLOGENETIC TREE SHAPE

Arne Ø. Mooers, Luke J. Harmon, Michaël G. B. Blum, Dennis H. J. Wong, and Stephen B. Heard

Abstract

As products of diversifying evolution, phylogenetic trees retain signatures of the evolutionary events and mechanisms that gave rise to them. Researchers have used a variety of theoretical models to represent different hypotheses about how diversification might proceed through the evolution of a clade. We outline two widely-used measures of phylogenetic tree shape, review a number of tree-generating models, and set out the predictions they make about tree shapes. The simplest of these models (the 'Yule' and 'Hey' models) are still used routinely, sometimes as if they provided good representations of diversification in nature; in fact, they do rather poorly when confronted with real data. More complex models that incorporate hypothesized macroevolutionary processes can in some cases provide a better fit to real data. We recommend further development of these more complex models—for instance, exploration of models that treat species as collections of individuals rather than as simple lineages. Much work remains to be done in estimating trees (especially waiting times), in exploring tree-generating models, and in assessing patterns in the shapes of real phylogenies.

5.1 Introduction

Phylogenetic trees represent the evolutionary histories of lineages and so bear the impression of the evolutionary forces that gave rise to those lineages. Advances in molecular and computational techniques continually increase the number and size of our phylogenetic estimates. In the 1990s, both we [41] and Purvis [52] surveyed the two main aspects of phylogenetic tree pattern: variation in realized diversification rate among contemporaneous lineages, and changes in realized diversification rates through time. The techniques highlighted in these reviews have been used very successfully (see, e.g. [4, 10, 11, 62, 63]).

In parallel, researchers have continued to present generating process models for phylogenetic trees, in the hopes of being able to compare these with the real things. We offer a biological perspective on some of these models here. Our general thesis is that these models should do more than mimic reasonable tree shapes: they should offer clear hypotheses that can be tested with the data as they become available. It is likely that real trees will be shaped by many factors and so these models should not be seen as mutually exclusive. All the models we survey are extensions of the simple birth–death process, in that evolving lineages have defined probabilities per unit time of giving birth to new lineages (causing a bifurcation) or terminating, and differ only in how these probabilities are assigned. We consider the strengths and problems of the models we survey and direct readers to some that we feel might show promise.

5.2 Background

We use the term 'tree shape' to refer generically to both the distribution of sizes of the groups defined by nodes (called 'clades' by evolutionary biologists) and the distribution of edge weights (called 'branch lengths' by evolutionary biologists) on a directional bifurcating acyclic graph (Fig. 5.1). Our choice of graph structure is motivated by the fact that evolution is directional and primarily diversifying, and that events leading to multifurcations (i.e. vertices of degree > 3) are rare [29]. We recognize that our formulation overlooks other interesting graph structures relevant to evolution (e.g. cycles produced by recombination in gene trees or by hybrid species formation in species trees; uncertainty expressed in unrooted trees or in graphs with multifurcations). We further restrict ourselves to ultrametric trees, and refer to edge lengths and waiting times using time units. This is because we are interested in the actual diversification process through time, rather than in the inference process. This glosses over some painful facts—very few inferred trees have a robust timeline, and rooting trees is very difficult.

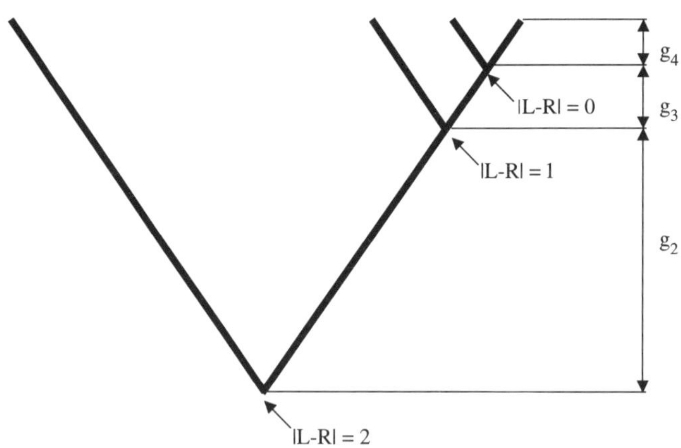

FIG. 5.1. A simple bifurcating tree highlighting the measures taken to summarize topology and waiting times. The sum of $|L - R|$ is used to give a measure of tree balance, while the waiting times g are used to create a measure of the relative placement of nodes between the root and the tips.

We concentrate on two aspects of tree shape. The first is the variation in subgroup size, captured very efficiently (see [2, 36]) by Colless' measure of imbalance [9, 23]. Colless' index I_c considers the number of leaves in the two partitions defined by each internal node (L and R) and is the sum of $|L - R|$ over all the $n - 1$ nodes in the tree, often normalized by the maximum possible value for a tree of size n (Fig. 5.1). Besides being the most commonly-used metric for bifurcating trees, it has a clear biological interpretation as an average measure of the realized differences in diversification rate of sister groups. Though $E(I_c)$ scales with n [23, 58], its distribution has been characterized under the pure birth model [5]. I_c also has the property of being most sensitive to variation in clade size nearest the root of the tree [2, 23].

The second aspect of tree shape is a measure of the distribution of nodes from the root to the leaves, or 'waiting times' (Fig. 5.1), as captured in lineage-through-time plots [46, 47]. Indices designed to summarize nodal distribution include stemminess [59] and γ [53] (or the closely related δ [54]). We return to these measures below.

As summary statistics, I_c and γ (or δ) do not capture all the variation within a sample of rooted trees [36]: for instance, two trees of size n with different topologies can nevertheless share the same I_c score [58]. They do, however, capture both differences in diversification rate among contemporaneous clades (I_c) and differences in diversification rate as one proceeds through time (γ). Importantly, the two axes are not expected to vary independently [2, 17, 41]; unfortunately, there is still little work that considers how reasonable diversification processes affect both aspects simultaneously.

5.3 Yule and Hey models

The simplest possible model of diversification also has the oldest pedigree, developing from a simple model of diversification presented in 1924 [77]. Though the original model had two parameters (one for the birth of species within genera and one for the birth of genera), the 'Yule Process' refers to a model where there is no death, and diversification is modeled as a Markov process with a single parameter λ, the instantaneous rate of birth [3]. The parameter λ can be thought of as the average number of speciation events that occur in one lineage per unit of time. The topologies of trees produced by this pure birth model can be described recursively. At a node that subtends n species, the number of species of the 'left' subtree is chosen uniformly over $\{1, \ldots, n-1\}$ and this process is continued in the left and right subtrees until the tips are reached [21, 70]. Blum and colleagues [5] have recently derived the asymptotic distribution of I_c under this model. The times during which there are i lineages, in the Yule model, are exponential random variables with parameter λi [28, 47], and the conditional probability distributions of the branching times given that n lineages are found after t units of time can be found in [44] and [76]. Pybus and Harvey [53] took advantage of constant λ to produce a standardized statistic γ such that $\gamma > 0$ with an increasing λ from the root to the tips, and $\gamma < 0$ with a slowdown in diversification through time:

$$\gamma = \frac{\frac{1}{n-2}\sum_{i=2}^{n-1}\sum_{j=2}^{i} jg_j - \frac{T}{2}}{\frac{T}{\sqrt{12(n-2)}}}, \tag{5.1}$$

where T is given by

$$T = \sum_{j=2}^{n} jg_j.$$

The expression for γ is obtained after modifications of a test statistic introduced by Cox [12] in the context of Poisson processes (see Appendix).

The Yule model was extended to include a constant death rate (μ) by Raup and colleagues in the early 1970s [56]. Because the rates are invariant across lineages, this addition does not change the expected distribution of topologies. However, because we will now sample lineages in the present time that are destined to go extinct [22], there are 'too many' lineages near the present, and $\gamma > 0$.

Another simple model for generating tree shapes was presented by Hey [28]. This model assumes that the total number of species N is fixed and that each lineage bifurcates at rate λN. More precisely, the time before speciation in each lineage is an exponential random variable with parameter λN. Furthermore, at each speciation event a lineage chosen uniformly among all lineages goes extinct, insuring that the total number of species remains constant. The same model is known in population genetics as the Moran model [42] [13, p. 18–23]. What is usually called the Hey model, though, is not the forward-in-time model that is described above, but the backward-in-time model that corresponds to the genealogy of n species sampled among the N extant species. Therefore it should be emphasized that Yule trees describe the genealogies of entire monophyletic groups, whereas Hey trees describe, within a monophyletic group, the genealogies of samples of species ($n \leq N$). The genealogy in the Hey model is equivalent to a well-characterized model known in population genetics as the coalescent [27] [Felsenstein, this volume]. The topology of the Hey (coalescent) process is simply described as follows: starting with n lineages, two pairs of lineages are chosen uniformly among all the possible pairs to coalesce and this coalescence process is continued until there is only one remaining lineage. The topologies of the Hey trees (and so its measure, I_c) are distributed identically to the topologies of the Yule trees [47]. In the Hey model, the time during which there are exactly i lineages is an exponential random variable with parameter $\lambda i(i-1)$ [28, 47]. Note that the expected values of coalescence times $1/i(i-1)$ (when $\lambda = 1$ as it is usually assumed in the Moran model) in the Hey model differ by a factor 2 to the expected values of coalescence times $2/i(i-1)$ in the coalescent model as it is usually used in population genetics [13, p. 23]. Under the Hey model, the statistic γ is expected to be large, with more nodes found nearer the leaves than under the Yule model. Pybus and colleagues [54] took advantage of the known waiting times for events under the coalescent to produce a new standardized measure denoted δ:

$$\delta = \frac{\dfrac{T'}{2} - \dfrac{1}{n-2}\sum_{i=n}^{3}\sum_{j=n}^{i} j(j-1)g_j}{\dfrac{T'}{\sqrt{12(n-2)}}}, \tag{5.2}$$

where T' is given by

$$T' = \sum_{j=n}^{2} j(j-1)g_j.$$

The expression of δ given by Pybus *et al.* [54, their equation (2)] results from our equation (5.2) after dividing the numerator and the denominator of our equation (5.2) by 2. The derivation of the statistic δ is given in the Appendix. We note that Pybus *et al.* did not apply δ to species trees.

Importantly, both these models do a remarkably poor job of capturing the distribution of tree shapes reported in the literature [2, 6, 30, 41, 69]: published trees are much more imbalanced (have higher I_c values) than expected. This is an important and perhaps still under-appreciated finding: if our published trees are unbiased with respect to shape, there are strong macroevolutionary forces at work that demand explanation. However, perhaps because of their convenience, these null models are still often used either explicitly [44, 45, 78] or implicitly (see, e.g. [7]).

5.4 λ = function(trait)

The core assumption of the models presented above, that all species have equal speciation rates at a given time, is an assumption that most evolutionary ecologists would always have rejected. Instead, at least since Darwin's time, an enormous amount of attention has been paid to the notion that some lineages might experience higher speciation rates (or lower extinction rates) than others, either due to intrinsic properties of the species, extrinsic factors having to do with the environment, or the interaction of the two [25, 43, 62]. Differences in diversification rates among related lineages have in fact been documented for a variety of clades (e.g. [7, 38, 67]), and analyses of branch-length distributions in phylogenies [61] have established that differences in diversification rate not only exist, but tend to be propagated along evolving lineages (such that high or low rates are 'heritable' from ancestral to descendent species). An important class of generating models [24] seeks to incorporate some of this biology by considering the case where the speciation rate λ is a (perhaps nonlinear) function of some variable x, where x takes on a value for each species that is determined by an evolutionary model over the phylogeny of an evolving clade. Most simply, x can be interpreted as any evolving trait (simple or complex) of the organisms, such as body size, dispersal rate, feeding strategy, or pollination syndrome [24], but it could equally represent a characteristic of the environment, so long as restricted dispersal by the organisms constrained the value of x for one species to resemble the value of x for its ancestor. In either case, λ varies among species in an evolving clade, but does so with non-zero heritability (there is a resemblance between ancestor and descendent) such that whole lineages are typified by higher or lower speciation rates.

Heard [24] explored a model belonging to this class, in which a trait value x evolved in a clade by a random walk, with changes either gradual (continuous in time) or punctuated (occurring only at speciation events). In this model, λ for each species was a simple function of the trait value x, plus a 'noise' term ϵ representing other influences on speciation rate. Heard [24] found that this model produced phylogenies with high I_c compared to the ERM, and that I_c values typical of real phylogenies could be produced—albeit with high rates of evolution in the trait value x (or, more generally, in the rate of evolution of the diversification rate parameter itself). Furthermore, speciation-rate variation arising through the addition of the 'noise' term increased I_c, but only when values of ϵ were persistent through time (that is, when ϵ changed only at speciation events, rather than continuously through time). This model drew attention to the importance of differences in diversification rates that are maintained by lineages through time (either through trait heritability or through other temporally persistent effects on λ) in generating phylogenies with high I_c. Efforts to demonstrate the existence of heritable diversification-rate variation [61] and to devise tests for correlates of diversification rate (see, e.g. [50]) were inspired directly by this generating model.

Heard [24] did not consider the nodal height distribution property of the trees produced by his model. Because clades in Heard's model become dominated by high-diversification-rate lineages [24] via species sorting [74], we would expect their phylogenies to have $\gamma > 0$ as more speciation events occur closer to the present. However, whether models of this type can produce trees with realistic values of γ (and do so for the same parameter values that produce realistic I_c) remains unknown.

5.5 λ = function(age)

In this class of generating models, λ varies among species only as a function of the time elapsed since a species's last speciation event (its age). One can imagine biological circumstances under which speciation rates might be either higher or lower for young species, and both cases have been modelled.

Models in which young species have smaller λ are biologically plausible when young species tend to have small population sizes or small geographic ranges. This is, in fact, a prediction of most models of speciation, most notably of the peripheral-isolate model [39]. Two slightly different models have been proposed. Losos and Adler [35] described a model in which speciation rate $\lambda = c$ for all lineages, except that following speciation, one daughter lineage has $\lambda = 0$ during a refractory period of length a^*. As an alternative, Chan and Moore [8] modelled λ as increasing linearly from zero to c over a period a^* for both daughters following a split. In either case, with a^* small to moderate compared to total tree height, these models produce phylogenies more balanced (lower I_c) than does the pure-birth model. (When a^* is a substantial fraction of total tree height, the resulting phylogenies have higher I_c than pure-birth, but such large values of a^* are probably not plausible in the biological context that inspired the models). Because these models, then, produce phylogenies even more unrealistic than the pure-birth model ('real' trees have higher I_c than pure-birth, not lower), they have not attracted much recent attention. Our preliminary work (SBH and DHJW) suggests that reasonable values of a^* have no effect on γ. Moderate refractory periods lower the effective speciation rate, but do not change the relative distribution of speciation events over the height of the tree. Much longer refractory periods do give rise to trees with negative γ, but again, such long a^* are probably unrealistic.

Models in which young species have larger λ are biologically plausible when speciation events are likely to occur in bursts—for instance, because lineages that are speciating have colonized a new region, and a new region with many open niches favours multiple speciation events [70]. Agapow and Purvis [2] considered a discrete time model in which λ increases after speciation, followed by decay back to $c : \lambda(a) = c + Ka^{-0.5}$, where a is age (time post-speciation, with both daughters of a speciation event beginning with age $a = 0$). Steel and McKenzie [70] examined a general class of models in which λ decreases monotonically with a (the Agapow and Purvis model is a special case), but developed in particular a subclass in which $\lambda(a) = 0$ for $a > m$, where m is a constant speciation window. A simple version of this model, essentially the converse of the Losos–Adler refractory period model, would have $\lambda(a) = c$ for $a \leq m$, and $\lambda(a) = 0$ for $a > m$. Both the Agapow–Purvis and the Steel–McKenzie models produce imbalanced phylogenies (high I_c, which is realistic), and distributions of nodal heights with more speciation events towards the root of the tree ($\gamma < 0$). However these models generally have been explored only by simulation; formal results establishing distributions of I_c or γ are known for only a few special cases (see, e.g. [5]).

There are (at least) two interesting questions one could ask about Agapow–Purvis and Steel–McKenzie models. The first of these is statistical, and concerns the ability of the models to produce trees with any given distribution of shapes. The second question is more biological, and concerns the fit of model results to real-world trees.

The Steel–McKenzie model was motivated by the Uniform distribution of phylogenies, a natural distribution of interest to many mathematicians whereby all labelled cladograms (rooted trees where the branch lengths are not considered) are equally likely. Under this distribution, trees are random guesses [68]. This model might be useful as a prior for Bayesian tree inference. However, despite its mathematical attractiveness, evolutionary biologists have largely failed to imagine plausible process models that produce such a distribution. Steel and McKenzie [70] proved that their model does produce the Uniform distribution when $\lambda(a) = 0$ for $a > m$, $m < \frac{T}{n}$, where T is an arbitrary time horizon. However, upon closer examination this result appears to be of primarily mathematical interest because the trees produced under these conditions are not biologically plausible. To see this, consider that any lineage that fails to speciate before a time m since its birth is a spinster that can never speciate again; and a tree in which all lineages are spinster lineages is a spinster clade that can never increase in size. But the condition for producing the Uniform distribution is $m < \frac{T}{n}$, or $T > mn$. Since each lineage must speciate within an interval m or become a spinster, after a period $T > mn$ the only trees of n lineages that can exist are spinster trees. We do not believe that many (if any) real clades are spinster clades in which the origin of new lineages is no longer possible; on the contrary, available evidence suggests that speciation continues today in many if not all clades (e.g. [62, 71]).

This does not mean, however, that the model should be discarded. Instead, one can ask a second question about the model: can it produce realistic tree shapes with plausible parameter values? Using the same approach we have described for other model classes, one could compare Steel–McKenzie model trees with collections of real estimated phylogenies, asking whether plausible values of model parameters (c, K, m, or others in more complex models of the class) can produce phylogenies with realistic I_c and γ. This is an open question, in part because what constitutes 'realistic' γ values is not well established, and in part because the biological or palaeontological data needed to assess plausibility of a particular choice of K or a^* are not obviously available. Analysis of this sort (as in [24]) is logically straightforward, at least, and could establish whether 'speciation-burst' biology is a good candidate as a contributor to the shapes of real trees.

5.6 λ = function(time)

There are several verbal models that make λ a declining function of absolute time rather than the age of the lineage; for instance, key innovations or new biogeographic opportunities may allow for an initial flourish of speciation that then settles down. However, the model that has received the most attention is that of adaptive radiation (AR [62]). Adaptive radiation is the evolution of phenotypic divergence in a rapidly multiplying lineage [62]; indeed, it is primarily the emphasis on phenotypic divergence that separates AR models from the models considered in the previous section. Some claim that adaptive radiation may account for much or even most of present day diversity (D. Schluter, pers. comm.). One expectation from AR theory is that speciation is rapid in its initial stages and then slows down (so, e.g. $\gamma < 0$; [19, 62]). This seems to be the case for some fossil [18] and some extant clades [46, 51, 60, 66]. One presumed underlying pattern has clades growing rapidly and then, as birth rates decline below extinction rates, shrinking. We note that this particular trajectory has been formally modelled for species numbers by Raup and colleagues [56] and Strathman and Slatkin [72] and presented as an example for waiting times on trees by Nee and colleagues [47].

More quantitative work on AR tree shape is needed. Gavrilets and Vose [19] have made a start with an individual-based simulation approach to AR, where sexual diploid individuals with complex genomes evolve on discrete patches arranged on an initially empty but heterogeneous grid. These individuals migrate, undergo selection, and eventually form populations that speciate. They found that speciation was vastly more common during the early stages of the diversification; resulting trees would have low γ values. They also often observed 'overshooting', where the clade size at the end of a run was smaller than the maximum reached during a run. Though they do not look at tree balance, Gavrilets and Vose [19] interpret some of their simulation results in light of a verbal model of a few generalist lineages rapidly speciating into slower-evolving specialists, which might give rise to imbalanced trees. The generalist to specialist pattern is, however, not strongly supported by available comparative data [49, 62].

5.7 The neutral model

Another rich, if controversial, approach to explaining biodiversity production is Hubbell's 'Unified Neutral Theory' or UNT [31]. The UNT has at its core a metacommunity landscape saturated with competitively identical individuals. This landscape is made up of patches that can be occupied by only one individual, regardless of its species. In this model, individuals in the metacommunity compete for space, with patches vacated by death filled by migration of a new individual from surrounding patches. This feature makes the UNT a null model for community organization and evolution, and it is widely agreed that at least some communities deviate strongly from the UNT. However, the extent to which this is true is currently under intensive debate (e.g. [20, 40]). Although much of the focus of this debate has been placed on the ability of this theory to explain relative abundances within communities (e.g. [16, 73]), the UNT also makes predictions about the shapes of phylogenetic trees. In the context of diversification, Hubbell's model has an unchanging per-individual speciation rate over the entire metacommunity, while extinction occurs whenever the population size of any species reaches zero individuals. As a consequence, per-species speciation and extinction rates are a function of population size. Critically, under the UNT, extant lineages differ in a predictable fashion in relative abundance, collectively approximating a truncated log-normal distribution. This means that at any time, extant lineages differ predictably in their propensities to speciate and to go extinct. Hubbell [31] was able to demonstrate by simulation that the UNT produces trees with a concentration of short branches near the tips (high γ values), because extinction is highest when there are many species at small population size.

The UNT is qualitatively different from AR in that lineages do not evolve to take advantage of heterogeneous resources. Also, because speciation is conceived of as a point mutation in one individual, its behaviour in terms of population size is punctuational [15]—the parent species is very similar in size before and after the speciation event, while the daughter lineage is made up of a single individual and so it initially has a very low probability of speciating and a very high probability of going extinct.

In order to address what sorts of trees this explicitly ecological model produces, we modeled the UNT for a metacommunity composed of 441 local communities arranged in a 21 × 21 grid. Each local community was made up of 100 individuals for a total metacommunity size (J_m) of 44,100. Hubbell [31] defines a 'fundamental biodiversity number' $\theta = 2JmV$, where v is the per-capita speciation rate. For our simulations, we used a value of $\theta = 5$. Following Hubbell [31], we ran simulations in discrete time and allowed one individual per local community to die and be replaced by a birth, migration, or speciation event in each generation. We limited migration to communities that were immediate neighbours in the grid [31]. We then simulated community drift and diversification under a range of migration rates. We started each simulation with a metacommunity completely filled with individuals of a single species, and ran them until both species-abundance distributions and phylogenetic tree shape reached a dynamic equilibrium. For this set of parameter values, tree shape equilibrium was reached at around 100,000 generations, but to ensure that our results represent tree shapes at metacommunity equilibrium, we ran simulations for 500,000 generations to produce phylogenetic trees. Increasing migration rates had a negative impact on metacommunity species richness (Fig. 5.2; [31]). As stated by Hubbell [31], we found that phylogenetic trees generated from these simulations show a concentration of short branches near the tips; as a consequence, γ values were consistently high over a range of migration rates. In fact, for most sets of simulations, over half of the produced phylogenies had $\gamma > 2$, and would constitute a significant deviation from the pure-birth expectation. This effect was most pronounced for very low migration rates, but that may be influenced by higher power of the γ statistic [53] for the larger trees such simulations produce. Phylogenies produced by this model were highly imbalanced; in fact, most phylogenies were completely pectinate trees (Fig. 5.2). The percentage of completely pectinate trees increased with higher migration rates. This is because under Hubbell's model, metacommunities with high migration rates have a steeper rank abundance curve [31]. Since variation in speciation rate in a metacommunity is related to the slope of the rank abundance distribution, communities dominated by a single abundant species will have more imbalanced phylogenetic trees than communities with a more even abundance distribution. This prediction is probably robust to many aspects of Hubbell's model, and follows from the mode of speciation and relationship of speciation rates to abundances.

Formal comparisons of UNT tree-shape predictions with the shapes of real phylogenies have not been conducted, but it seems fairly clear that the UNT

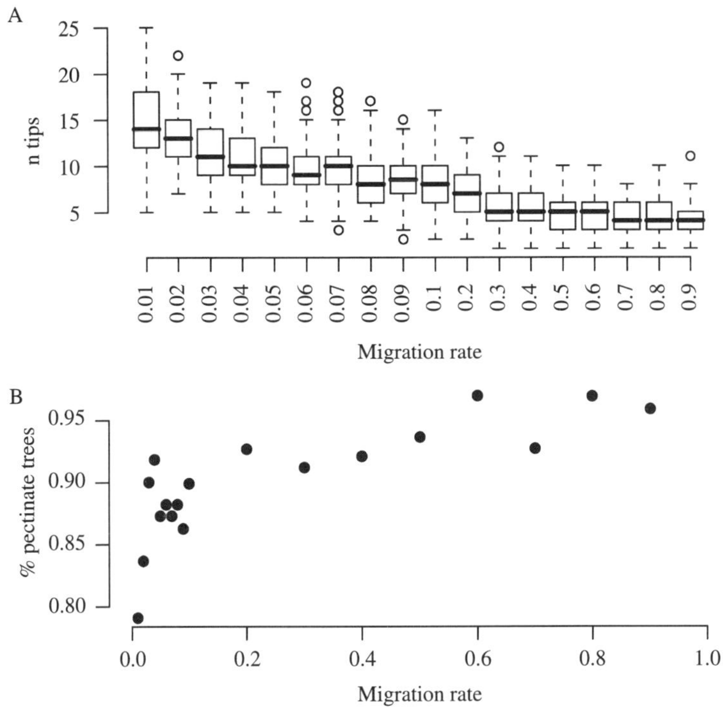

FIG. 5.2. The behaviour of diversification under the Hubbell's Unified Neutral Model. (A) The average size of trees with increasing migration rate among patches in the metacommunity. (B) The proportion of fully pectinate trees at equilibrium for communities with different migration rates among patches in the metacommunity. Because many of the trees at high migration rates are small, this is a better measure of tree shape than standardized I_c. For all runs, $J_m = 44,100$ and $\theta = 5$.

as we implement it above produces trees that are much too imbalanced to be realistic (it is more difficult to assess predictions for γ, since the distribution of γ for real trees is not known). This is an interesting result, since other modelling efforts have found it difficult to produce trees that are imbalanced enough for realism [24]. It is unknown whether elements of the UNT assumptions (such as population-size dependence), might in a more sophisticated model be able to produce realistic distributions of I_c and γ; this is an area for further study. Hubbell [31] shows, in addition, that species abundance distributions are much more even under a 'fission' model of speciation, where speciation events involve randomly dividing the ancestral population into two parts; we predict that this mode of speciation will result in more balanced phylogenetic trees than those produced in our implementation (Fig. 5.2). However, there will still be some relationship between ancestral and daughter population sizes, and trees will likely be more imbalanced than those produced under the Yule model.

5.8 λ = function (N)

One feature missing from all models discussed so far is any tendency for diversity to be limited—that is, for diversity to reach an equilibrium N^* analogous to carrying capacity in the logistic model of population growth. Such an equilibrium will result if per-capita extinction rates increase, or per-capita speciation rates decrease, with standing diversity. Such effects are plausible for a variety of biological reasons—for instance, if high diversity means smaller population sizes for each species, raising extinction risk. However, whether such limits to diversity are ever reached in nature is an open question. Paleontologists have modelled diversity in the marine fossil record with logistic-like functions that assume limits to diversity, with some success for Paleozoic faunas but more debatable results for Mesozoic and Quaternary faunas [64, 65]. Ecologists have also devoted considerable theoretical and empirical attention to the idea of 'limiting similarity' in communities (and by extension, clades), which would impose limits to diversity by setting a maximum number of niches available for occupation [1, 32, 34, 37]. A half-century of research, though, has produced no consensus on whether such models explain much about real communities. Indeed, some models of diversification either assume or imply that diversification is more likely to proceed with positive feedback than with negative: for instance, escape-and-radiation [14] and cascading host-race formation [71].

Surprisingly, little is known about tree shapes under models of limited diversification. Harvey *et al.* [22] considered a model in which extinction rate increases with diversity, but speciation rate is constant. However, they did not report balance for their model, and (considering only the extant species) only report that nodal height distributions are similar to those from a mass-extinction model. More complex models, with both speciation and extinction rates responding to diversity, show more complex behaviour (DHJW and SBH, unpublished data), for instance with γ depending strongly on the ratio of speciation to extinction rates at half of N^*. Few studies have yet asked whether limited-diversity models produce tree shapes typical of real clades, although Nee *et al.* [46] interpreted the shape of a compound bird-phylogeny as consistent with niche-filling model (though one with diversification rate decreasing to zero only as N^* approaches infinity).

A rather different approach to modelling limited diversity is implicit in the simple Hey model [28]. In Hey's model, a clade reaches size N^* and subsequently each speciation event (as speciation continues with constant rate) is balanced by a randomly imposed extinction event. Notably, the Hey model is mute with regard to how a clade reaches size N^* [47]. So, for instance, Zhaxybayeva and Gogarten [78], who recently used the model to simulate the early tree of life, simply start with N^* unrelated lineages and allow the model to run until all the extant individuals have a single common ancestor (all other $N^* - 1$ lineages having died out). Another approach that better mimics radiations is to consider a two-phase process: a tree first grows to size N^* ('growth phase'), followed by some time spent at size N^* ('Hey phase'). So long as the tree's Hey phase is long enough to reach the stationary distribution of tree shapes (that is, for any signature of the growth phase to be erased), the growth phase model doesn't matter. But how long a Hey phase might be required to reach the stationary distribution, and is this plausible for real trees?

This question has not been addressed for any growth phase model, but we can make a start by examining one simple possibility: growth phase diversification under the Yule model. We implemented a simulation model (following [23, 24]) of tree growth under the Yule model, followed by speciation and extinction (still at a constant rate for all lineages) in a Hey phase of variable length. We measure the length of a Hey phase in terms of species turnover: if there are N^* species when the Hey phase begins, then a Hey phase of length 1 has N^* speciation (and N^* extinction) events; the average species is replaced in the phylogeny once. We generated 500 replicate trees of $N^* = 10, 20, 50, 100$, and 500, with Hey phases of length 1, 5, 10, 25, and 50. We consider a Hey phase of even length 10 to be extraordinarily long, as it implies that since the clade reached its equilibrium diversity N^*, each species has (on average) been replaced 10 times over; or alternatively, over 90% of the history of the clade has been spent at equilibrium diversity. Since our Yule trees start with the same distribution of I_c as expected following the Hey phase [47], there is no change in this attribute of

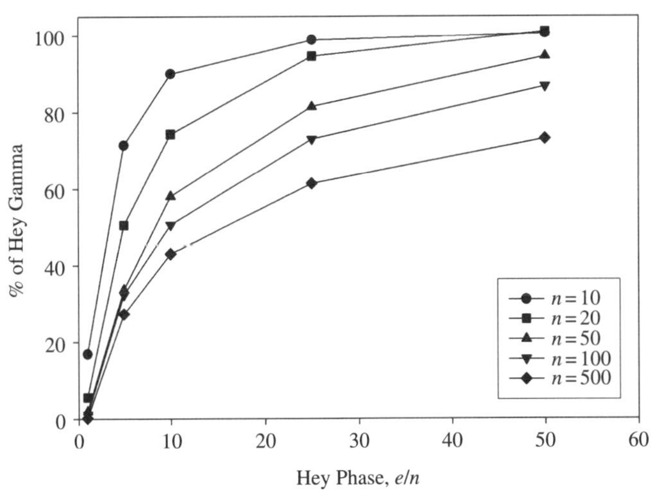

FIG. 5.3. Approach to stationary γ distribution for trees grown to size n under a Yule model, followed by balanced speciation and extinction under Hey's [28] model. The length of the Hey phase is measured as number of speciation/extinction events (e) as a multiple of the number of species in the tree (n), and γ expressed as a percentage of that expected under the Hey model.

tree shape (as there might be under other growth-phase generating models). The nodal height distribution does, however, change: the Yule trees that enter the Hey phase have growth-phase $\gamma = 0$ [53], (we confirm this in our simulations), whereas Hey trees will have large, positive γ. Importantly, for trees of moderate to large size, the approach to stationary Hey-phase γ is quite slow (Fig. 5.3): for instance, a Hey phase of length 10 brings trees of $n = 50$ and $n = 500$ trees just 58% and 43% respectively of the way from the growth-phase γ to stationary Hey-phase γ.

Since we have little evidence that modern clades are at an equilibrium diversity (N^*) at all, let alone that clades spend much time at N^*, we conclude that the Hey model is probably not very relevant to the shapes of real phylogenies. Of course, our use of the Yule model for the growth phase can (and should) be criticized, but we do not expect this to change the overall picture much. Indeed, because the Hey model produces trees with the Yule distribution of topologies, it does not mimic the trees we infer from nature.

5.9 Concluding remarks

Since the 1980s, we have known that the simplest models do a poor job of modelling the shapes of published phylogenetic trees. Tree reconstruction methods may be biased with respect to shape [41], but current surveys suggest the problem may not be grave for trees $<\sim 50$ tips [69], which is good news. However, Wilkinson *et al.* [75] point to some clear biases with supertree construction methods, and now that we are routinely reconstructing very large trees ($N \gg 100$ tips) using heuristic methods, we continue to urge caution. A related problem is taxon choice: sampling methods that inadvertently lead to unbalanced trees are easy to conceive of, but little empirical or theoretical work has been done in this area. However, our intuition is that the tree of life is very imbalanced at all levels. There may be different processes occurring at different temporal scales; if so, we must be careful what we infer from our biologically-motivated models. Even at any one scale, different processes likely co-occur in nature, and at different times during the history of a lineage. Also, the same model applied at different scales can produce different patterns. For instance, the last two sections above consider two models (UNT and Hey) that are each based on the coalescent. However, the expected shapes are very different because the first applies the coalescent to individuals that are then aggregated into species, and the second applies the coalescent to the species themselves. This scale issue may be interesting to pursue.

We know of no general survey of waiting times (e.g. looking at γ or δ) for ultrametric trees; Hey [28] considered eight small trees but could reject neither the Yule nor the Hey models for any, suggesting very low power. Getting branch lengths that are proportional to time is very difficult (see, e.g. [60]), and advances in this area are needed. However, from the modelling perspective, there is room for much more work. As an example, the Adaptive Radiation and Hey models both have aspects of diversity-dependent cladogenesis [56], but they make strongly contrasting predictions about waiting time: AR models should produce many longer terminal branches than the Hey model.

Our approach in this chapter has been to consider simple tree measures as reflections of the underlying macroevolutionary models and to compare these to expectations under various null models. A more powerful approach would use maximum likelihood and information criteria to select among alternative models directly. Building on the pioneering work on birth–death models by Nee and colleagues [47], such tools have been created [55], but more work is needed to confront a wider range of possible models of diversification [54].

In the introduction, we stated that tree shape is the record of the tempo and mode of diversification. For evolutionary biologists, being able to explain these patterns is a major intellectual goal. However, we would like to end this chapter by pointing out that the study of tree shape resonates at a more immediate level. Rauch and Bar-Yam [57] point to the fact that reasonable models of gene genealogies imply that genetic diversity within species is highly skewed. Non-random loss of the few individuals bearing divergent genes can greatly decrease a species's overall genetic diversity. The same holds true at larger scales: to the extent that phylogenies are highly imbalanced and have long terminal branches, a few lineages will represent much of the tree's total diversity and their extinctions would represent disproportionate loss to the tree of life [26][Hartmann and Steel, this volume]. Given the rate at which we are losing species through anthropogenic extinctions, understanding this may help us in efforts to preserve the products of evolution for the distant future.

Acknowledgements

AOM thanks Olivier Gascuel, Mike Steel, and the organizers of the MEP2005 conference for the opportunity to present some ideas on tree shape to a perspicacious audience—most of which are not in this chapter. We thank our various universities, NSERC Canada (SBH, AOM), the U.S. National Science Foundation (SBH), and the Wissenschaftskolleg zu Berlin (AOM) for facilitating collaboration; Andrew Rambaut for ongoing technical help of various kinds; and Olivier Francois, Olivier Gascuel, Klaas Hartmann, Oliver Pybus, and the fab*-lab at SFU for useful feedback on some of the ideas presented here.

5.10 Appendix

The statistics γ and δ that have been introduced by Zink and Slowinski [79] and Pybus and colleagues [53, 54] in order to detect trends in speciation rates have been derived from a test statistic that can be found in [12]. In [12], the null model is a simple homogeneous Poisson process and the alternative hypothesis corresponds to a model where the instantaneous rate of occurrence λ is not constant anymore but varies with time. In a homogeneous Poisson process, the times between successive events are exponentially distributed with the same parameter λ.

FIG. 5.4. An illustration of a Poisson process. The $t_i's$ denote the time at which events occurred. In a homogeneous Poisson process, the $(t_{i+1} - t_i)$'s are exponentially distributed with the same parameter λ.

An important result concerning homogeneous Poisson processes is that the joint probability distribution of $(t_1, \ldots, t_n)$ conditional on observing n events between time 0 and time t_0 is the same as the joint probability distribution of an ordered vector of independent and uniform (over $(0, t_0)$) random variables (Fig. 5.4).

Thus, because the value of the sum of the t_i's is invariant by permutation, we have

$$E\left[\sum_{i=1}^{n} t_i\right] = n\frac{t_0}{2}. \tag{5.3}$$

Here we used the fact that the expected value of a uniform random variable over $(0, t_0)$ is $t_0/2$. Using equation (5.3) and the fact that $\mathrm{Var}[\sum t_i] = nt_0^2/12$, Cox [12] introduced the following test statistic

$$S = \frac{\sum_{i=1}^{n} t_i - n\frac{t_0}{2}}{t_0\sqrt{\frac{n}{12}}}. \tag{5.4}$$

This statistic captures the fact that the mean of the t_i's should be around $t_0/2$ if there is no trend (i.e. $\lambda =$ constant), larger than $t_0/2$ if the rate increases with time ($t_i's$ will be nearer t_0), and smaller than $t_0/2$ if the rate decreases with time ($t_i's$ nearer to 0) .

Sometimes, events are not observed until a given point of time but until, let's say, the nth event. In that case $t_n = t_0$ and n should be replaced by $n - 1$ in equation (5.4) (see [12])

$$S = \frac{\sum_{i=1}^{n-1} t_i - (n-1)\frac{t_0}{2}}{t_0\sqrt{\frac{n-1}{12}}}. \tag{5.5}$$

In the Yule process, the inter-speciation times are still exponentially distributed but the parameter of the exponential random variables varies with the number of species. If we denote by g_i's the inter-speciation time corresponding to the time during which there are exactly i species (see Fig. 5.1), the random variable g_i is exponentially distributed with parameter λi.

In order to build a test statistic, the trick consists of considering the ig_i's rather than the g_i's because the ig_i's are exponentially distributed with the same parameter λ. Equation (5.4) can then be used by first noting that the first event (at the root) gives no information (i.e., it only defines $t = 0$). We therefore have only $n - 2$ observations. The test statistic is then obtained from equation (5.4) after replacing n by $n - 2$, t_i by $\sum_{j=2}^{i+1} jg_j$, t_0 by $\sum_{j=2}^{n} jg_j$ and simplifying the summands in the first term:

$$\gamma = \frac{\sum_{i=2}^{n-1} \sum_{j=2}^{i} jg_j - \left(\frac{n-2}{2}\right) \sum_{j=2}^{n} jg_j}{\sum_{j=2}^{n} jg_j \sqrt{\frac{n-2}{12}}}. \tag{5.6}$$

In the Hey process, speciation events should be viewed backwards. The g_i's are exponentially distributed with parameter $i(i-1)\lambda$ and the root corresponds to the last (viewed backward) event. Because the end of the time interval during which the process is observed corresponds to a speciation event (the first one at the root), equation (5.5) should be used instead of equation (5.4) and the test statistic becomes

$$\delta = -\frac{\sum_{i=n}^{3} \sum_{j=n}^{i} j(j-1)g_j - \left(\frac{n-2}{2}\right) \sum_{j=n}^{2} j(j-1)g_j}{\sum_{j=n}^{2} j(j-1)g_j \sqrt{\frac{n-2}{12}}}. \tag{5.7}$$

The minus sign is introduced so that that the statistic is positive when nodes are closer to the tips than expected, and negative conversely. The reason why the sum ranges from n to 2 or 3 is that the speciation process should be viewed backwards in the Hey (coalescent) process. Thus, the 'first' speciation event occured g_n units of time before the present, the 'second' speciation event occured g_{n-1} units of time before the 'first' speciation event, and so on. The statistics introduced by Pybus [53, 54] (our equations (5.1) and (5.2)) are then given by equation (5.6) and (5.7) after dividing their numerators and their denominators by $n - 2$.

References

[1] Abrams, P. A. (1983). The theory of limiting similarity. *Annual Review of Ecology and Systematics*, **14**, 359–376.

[2] Agapow, P. M. and Purvis, A. (2002). Power of eight tree shape statistics to detect nonrandom diversification: A comparison by simulation of two models of cladogenesis. *Systematic Biology*, **51**, 866–872.

[3] Aldous, D. J. (2001). Stochastic models and descriptive statistics for phylogenetic trees, from Yule to today. *Statistical Science*, **16**, 16–34.

[4] Baldwin, B. G and Sanderson, M. J. (1998). Age and rate of diversification of the Hawaiian silversword alliance (Compositae). *Proceedings of the National Academy of Sciences (USA)*, **95**, 9402–9406.

[5] Blum, M. G. B., François, O., and Janson, S. (2006). The mean, variance and limiting distribution of two statistics sensitive to phylogenetic tree balance. *Annals of Applied Probability*, **16**, 2195–2214.

[6] Blum, M. G. B. and François, O. (2006). Which random processes describe the Tree of Life? A large-scale study of phylogenetic tree imbalance. *Systematic Biology*, **55**, 685–691.

[7] Cardillo, M., Orme, C. D. L., and Owens, I. P. F. (2005). Testing for latitudinal bias in diversification rates: an example using New World birds. *Ecology*, **86**, 2278–2287.

[8] Chan, K. M. A. and Moore, B. R. (1999). Accounting for mode of speciation increases power and realism of tests of phylogenetic asymmetry. *American Naturalist*, **153**, 332–346.

[9] Colless, D. H. (1982). Phylogenetics: the theory and practise of phylogenetic systematics II (book review). *Systematic Zoology*, **31**, 100–104.

[10] Cotton, J. A. and Page, R. D. M. (2005). Rates and patterns of gene duplication and loss in the human genome. *Proceedings of the Royal Society of London Series B—Biological Sciences*, **272**, 277–283.

[11] Cotton, J. A. and Page, R. D. M. (2006). The shape of human gene family phylogenies. *BMC Evolutionary Biology*, **6**, 66.

[12] Cox, D. R. and Lewis, P. A. W. L. (1966). *The Statistical Analysis of a Series of Events*. Chapman and Hall, London.

[13] Durrett, R. (2002). *Probability Models for DNA Sequence Evolution*. Springer-Verlag, New York.

[14] Ehrlich, P. R. and Raven, P. H. (1964). Butterflies and plants: a study in coevolution. *Evolution*, **18**, 586–608.

[15] Eldredge, N. and Gould, S. J. (1972). Punctuated equilibira: an alternative to phyletic gradualism. In *Models in Paleobiology* (ed. T. J. M. Schopf and J. M. Thomas), pp. 305–322, Freeman, Cooper, San Francisco.

[16] Etienne, R. S. and Olff, H. (2005). Confronting different models of community structure to species-abundance data: a Bayesian model comparison. *Ecology Letters*, **8**, 493–504.

[17] Felsenstein, J. (2004). *Inferring Phylogenies*. Sinauer Associates, Sunderland.

[18] Foote, M. (1993). Discordance and concordance between morphological and taxonomic disparity. *Paleobiology*, **19**, 185–204.

[19] Gavrilets, S. and Vose, A. (2005). Dynamics of adaptive radiation. *Proceedings of the National Academy of Sciences (USA)*, **102**, 18040–18045.

[20] Gilbert, B. and Lechowicz, M. J. (2004). Neutrality, niches, and dispersal in a temperate forest understory. *Proceedings of the National Academy of Sciences (USA)*, **101**, 7651–7656.

[21] Harding, E. (1971). The probabilities of rooted tree-shapes generated by random bifurcation. *Advances in Applied Probability*, **3**, 44–77.

[22] Harvey, P. H., May, R. M., and Nee, S. (1994). Phylogenies without fossils. *Evolution*, **48**: 523–529.

[23] Heard, S. B. (1992). Patterns in tree balance among cladistic, phenetic, and randomly generated phylogenetic trees. *Evolution*, **46**, 1818–1826.

[24] Heard, S. B. (1996). Patterns in phylogenetic tree balance with variable and evolving speciation rates. *Evolution*, **50**, 2141–2148.

[25] Heard, S. B. and Hauser, D. L. (1995). Key evolutionary innovations and their ecological mechanisms. *Historical Biology*, **10**, 151–173.

[26] Heard, S. B. and Mooers, A.Ø. (2000). Phylogenetically patterned speciation rates and extinction risks change the loss of evolutionary history during extinctions. *Proceedings of the Royal Society of London Series B—Biological Sciences*, **267**, 613–620.

[27] Hein, J., Schierup, M. H., and Wiuf, C. (2005). *Gene Genealogies, Variation and Evolution.* Oxford University Press, Oxford.

[28] Hey, J. (1992). Using phylogenetic trees to study speciation and extinction. *Evolution*, **46**, 627–640.

[29] Hoelzer, G. A. and Melnick, D. J. (1994). Patterns of speciation and limits to phylogenetic resolution. *Trends in Ecology and Evolution*, **9**, 105–107.

[30] Holman, E. (2005). Nodes in phylogenetic trees: the relation between imbalance and number of descendent species. *Systematic Biology*, **54**, 895–899.

[31] Hubbell, S. P. (2001). *The Unified Neutral Theory of Biogeography.* Princeton University Press, Princeton.

[32] Hutchinson, G. E. (1959). Homage to Santa Rosalia, or why are there so many kinds of animals? *American Naturalist*, **93**, 145–159.

[33] Kingman, J. F. C. (1982). On the genealogy of large populations. *Journal of Applied Probability*, **19**, 27–43.

[34] Kinzig, A. P., Levin, S. A., Dushoff, J., and Pacala, S. (1999). Limiting similarity, species packing, and system stability for hierarchical competition-colonization models. *American Naturalist*, **153**, 371–383.

[35] Losos, J. B. and Adler, F. R. (1995). Stumped by trees? A generalized null model for patterns of organismal diversity. *American Naturalist*, **145**, 329–342.

[36] Matsen, F. A. (2006). A geometric approach to tree shape statistics. *Systematic Biology*, **55**, 652–661.

[37] May, R. M. and MacArthur, R. H. (1972). Niche overlap as a function of environmental variability. *Proceedings of the National Academy of Sciences (USA)*, **69**, 1109–1113.

[38] Mayhew, P. J. (2002). Shifts in hexapod diversification and what Haldane could have said. *Proceedings of the Royal Society of London Series B—Biological Sciences*, **269**, 969–974.

[39] Mayr, E. (1954). Change of genetic environment and evolution. In *Evolution as a Process* (ed. J. Huxley, A. C. Hardy, and E. B. Ford), pp. 157–180. Allen and Unwin, London.

[40] McGill, B. J., Hadley, E. A., and Maurer, B. A. (2005). Community inertia of Quaternary small mammal assemblages in North America. *Proceedings of the National Academy of Sciences (USA)*, **102**, 16701–16706.

[41] Mooers, A. Ø. and Heard, S. B. (1997). Inferring evolutionary process from phylogenetic tree shape. *Quarterly Review of Biology*, **72**, 31–54.

[42] Moran, P. A. P. (1958). A general theory of the distribution of gene frequencies. 1. Overlapping generations. *Proceedings of the Royal Society of London Series B—Biological Sciences*, **149**, 102–112.

[43] Mitter, C., Farrell, B., and Wiegmann, B. (1988). The phylogenetic study of adaptive zones: Has phytophagy promoted insect diversification? *American Naturalist*, **132**, 107–128.

[44] Nee, S. (2001). Inferring speciation rates from phylogenies. *Evolution*, **55**, 661–668.

[45] Nee, S. and May, R. M. (1997). Extinction and the loss of evolutionary history. *Science*, **278**, 692–694.

[46] Nee, S., Mooers, A. Ø., and Harvey, P. H. (1992). Tempo and mode of evolution revealed from molecular phylogenies. *Proceedings of the National Academy of Sciences (USA)*, **89**, 8322–8326.

[47] Nee, S., May, R. M., and Harvey, P. H. (1994). The reconstructed evolutionary process. *Philosophical Transactions of the Royal Society Series B—Biological Sciences*, **344**, 305–311.

[48] Nee, S., Holmes, E. C., May, R. M., and Harvey, P. H. (1995). Estimating extinction from molecular phylogenies. In *Estimating Extinction Rates* (ed. Lawton, J. L. and R. M. May), pp. 164–182. Oxford University Press, Oxford.

[49] Nosil, P. and Mooers, A. Ø. (2005). Testing hypotheses about ecological specialization using phylogenetic trees. *Evolution*, **59**, 2256–2263.

[50] Paradis, E. (2005). Statistical analysis of diversification with species traits. *Evolution*, **59**, 1–12.

[51] Price T., Gibbs, H. L., de Sousa, L., and Richman, A. D. (1998). Different timing of the adaptive radiations of North American and Asian warblers. *Proceedings of the Royal Society of London Series B—Biological Sciences*, **265**, 1969–1975.

[52] Purvis, A. (1996). Using interspecies phylogenies to test macroevolutionary hypotheses. In *New Uses for New Phylogenies.* (ed. P. H. Harvey, A. J. L. Brown, J. M. Smith and S. Nee), pp. 153–168. Oxford University Press, Oxford.

[53] Pybus, O. G and Harvey, P. H. (2000). Testing macro-evolutionary models using incomplete molecular phylogenies. *Proceedings of the Royal Society Series B—Biological Sciences*, **267**, 2267–2272.

[54] Pybus, O. G., Rambaut, A., Holmes, E. C., and Harvey, P. H. (2002). New inferences from tree shape: Numbers of missing taxa and population growth rates. *Systematic Biology*, **51**, 881–888.

[55] Rabosky, D. L. (2006). Likelihood methods for detecting temporal shifts in diversification rates. *Evolution*, **60**, 1152–1164.

[56] Raup, D. M., Gould, S. J., Schopf, T. J. M., and Simberloff, D. S. (1973). Stochastic models of phylogeny and the evolution of diversity. *Journal of Geology*, **81**, 525–542.

[57] Rauch, E. M. and Bar-Yam, Y. (2004). Theory predicts the uneven distribution of genetic diversity within species. *Nature*, **431**, 449–452.

[58] Rogers, J. S. (1994). Central moments and probability-distribution of Colless' Coefficent of tree imbalance. *Evolution*, **48**, 2026–2036.

[59] Rohlf, F. J., Chang, W. S., Sokal, R. R., and Kim, J. (1990). Accuracy of estimated phylogenies: effects of tree topology and evolutionary model. *Evolution*, **44**, 1671–1684.

[60] Rüber, L., and Zardoya, R. (2005). Rapid cladogenesis in marine fishes revisited. *Evolution*, **59**, 1119–1127.

[61] Savolainen, V., Heard, S. B., Powell, M. P., Davies, T. J., and Mooers A. Ø. (2002). Is cladogenesis heritable? *Systematic Biology*, **51**, 835–843.

[62] Schluter, D. (2000). *Ecology of Adaptive Radiation*. Oxford University Press, Oxford.

[63] Schneider H., Schuettpelz E., Pryer K. M., Cranfill R., Magallon S., and Lupia R. (2004). Ferns diversified in the shadow of angiosperms. *Nature*, **428**, 553–557.

[64] Sepkoski, J. J. Jr. (1979). A kinetic model of Phanerozoic taxonomic diversity. II. Early Phanerozoic families and multiple equilibria. *Paleobiology*, **5**, 22–251.

[65] Sepkoski, J. J. Jr. (1984). A kinetic model of Phanerozoic taxonomic diversity. III. Post-Paleozoic families and mass extinction. *Paleobiology*, **10**, 246–267.

[66] Shaw, A. J., Cox, C. J., Goffinet, B., Buck, W. R., and Boles, S. B. (2003). Phylogenetic evidence of a rapid radiation of pleurocarpous mosses (Bryophyta). *Evolution*, **57**, 2226–2241.

[67] Sims, H. J. and McConway, K. J. (2003). Nonstochastic variation of species-level diversification rates within angiosperms. *Evolution*, **57**, 460–479.

[68] Simberloff, D. S., Hecht, K. L., McCoy, E. D., and Conner, E. F. (1981). There have been no statistical tests of cladistic biogeographic hypotheses. In *Vicariance Biogeography: A Critique* (ed. G. Nelson and D. E. Rosen), pp. 40–63. Columbia University Press, New York.

[69] Stam, E. (2002). Does imbalance in phylogenies reflect only bias? *Evolution*, **56**, 1292–1295.

[70] Steel, M. and McKenzie, A. (2001). Properties of phylogenetic trees generated by Yule-type speciation models. *Mathematical Biosciences*, **170**, 91–112.

[71] Stireman, J. O. III., Nason, J. D., Heard, S. B., and Seehawer, J. M. (2006). Cascading host-associated genetic differentiation in parasitoids of phytophagous insects. *Proceedings of the Royal Society of London Series B—Biological Sciences*, **273**, 523–530.

[72] Strathman, R. R. and Slatkin, M. (1983). The improbability of animal phyla with few species. *Paleobiology*, **9**, 97–106.

[73] Volkov, I., Banavar, J. R., Hubbell, S. P., and Maritan, A. (2003). Neutral theory and relative species abundance in ecology. *Nature*, **424**, 1035–1037.

[74] Vrba, E. S. and Eldredge, N. (1984). Individuals, hierarchies, and processes: towards a more complete evolutionary theory. *Paleobiology*, **10**, 146–171.

[75] Wilkinson, M., Cotton, J. A., Creevey, C., Eulenstein, O., Harris, S. R., Lapointe, F. J., Levasseur, C., McInerney, J. O., Pisani, D., and Thorley, J. L. (2005). The shape of supertrees to come: Tree shape related properties of fourteen supertree methods. *Systematic Biology*, **54**, 419–431.

[76] Yang, Z. and Rannala, B. (1997). Bayesian phylogenetic inference using DNA sequences: a Markov Chain Monte Carlo Method. *Molecular Biology and Evolution*, **14**, 717–724.

[77] Yule, G. U. (1924). A mathematical theory of evolution based on the conclusions of Dr J. C. Willis. *Philosophical Transactions of the Royal Society (London) Series B—Biological Sciences*, **213**, 21–87.

[78] Zhaxybayeva, O. and Gogarten, J. P. (2004). Cladogenesis, coalescence and the evolution of the three domains of life. *Trends in Genetics*, **20**, 182–187.

[79] Zink, R. M. and Slowinski, J. B. (1995). Evidence from Molecular Systematics for Decreased Avian Diversification in the Pleistocene Epoch. *Proceedings of the National Academy of Sciences (USA)*, **92**, 5832–5835.

6

PHYLOGENETIC DIVERSITY: FROM COMBINATORICS TO ECOLOGY

Klaas Hartmann and Mike Steel

Abstract

The phylogenetic diversity (PD) of a set of taxa contained within a phylogenetic tree is a measure of the biodiversity of that set. PD has been widely used for prioritizing taxa for conservation and is the basis of the 'Noah's Ark Problem' in biodiversity management. In this chapter we describe some new and recent algorithmic, mathematical, and stochastic results concerning PD. Our results highlight the importance of considering time scales and survival probabilities when making conservation decisions. The loss of PD under a simple extinction process is also described for any given tree—this provides contrasting results depending on whether extinction is measured as function of time or of the number of lost species. Lastly we explore a very different application of PD, its use for reconstructing trees and the associated mathematical properties. The wide range of applications in this chapter shows the usefulness of PD for exploring phylogenetic tree structure with further applications sure to follow.

6.1 Introduction and terminology

Phylogenetic diversity (PD) is a measure of the evolutionary history spanned by a set of taxa within a larger phylogenetic tree. Briefly, the PD score of a subset of taxa, S, is the sum of the lengths of those edges of the tree that span S (a more precise definition follows later). It has been used as a comparative measure in biodiversity conservation, following its introduction by Dan Faith in 1992 [14]. Subsequent authors (see, for example [2, 10, 29, 36, 52] and the references therein) have further explored its application to biodiversity assessment and conservation. Properties of phylogenetic diversity have also been applied recently by Pardi and Goldman [34] to shed light on the relative merits of cooperative versus greedy strategies for taxa sampling for genomic sequencing.

In this chapter we explore some mathematical and stochastic properties of phylogenetic diversity in three different settings: biodiversity conservation, patterns in biodiversity loss, and its relevance for tree reconstruction.

First we explain how PD satisfies two combinatorial properties, including a certain greedoid-type inequality (from [46]) which is algorithmically useful for selecting sets of taxa to optimize PD. We also consider some taxon-specific indices based on PD that give an indication of the relative distinctiveness of each taxon in a tree. We show that these indices have some shortcomings when used to guide biodiversity conservation and consider a framework that overcomes some of these limitations. This framework is the 'Noah's Ark Problem' (NAP) introduced by Weitzman [52]. In the NAP each taxon has survival probabilities and conservation costs. The NAP seeks the optimal allocation of limited funds to conserving taxa such that the expected remaining future PD is maximized. Under certain restrictions a fast, 'greedy' algorithm provides a solution to this problem. We extend one such result from [46] and use this extension to investigate the management time scale that is implicit in the NAP.

We then investigate the loss of PD as taxa randomly become extinct. Nee and May [30] investigated this process for randomly generated trees. They found a characteristic concave shape in the relationship between expected PD and the proportion of taxa deleted. We describe a result that shows how this is the expected behaviour for any given tree (with any given branch lengths). This indicates that most of the loss of PD comes near the end of an extinction process. However, if one examines the behaviour of expected PD as a function of time, then a contrasting (partially convex, rather than concave) relationship emerges.

Finally we examine the role of PD in the reconstruction of phylogenetic trees. PD estimates for triples (or larger numbers) of taxa have recently been investigated as a way to refine the popular Neighbor-Joining algorithm; it is also possible to consider PD over any abelian group. We describe the mathematical properties of PD in these two settings.

6.2 Definitions and combinatorial properties

Mostly we follow the notation used by Semple and Steel [42]. We let $\mathcal{T}$ denote a *phylogenetic X–tree*, that is, a tree whose leaves comprise the set X of taxa (generally species or populations) under study, and whose remaining vertices (nodes) are of degree at least 3 (the degree of a vertex is the number of edges that are incident with it). The vertices at the tips are called *leaves.* If all the non-leaf vertices in a tree have three incident edges the tree is said to be *fully resolved* (sometimes called 'binary'—these are the trees without polytomies, and so are maximally informative). We also deal with *rooted trees* which have some vertex (often the mid-point of an edge) distinguished as a root vertex. If we direct all the edges of the tree away from the root (i.e. so they are consistent with a time direction if the root is the ancestral taxon) then we can talk about the *clusters* of the tree—the subsets of X that lie below the different vertices of the tree. It is a classical result that any rooted phylogenetic tree can be uniquely reconstructed from its set of clusters. Often the edges of the trees (rooted or unrooted) will have a *(branch) length*—corresponding perhaps to the expected amount of evolutionary change on that edge.

Given a (rooted or unrooted) phylogenetic X–tree, $\mathcal{T}$, with branch lengths, and given a subset Y of X, the *phylogenetic diversity* (PD) of Y, denoted $PD(Y)$

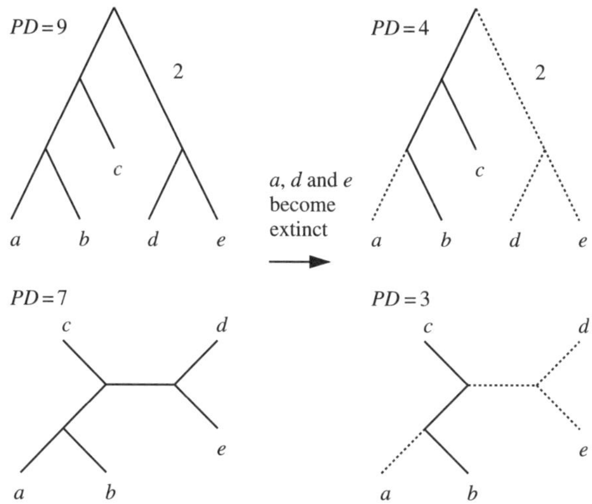

FIG. 6.1. The PD of the trees on the left is calculated by summing the edge lengths. All edges are length 1 except for the long edge on the rooted tree which has length 2. The trees on the right show which edges are considered to remain (solid lines) after taxa a, d, and e become extinct. The PD of these trees is the sum of the remaining edges.

is the sum of the lengths of all the branches that connect the leaves in Y (and also the root of $\mathcal{T}$ if $\mathcal{T}$ is a rooted tree). That is, if we denote the length of an edge e of $\mathcal{T}$ by λ_e we have:

$$PD(Y) = \sum_{e} \lambda_e,$$

where the summation is over all edges e in $\mathcal{T}$ that lie on the minimal subtree of $\mathcal{T}$ connecting the taxa in Y (and if $\mathcal{T}$ is rooted, also connecting the root). There has been some debate about whether the root should be included, however the original definition in [14] and prevailing usage include the root (see [10], [15] and [9] for further discussion). Figure 6.1 illustrates the various PD measures we have discussed here.

Depending on the data from which a tree is derived, the branch lengths may have different interpretations. Branch lengths may correspond to an evolutionary time-scale (i.e. the number of millions of years between speciation events), or to genetic distance, or to the extent of morphological differences, or perhaps some combination of these (or other) measures of evolutionary distance. Throughout this chapter, no particular interpretation is assumed, so as to allow the greatest degree of generality for applications; in particular, unless we state so explicitly, we do not assume that the tree is ultrametric (an *ultrametric tree* is one for which the distance from the root to any leaf is the same, as would occur for (a) genetic distance under a 'molecular clock', or (b) an evolutionary time-scale).

The PD measure has two basic combinatorial properties which we now describe.

6.2.1 *The strong exchange property*

For any function f defined from the collection of subsets of X of size at least r into the real numbers, we say that f satisfies the *strong exchange property* if for any two subsets Y and Z with $r \leq |Y| < |Z|$ there exists some taxon $z \in Z - Y$ such that:

$$f(Z - \{z\}) - f(Z) + f(Y \cup \{z\}) - f(Y) \geq 0. \tag{6.1}$$

This condition is a sufficient condition for the greedy algorithm to construct subsets of any given size ($\geq r$) that maximize f starting from any given set of size r that maximizes f. This follows by standard arguments from 'greedoid' theory (see [23]). To construct such a subset the greedy algorithm iteratively adds the element (taxon) that gives a maximal increase in f until the subset contains r elements.

The strong exchange property was established for $f = PD$ (and $r = 2$ in the case of unrooted phylogenetic trees) in [46]; its interpretation in this setting is that for any two of the subsets, the larger one contains some taxon (z) that would contribute at least as much to the PD value of the smaller subset than it adds to that of the larger one.

Consider both trees in Fig. 6.1 and the situation where the two subsets Y and Z are $\{a, c\}$ and $\{b, d, e\}$ respectively; clearly $|Y| < |Z|$. Deleting taxon b from subset Z and adding it to Y results in a loss of the combined PD of Y and Z in both trees, hence b does not satisfy the strong exchange property. However the combined PD of Y and Z is increased if taxon d is removed from Z and added to Y, thus satisfying the strong exchange property.

Note that the strong exchange property for PD fails for $r = 1$ and $r = 0$ for unrooted trees, but holds for rooted trees. Moreover, as demonstrated in [34], for any given set of taxa W of size at least 2 (or 1 in case of rooted trees) the strong exchange property also ensures that amongst the collection of all subsets of size k containing W, the one(s) of maximal PD value can be constructed from W by the greedy algorithm (even though W itself may not have optimal PD score for its cardinality).

6.2.2 *Generalized Pauplin formula*

The second combinatorial property of PD is that it can be written canonically as a linear combination of pairwise distances within the tree. That is, if $d(x, y)$ denotes the distance between x and y in $\mathcal{T}$, the PD of a set W can be written as

$$PD(W) = \sum_{x,y \subseteq W} \mu_{\mathcal{T},W}(x, y) d(x, y) \tag{6.2}$$

where $\mu_{\mathcal{T},W}$ is a function that depends on $\mathcal{T}$ and W but not the branch lengths. Actually there are many possible choices of $\mu_{\mathcal{T},W}$ but there is one that is particularly natural and which is defined as follows. Let $\mathcal{T}_W$ denote the subtree of $\mathcal{T}$

connecting W and let $p(\mathcal{T}_W, x, y)$ be the set of non-leaf vertices of $\mathcal{T}_W$ that lie on the path connecting x and y. Then set

$$\mu_{\mathcal{T},W}(x,y) = \prod_{v \in p(\mathcal{T}_W, x, y)} (d(v)-1)^{-1}$$

where $d(v)$ is the degree of vertex v in $\mathcal{T}_W$. The validity of equation (6.2) for this choice of $\mu_{\mathcal{T},W}$ was described (for $W = X$) for binary phylogenetic X–trees by Pauplin [35], and generalized to arbitrary phylogenetic X–trees in Semple and Steel [43]. The Pauplin formula also provides an interesting starting point for forming species specific indices of biodiversity such as the Equal-Splits index (Section 6.3.1).

6.2.3 *Exclusive molecular phylodiversity*

We end this section by noting that Lewis and Lewis [26] recently investigated a related measure they called the 'exclusive molecular phylodiversity' of a set Y, defined by:

$$E(Y) = PD(X) - PD(X - Y).$$

This measure has also been used by [41] to assess the evolutionary history of endemic species in biodiversity hotspots. The benefit of exclusive molecular phylodiversity in that context is that it avoids the need for any information about non-endemic species, effectively assuming that these are well represented elsewhere. It is easy to show that this measure does not satisfy the strong exchange property (equation (6.1)) and that greedy algorithms cannot be guaranteed to produce an optimal subset, Y.

6.3 Biodiversity conservation

Ross Crozier summarizes the rationales for conserving biodiversity into three categories: 'moral (other species have a right to exist), esthetic (species are like works of art, and it would be foolish to destroy them), and utilitarian (humans derive material benefit from the existence of other species)' [8]; these motivations are further explored in [31]. Given unlimited resources for conservation all three motivations dictate the same action—conserving all taxa. In a realistic setting where there are limited resources for conservation the taxa must be prioritized in some manner. In this case the three categories of motivation may dictate different prioritizations.

If conservation is motivated by moral considerations, as many taxa as possible should be conserved. A conservation scheme should therefore allocate its resources so that the net survival increase of all taxa is as high as possible.

If the motivation for conservation is utilitarian, the distinctiveness of the remaining taxa is of great importance. For example, protecting the sole remaining taxon from a clade has greater utilitarian benefits than protecting a taxon from a well represented clade as the former has greater unique genetic potential for further evolution and bio-prospecting [8].

Lastly, if conservation is motivated by aesthetic reasons the role that distinctiveness should play is dependent on the uncertain definition of aesthetic value. However, given the choice of saving either a taxon from a well represented clade or a taxon that is the 'last of its kind' it seems difficult to find a general justification for not choosing the latter.

Most biodiversity conservation approaches aim to conserve as many taxa as possible [18], but the reasons used to motivate conservation are often utilitarian in nature (e.g. Chapter 1, [37]) and should therefore take taxon distinctiveness into account.

In this subsection we discuss several methods for prioritizing taxa to conserve that allow distinctiveness to be taken into consideration. Throughout this chapter we apply various methods and indices to the tree depicted in Fig. 6.2. This tree shows the phylogenetic relationship of Crested penguins (*Eudyptes*) as produced by Sara Bertelli and Norberto Giannini [3],[19]. Note that no edge lengths were given in the original tree, and so here we have assumed that the tree is ultrametric, as shown in Fig. 6.2. We are using this example for illustrative purposes only, but it is of interest to note that according to the 2004 IUCN Red

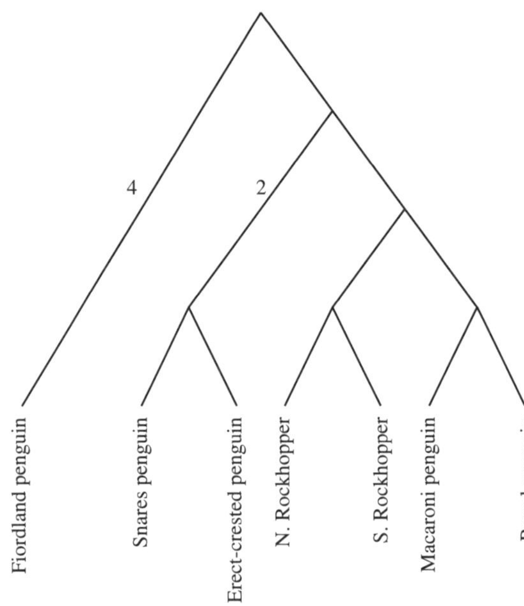

FIG. 6.2. The phylogenetic tree for Crested penguins. This tree was derived from the tree in [3] and [19] which had no branch lengths. For illustrative purposes each level in the original tree was assumed to be separated by the same distance such that all edges in this tree are of length 1 except for the two marked edges.

List [22] all of the species are vulnerable except the Erect-crested penguin, which is endangered.

6.3.1 *Simple indices*

Many indices for measuring or ranking the distinctives of a single taxon have been proposed. These indices have the advantage of being easy to compute and, as each taxon is assigned some value, they can be readily combined with other information for decision making (such as the conservation cost or economic importance of that taxon). The disadvantage of these indices is that they do not take into account the complexities of conserving multiple taxa. For example, if one taxon is conserved the relative importance of conserving closely related taxa may decrease as we wish to conserve as distinctive a *set* of taxa as possible.

Here we consider some simple indices with a particular focus on those that are based on the concept of phylogenetic diversity. For the interested reader some notable work not considered here (some of which is not based on PD) is contained in [2], [6], [7], [20], and [51].

One of the conceptually simplest indices is the 'Pendant Edge' (PE) measure introduced by Stephen Altschul and David Lipman [1] where each taxon is assigned a value equal to the length of its pendant edge. Strictly speaking the PE measure is not based on PD but has been included here for illustrative purposes.

The PE value of each species in Fig. 6.2, is easily determined: in this case the Fiordland penguin has the highest PE with the other taxa having an equal second highest value. PE suggests that the Fiordland penguin is the most important to conserve but does not differentiate between the other taxa. It seems logical, though, that if some of the other taxa were to be conserved, we should not choose the most closely related of these.

A conceptually appealing family of indices divides the total phylogenetic diversity of a tree amongst the taxa corresponding to the leaves of that tree. An example of these indices is the Equal-Splits (ES) index [38] which is closely related to the previous discussed Pauplin formula. This index splits the PD value of an edge equally between it's daughter trees. Denoting the edge length between a node, j, and its direct ancestor by λ_j, the equal splits index for a taxon, i, can be calculated by summation over all the nodes between i and the root (including i):

$$ES(i) = \sum_j \frac{\lambda_j}{2^{d'(i,j)}}$$

where $d'(i,j)$ is the number of edges between the taxon (node i) and node j. Applying the ES index to the tree in Fig. 6.2 again suggests that the Fiordland penguins are the most important species to conserve with an index value of 4. The Snares and Erect-crested penguins have an index equal value of $\frac{9}{4}$ whilst the remaining species have a value of $\frac{15}{8}$; if, for example, three species could be conserved, this suggests that the Fiordland, Snares, and Erect-crested penguins should be chosen. Intuitively, however, it seems more beneficial to conserve one of the other species instead of the Snares or Erect-crested penguins and thus protect more of the internal edges. The problem with ES (and other simple indices) is that the decision to conserve one taxon does not affect the importance assigned to conserving the remaining taxa.

One can also consider the expected contribution to PD that a taxon will make at some time in the future if the survival of all other taxa is uncertain. To make this idea precise, for each subset S of X, and each taxon $i \in X - S$ let

$$\Delta_{PD}(S,i) = PD(S \cup \{i\}) - PD(S);$$

$\Delta_{PD}(S,i)$ is the increase in PD that taxon i provides when added to S. Now, suppose that each taxon $j \in X - \{i\}$ has a probability a_j that it is not extinct at some time t in the future. If we assume that extinction events are independent between taxa, and let E be the (random) set of taxa that are extant at time t then we can ask how much we expect taxon i to contribute to the PD at time t—this value, ψ_i is simply the expected value of $\Delta_{PD}(E,i)$, given formally by

$$\psi_i = \sum_{S \subseteq X - \{i\}} \mathbb{P}[E = S]\Delta_{PD}(S,i).$$

Note that $\mathbb{P}[E = S]$ is the probability that the set of extant taxa at time t will be S; this depends on the survival probabilities (a_j's).

Although this last equation involves a summation over an exponential number of terms, it has an equivalent description that allows for its rapid (polynomial-time) calculation (Steel, M., A. Mimoto and A. O. Mooers, submitted). A related but different index to ψ_i is the Shapley value which has been considered in detail elsewhere [20].

6.3.2 *Noah's Ark Problem*

Martin Weitzman introduced the 'Noah's Ark Problem' (NAP) [52], a comprehensive framework for allocating limited funding for biodiversity conservation that overcomes some of the problems associated with the simple indices discussed previously. In the NAP framework each taxon, j, has some probability, a_j, of remaining extant. If some conservation intervention of cost c_j is applied to this taxon, then this survival probability can be increased from a_j to b_j. The aim is to identify the subset of taxa to conserve: S, that maximizes the future expected phylogenetic diversity $\mathbb{E}(PD|S)$ subject to the budgetary constraint, B. The notation '$|S$' indicates that the expected value is conditional on the set S of taxa being conserved. The formulation of the NAP as used in this chapter is:

Given an edge-weighted phylogenetic tree, and values (a_j, b_j, c_j) for each taxon j, maximize $\mathbb{E}(PD|S)$ over all subsets S of taxa, subject to the constraint: $\sum_{j \in S} c_j \leq B$.

The original formulation used by Weitzman [52] allowed the inclusion of an intrinsic value for the taxa (for example the tourism value of a species of whale), however this value can readily be included here by increasing the length of each taxon's pendant edge appropriately. There is of course an inherent difficulty in combining phylogenetic diversity and other socio-economic values; this is also the case in the original formulation of the NAP.

$\mathbb{E}(PD|S)$ is calculated by summing all the edge lengths, λ_e, in the tree, weighting each edge by the probability that it will be spanned by the surviving taxa:

$$\mathbb{E}(PD|S) = \sum_e \lambda_e p(e|S).$$

For rooted trees the probability that an edge is spanned, $p(e|S)$, is simply the probability that at least one of the taxa in the tree subtended by edge e will remain extant.

Variations of the NAP have been used in a variety of applications including biodiversity conservation (e.g. [10], [29], and [45]) and prioritizing taxa for genomic sequencing [34]. Additional intrinsic values for the taxa can be incorporated in this version of the NAP by adding the intrinsic value of each taxon to its pendant edge.

A problem with the NAP is that no efficient algorithm has been found for producing solutions to it. To find an optimal solution it may be necessary to consider many of the possible subsets of taxa. The number of subsets increases at rate $2^{|X|}$, therefore considering a large proportion of these is infeasible for more than a few dozen taxa. For example, if one has a tree with (say) 1,000 taxa, and one wishes to find a subset of (say) 100 taxa that maximizes $\mathbb{E}(PD|S)$ then it is impossible for any computer to search all subsets of size 100 from the 1,000. Having efficient algorithms for solving the NAP is therefore essential for applying the NAP to large trees.

Several variations of the NAP where additional constraints are imposed have been shown to be solvable using simple 'greedy' algorithms [21], [46]. These algorithms allow the optimal solutions for a particular problem to be found quickly. Here we provide a further extension to the scenario considered in [46].

First consider the class of NAPs where taxa become extinct unless they are conserved, all taxa cost the same to conserve and conserved taxa survive with certainty; this corresponds to $a_j = 0, b_j = 1$, and $c_j = c$ (where $c > 0$ is some constant) for each taxon j. We will call this type of NAP *Scenario 1.*

In this scenario, the expected remaining phylogenetic diversity ($\mathbb{E}(PD|S)$) is simply the phylogenetic diversity of the conserved taxa ($PD(S)$), since all other taxa become extinct with certainty. Solving the NAP is therefore equivalent to finding the subset S of X of size at most $\frac{B}{c}$ with maximal PD. This problem was shown to be solvable using a simple greedy algorithm in [46], from which we have the following result:

Theorem 6.1 *For a NAP under Scenario 1, the following greedy algorithm produces the optimal solution(s). For rooted trees the algorithm begins with an* *empty set S, and for unrooted trees it begins with a set S containing the two taxa that are furthest apart. The algorithm sequentially adds the taxon that provides the greatest increase in $\mathbb{E}(PD|S)$ until S contains as many taxa as the budget permits to be conserved. Where more than one taxon provides an equal increase in $\mathbb{E}(PD|S)$ one is chosen at random. Upon completion S contains an optimal solution, other optimal solutions (if they exist) are obtained by making different choices where a taxon was chosen at random.*

We will now extend Scenario 1 to allow non-zero survival probabilities in the absence of conservation ($a_j \neq 0$), as follows. We will refer to this extension as *Scenario 2* which has the remaining constraints that $b_j = 1$, c_j is constant and the tree is rooted. The following result was independently derived here and in [33].

Theorem 6.2 *For a NAP under Scenario 2, the greedy algorithm described in Theorem 6.1 produces the optimal solution(s) when applied to a rooted tree with suitably adjusted edge lengths, λ'_e. Denoting the set of children of edge e (the leaves/taxa separated from the root by e) by C_e the adjusted edge lengths are:*

$$\lambda'_e = \lambda_e \prod_{j \in C_e} (1 - a_j). \tag{6.3}$$

Proof Instead of maximizing $\mathbb{E}(PD|S)$ we can seek to maximize $\mathbb{E}(PD|S) - \mathbb{E}(PD|\emptyset)$, the increase in the expected PD that conservation of the taxa in S will provide. For a Scenario 2 problem the increase in the probability that a particular edge is spanned when the set, S, of taxa is conserved is:

$$\begin{aligned} p(e|S) - p(e|\emptyset) &= \begin{cases} 1 - (1 - \prod_{j \in C_e}(1 - a_j)), & \text{if } |C_e \cap S| > 0; \\ 0, & \text{if } |C_e \cap S| = 0; \end{cases} \\ &= \prod_{j \in C_e}(1 - a_j) \times \begin{cases} 1, & |C_e \cap S| > 0; \\ 0, & |C_e \cap S| = 0. \end{cases} \end{aligned}$$

The expected increase in the PD is simply the sum over all edges with each edge weighted by the increased probability:

$$\begin{aligned} \mathbb{E}(PD|S) - \mathbb{E}(PD|\emptyset) &= \sum_e \lambda_e (p(e|S) - p(e|\emptyset)) \\ &= \sum_e \lambda_e \prod_{j \in C_e}(1 - a_j) \times \begin{cases} 1, & \text{if } |C_e \cap S| > 0; \\ 0, & \text{if } |C_e \cap S| = 0; \end{cases} \\ &= \sum_e \lambda'_e \times \begin{cases} 1, & \text{if } |C_e \cap S| > 0; \\ 0, & \text{if } |C_e \cap S| = 0. \end{cases} \end{aligned}$$

This final expression for $\mathbb{E}(PD|S) - \mathbb{E}(PD|\emptyset)$ is equal to the objective, $\mathbb{E}(PD|S)$, for a Scenario 1 problem with branch lengths λ'_e as required. □

6.3.3 *Conservation time scale*

The survival probabilities (a_j) contain an implicit time scale as they represent the probability that a taxon will survive to some future time, t; in the absence of conservation the expected number of taxa surviving to t is $\sum_j a_j$. If the time t is in the distant future (a long time scale) the survival probability of unprotected taxa will be close to zero due to background extinction, for shorter time scales (t closer to the present) the survival probabilities will be closer to one. This choice of time scale affects solutions to the NAP as management strategies corresponding to longer time scales will place greater emphasis on internal edges. Note that Scenario 1 corresponds to long term management where only those taxa that were conserved remain, whereas in Scenario 2 the time scale can be freely chosen by selecting values for a_j that are of appropriate magnitude.

To illustrate the importance of selecting an appropriate time scale consider the tree in Fig. 6.3, where each taxon is equally likely to remain extant at any future time. Panel A corresponds to the situation where all taxa that are not conserved become extinct (a long time scale). If two taxa can be conserved, the optimal choice consists of one taxon from each branch of the tree. This optimal choice is found either by application of the greedy algorithm (Theorem 6.1) or by an exhaustive search.

Consider increasing the survival probability of unconserved taxa (a_j) so that all taxa have a $\frac{1}{4}$ chance of surviving; this represents a move to a shorter management time scale. To find the optimal solutions for this problem the transformation outlined in Theorem 6.2 is applied to the original tree (Panel A in Fig. 6.3) yielding the tree in Panel B. As expected from equation (6.3) the interior edges have had a greater reduction in length than the pendant edges; application of the greedy algorithm can now be used to obtain the optimal solutions. The pendant edge lengths of taxa a and b are now equal to the distance between the root and taxa c or d. Consequently conserving both taxa a and b is now also an equally good solution.

If the survival probabilities (a_j) are further increased (to, say, $\frac{3}{8}$), the interior edges of the transformed tree decrease in length to such an extent that the optimal set of taxa to conserve becomes $\{a, b\}$ (see Panel C).

We have illustrated that the optimal set of taxa to conserve is dependent on the management time scale. As the management time scale shifts from long term to short term, less emphasis is placed on interior edges as these are more likely to remain extant anyway.

A discussion of the merits of conservation time scales is beyond the scope of this work (see [4] and [25] for more details). However the optimal time scale will be highly dependent on the application. Of particular importance will be the time scale on which conservation focus can be shifted from one taxon to another. If this can occur rapidly, planning for the short term would be optimal and the conservation strategy should be reevaluated as taxa become extinct. For

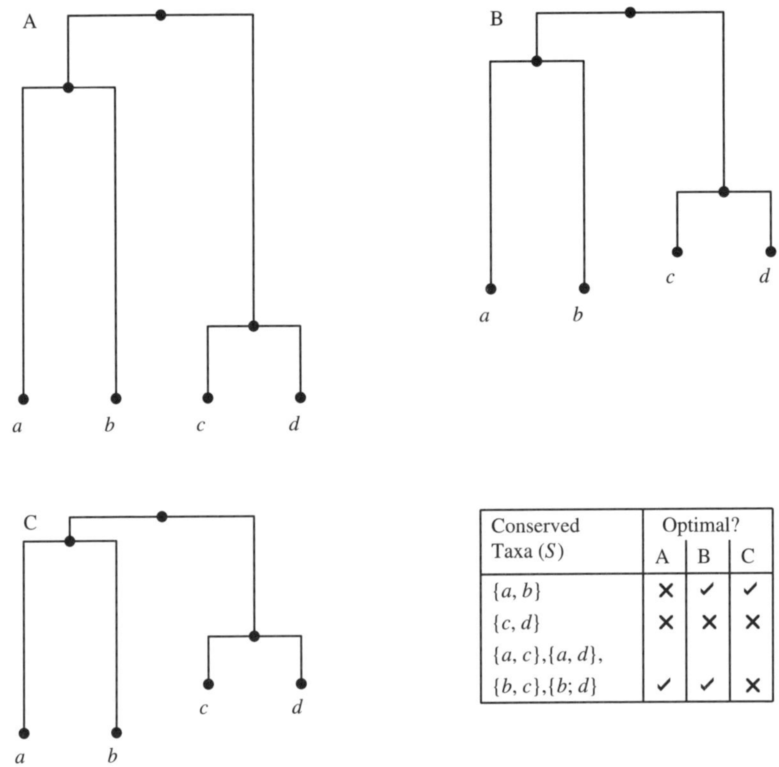

Conserved Taxa (S)	Optimal? A	B	C
$\{a, b\}$	✗	✓	✓
$\{c, d\}$	✗	✗	✗
$\{a, c\}$,$\{a, d\}$, $\{b, c\}$,$\{b; d\}$	✓	✓	✗

FIG. 6.3. Panel A depicts a tree where unconserved species become extinct with certainty ($a_j = 0$). Panels B and C depict the transformed tree as this survival probability is increased to 0.25 and 0.375 respectively. Optimal subsets of size 2 can be found by applying the greedy algorithm to these trees. The optimality of each subset for each panel is indicated in the table.

many taxa, conservation programmes are long term investments. In these cases, a longer time scale should be investigated when the taxa to be conserved are initially selected.

6.3.4 *Further algorithmic results*

For problems where a greedy algorithm is known to produce optimal solutions, a naïve implementation of the algorithm may be unnecessarily slow. An efficient implementation of the greedy algorithm for Scenario 1 is provided in [28], which in their simulations took 1/100th of the time of a naïve implementation. In [28] an alternative pruning algorithm is also provided, this algorithm begins with all the taxa and removes the least important taxon sequentially until a subset of the desired size is obtained. As expected if a large proportion of the taxa are to be included in the subset, the pruning algorithm is more efficient.

Two further variations of the NAP for which greedy algorithms produce optimal solutions were considered by the authors in [21]. The first variation permits the survival probability for conserved and unconserved taxa (a_j and b_j) to be varied, but these must be related by a particular relationship. The second variation permits variable conservation costs (c_j) but requires that taxa only survive if they are conserved ($a_j = 0$, $b_j = 1$). Additionally, for the greedy algorithm to produce optimal solutions, the tree must be ultrametric (satisfy a molecular clock).

A dynamic programming algorithm has also been produced for a less restrictive variation of the NAP with the sole restriction that conserved taxa survive with certainty ($b_j = 1$) [33].

6.3.5 *Extensions to the NAP*

The Noah's Ark Problem provides a satisfying framework for biodiversity resource allocation problems. It is, however, still a simplification of reality and some extensions to it have been suggested.

The NAP as presented here does not consider the possibility of partially conserving taxa and therefore being able to spread resources more thinly across a greater number of taxa. Weitzman [52] assumed that the survival probability of a taxon increases linearly with the conservation funding allocated to that taxon. Under this assumption optimal solutions to the NAP are extreme and allocate the maximum possible amount to a few taxa instead of partially conserving a greater number. An extension of the NAP to more realistic relationships between survival probability and expenditure was considered in [44], with an application to conservation of breed diversity in African cattle. A greedy algorithm was presented in that paper that the authors suggested would provide optimal solutions to all problems of this type. However, it was shown in [21] that this cannot be the case. This was extended further in [39] to allow for discontinuous relationships produced by multiple possible conservation schemes, necessitating a two step optimization procedure (which they state is not guaranteed to produce the global optimum).

Another implicit assumption in the NAP is that the survival probabilities are independent. That is, conserving one taxon does not raise or lower the survival probabilities of any others, and this may be unrealistic. For example, conserving the prey of one taxon may raise the survival probability of that taxon as well. This effect was considered in [50] where it was shown that failure to consider interdependent survival probabilities may result in an incorrect suggestion as to which species should be protected. The authors in this study stress the importance of their findings as 'more significant losses of biodiversity are exactly those in which ecological impacts are severe, that is, where the loss of one species affects the survival of others'.

In summary, whilst the NAP provides a good starting point, there are other important factors that influence which taxa should be conserved. Inclusion of some of these may prove more difficult than others and adding these factors will further complicate the problem of finding optimal solutions. For example, consider the following problem which is relevant to biodiversity conservation. We have a collection C of locations, where each location $l \in C$ contains some subset $S(l)$ of taxa from a set X of taxa; also we have a phylogenetic X–tree $\mathcal{T}$ with branch lengths. We wish to select k locations so as to maximize the PD of the set of taxa that occur in at least one selected location. If no taxon occurs in more than one location this problem is easily solved, by transforming it to the standard PD optimization problem and applying the greedy algorithm. In general, however, the problem is NP-hard. The proof consists of showing that one can transform the NP-complete problem 'Minimum cover' [16] to this problem, by selecting branch lengths for $\mathcal{T}$ that are 1 on all the pendant edges, and 0 on all the interior edges. For various approaches to solving this and related problems see [40], [5] and [53].

6.4 Loss of phylogenetic diversity under extinction models

We turn now to the statistical properties of PD as taxa go extinct, beginning with a recent result from [47]. Nee and May [30] investigated the loss of PD as taxa are randomly deleted from random trees under a simple model: each taxon is equally likely to be the next to become extinct (the 'field of bullets' model). The trees were ultrametric trees as generated by a random-birth model. They found a characteristic concave shape in the relationship between the expected remaining PD and the proportion of taxa deleted. This relationship is illustrated for the Crested penguins tree (Fig. 6.2) by the upper curve in Fig. 6.4.

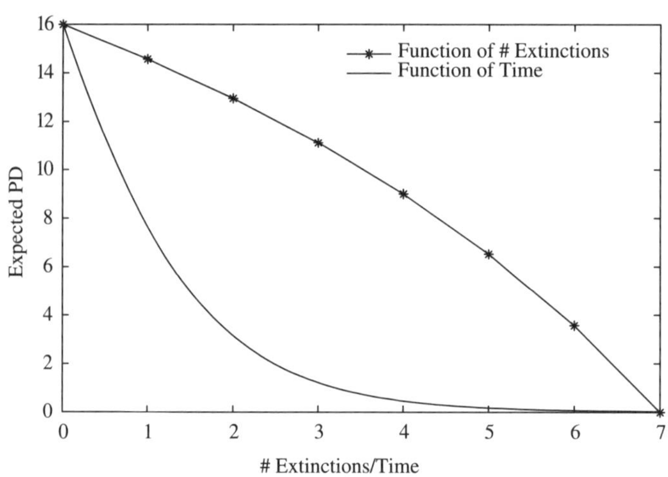

FIG. 6.4. The expected remaining PD after extinctions have occurred among the Crested penguins depicted in Fig. 6.2. This loss in PD is viewed as a function of both the number of extinctions that have occurred and the time that has elapsed since extinctions have been allowed to occur.

This relationship was further investigated recently in [45], which studied random deletion of taxa from certain biological trees. Once again the relationship between taxa deleted and remaining PD was concave. Recall that a sequence $x = (x_1, x_2, \ldots, x_n)$ of real numbers is *concave* if, when we let $\Delta x_r = x_r - x_{r-1}$ the following inequality holds for all r:

$$\Delta x_r - \Delta x_{r+1} \geq 0$$

and the sequence is *strictly concave* if the inequality is strict for all r. Geometrically this means that the slope of the line joining adjacent points in the graph of x_r versus r is decreasing. Note that x_r is concave precisely if the complementary (reverse) sequence $y_r = x_{n-r}$ is concave. The significance of (strict) concavity for PD is that it says (informally) that most PD loss comes near the end of an extinction process.

In this section we first describe a generic concave relationship observed between the average PD and the number of taxa deleted. This makes intuitive sense, because each interior branch survives until the point where there is no taxon below it and this is likely to occur towards the end of a random extinction process.

Consider a rooted phylogenetic tree having a leaf set X of size n. Let W be a random subset of taxa of size r sampled uniformly from X (for example, by selecting uniformly at random a set S of $n - r \geq 0$ elements of X and deleting them, in which case $W = X - S$). For $r \in \{1, \ldots, n\}$ let $\mu_r = \mathbb{E}[PD|r]$, the expected value of $PD(W)$ over all such choices of W. Equivalently, we can write $\mu_r = \binom{n}{r}^{-1} \sum_{W \subseteq X : |W| = r} PD(W)$, where $\binom{n}{r}$ is the binomial coefficient $(= \frac{n!}{r!(n-r)!})$, which is the number of ways of selecting r elements from a set of size n. For brevity we adopt the usual convention that $\binom{n}{r} = 0$ if r is greater than n or less than 0.

Clearly $\mu_n = PD(X)$. For $r \in \{1, \ldots, n\}$, let $\Delta\mu_r = \mu_r - \mu_{r-1}$. Note that, since $\mu_0 = 0$, we have $\Delta\mu_1 = \mu_1$. For an edge e of $\mathcal{T}$, and $r \in \{1, \ldots, n-1\}$ let

$$\psi(e, r) := \frac{n_e(n_e - 1)}{r(r+1)} \cdot \frac{\binom{n-n_e}{r-1}}{\binom{n}{r+1}}$$

where n_e denotes the number of leaves of $\mathcal{T}$ that lie 'below' e (i.e. separated from the root by e).

The proof of the following result is given in [47]. It shows that for any fully resolved tree, PD decays in a strictly concave fashion as taxa are randomly deleted, and the only trees for which the decay of PD is linear are fully unresolved 'star' trees. In the following theorem a *cherry* is a pair of leaves that are adjacent to the same vertex.

Theorem 6.3 *Consider a phylogenetic tree $\mathcal{T}$ with an assignment λ of positive branch lengths. Then, for each $r \in \{1, \ldots, n-1\}$,*

$$\Delta\mu_r - \Delta\mu_{r+1} = \sum_e \lambda_e \psi(e, r)$$

where the summation is over all edges of $\mathcal{T}$. In particular, μ is concave over this domain, and μ is strictly concave if and only if $\mathcal{T}$ has a cherry, while μ is linear if and only if $\mathcal{T}$ has no interior edges (i.e. is an unresolved 'star' tree).

Consider the tree for Crested penguins to which we have previously referred (Fig. 6.2). Figure 6.4 shows the expected PD as a function of the number of extinctions. As expected from the above theorem, the relationship depicted in this figure is strictly concave.

6.4.1 *Relationship between PD and time under an extinction process*

We have investigated the expected PD as a function of the number of extinctions that have occurred. So far each taxon has been considered as equally likely to be the next to become extinct. However, no consideration has been given to the timing of these extinctions. Here we consider the situation where each taxon has the same probability of becoming extinct at any point in time (the time to extinction for an individual taxon has an exponential distribution) and consider the expected PD as a function of the time instead of the number of extinctions that have occurred. We will show that the decline in expected PD does not in general have a concave shape and in fact after a specific time (dependent on the tree shape) the decline will become convex. Note that this is not a contradiction with the previous result; it is simply due to the fact that the number of extinctions decreases over time as there are fewer species left that could become extinct.

The probability that an edge, e, will be spanned by the taxa remaining at some time t, depends only on the number of children ($|C_e| = n_e$) of that edge. Denoting this probability by $p_e(t)$ we have:

$$p_e(t) = 1 - \left(1 - e^{-rt}\right)^{n_e}$$

where r is the rate of extinction. The expected PD at time t, $\mathbb{E}_t(PD)$ is easily found using these probabilities:

$$\mathbb{E}_t(PD) = \sum_e \lambda_e p_e(t).$$

Observe that $\mathbb{E}_t(PD)$ depends only on the sums of the edges with the same number of leaves attached, not on the individual edges themselves:

$$\mathbb{E}_t(PD) = \sum_{j=1}^{m} \alpha_j \left[1 - \left(1 - e^{-rt}\right)^j\right],$$

where $\alpha_j = \sum_{e,n_e=j} \lambda_e$, and m is the highest number of leaves below any edge—this corresponds to the edge(s) at the root with the most leaves descendant from them. To investigate the shape of $\mathbb{E}_t(PD)$ the second derivative is easily obtained:

$$\frac{d^2\mathbb{E}_t(PD)}{dt^2} = r^2 e^{-rt}\left(\alpha_1 + \sum_{j=2}^{m} \alpha_j j\left(1 - je^{-rt}\right)\left(1 - e^{-rt}\right)^{j-2}\right). \tag{6.4}$$

For convexity, the second derivative must be positive. The term corresponding to α_1 is clearly positive, but the sign corresponding to the other α-values depends on t. The term corresponding to a particular α_j is positive if $1 - je^{-rt} > 0$ which holds when

$$t > \frac{\ln(j)}{r}.$$

A sum of convex functions is convex, therefore once the above condition is satisfied for all j, $\mathbb{E}_t(PD)$ will be convex. The term that becomes convex the latest is the term with the highest value of j (namely m). Convexity is therefore guaranteed after $\hat{t} = \ln(m)/r$. In the limit as $\sum_{j<m} \alpha_j/\alpha_m \to 0$, $PD(t)$ will become convex exactly at $\hat{t}$, however $PD(t)$ will generally become convex earlier due to the other terms.

The terms corresponding to edges with high values of j are the last to become positive; as more weight is assigned to these the time to convexity lengthens. Variation in diversification rates through time and/or among clades can therefore affect the time to convexity.

The amount of PD loss that has occurred by the time that convexity is guaranteed ($\hat{t} = \ln(m)/r$) is difficult to characterize, but the number of taxa remaining at this time can be readily found. The probability of an individual taxon persisting to time t is e^{-rt}, so at $t = \hat{t}$ each taxon is extant with probability $1/m$. The total number of taxa is between $m + 1$ and $2m$ (depending on the imbalance of the tree at the root) and the expected number of extant taxa at $t = \hat{t}$ is therefore between 1 and 2. Accordingly, the convexity result may appear to be of limited biological interest, however, given a real tree, the expected number of taxa remaining by the time convexity is reached will usually be much higher.

Another interesting behaviour that can readily be examined and may be of more practical interest is the initial shape of the PD decline (that is at and just after $t = 0$). Substituting $t = 0$ in equation (6.4) we obtain:

$$\begin{aligned} \frac{d^2\mathbb{E}_t(PD)}{dt^2}|_{t=0} &= r^2 \left(\alpha_1 + \sum_{j=2}^{m} \left(\alpha_j j \,(1-j)\, 0^{j-2} \right) \right) \\ &= r^2 \left(\alpha_1 - 2\alpha_2 \right). \end{aligned} \tag{6.5}$$

Initial convexity requires $\alpha_1 > 2\alpha_2$ and concavity requires $\alpha_1 < 2\alpha_2$. The edges that contribute to α_1 are the pendant edges and those contributing to α_2 are edges above cherries. Any tree can have at most half as many 'above cherry' edges as pendant edges, so if pendant edges have similar lengths as the 'above cherry' edges then that tree will therefore exhibit initial convexity (as for the Crested penguins tree Fig. 6.2 and 6.4). It should be noted that even if the PD loss curve for a tree is convex at $t = 0$ and after $t = \hat{t}$ there is no guarantee that it will be convex between these two times due to the complexity of equation (6.4).

6.5 Tree reconstruction using PD

The simplest form of PD (on unrooted trees) considers subsets of taxa of size 2, in which case the PD value is just the path distance in the tree connecting the two taxa. Such pairwise distances suffice to reconstruct any tree (and indeed also the branch lengths). This is a classic result dating back to the mid-1960s [54], and it forms the basis of many fast and popular tree-building methods, such as Neighbor-Joining and BioNJ. However, despite their usefulness, pairwise distances have some drawbacks, and in this section we explore some of the ways in which PD–values on subsets of m–taxa (for $m > 2$) may provide a promising approach in future.

One (statistical) concern with using pairwise distance data is that converting sequence data to pairwise distances is a highly reductive transformation. That is, each distance matrix typically can be obtained from a huge number of different sets of aligned sequences, even under the usual Hamming distance measure (and even if we just count the frequencies of site patterns, not the order they occur in, [48]). Whether this extensive 'loss of information' is important for phylogeny reconstruction is a tantalizing question, though it is tempting to conjecture that it is. Phylogenetic diversity is one way of generalizing the idea of a distance in a tree—from pairs of leaves, to m-tuples of leaves—and this measure suggests a natural way of refining distance-based approaches, so that less information is lost in using sequences to build trees.

To illustrate this idea, consider a model-based approach to phylogeny reconstruction. Given a model of sequence evolution, one can generally compute the maximum-likelihood estimate of an 'evolutionary distance' $d(x, y)$ between any two sequences x, y. This 'evolutionary distance' is some quantity that is assumed to be additive on the underlying evolutionary tree. For example, for a stationary reversible Markov process of site substitution, the 'evolutionary distance' between x and y is usually understood as the expected number of substitutions occurring on the path separating x and y. Thus $d(x, y)$ can be viewed as an estimate of $PD(\{x, y\})$ for a suitable edge weighting of $\mathcal{T}$.

Notice that the PD values on subsets of X of size 3 are determined by the pairwise PD values, according to the following 3–point condition:

$$2PD(\{x, y, z\}) = PD(\{x, y\}) + PD(\{y, z\}) + PD(\{z, x\}). \tag{6.6}$$

Thus one could estimate $PD(\{x, y, z\})$ by using the pairwise distance estimates d, but again this results in a loss of information in reducing triplewise data to three pairwise marginals. Thus it may be more appropriate to estimate PD on m-element subsets by direct analysis of sequence data. For example, the PD score for three sequences might be estimated as the sum of the three branch lengths that maximize the likelihood score of the three sequences under a Markov process of site substitution (and perhaps also insertion and deletion). For certain models, the PD value when $m = 3$ can also be calculated explicitly (i.e. without optimizing branch lengths to maximize likelihood) by the 'tangle' triplewise distance described in [49].

When $m = 3$, estimation of PD values does not require estimating the tree structure connecting the m taxa. However, for any value $m > 3$, consideration of different trees connecting the m taxa is necessary.

Suppose that one was able to exactly calculate the true PD values for all m-elements subsets of X. A natural question is whether this information uniquely determines the underlying phylogenetic X-tree $\mathcal{T}$. It is clear that in general the answer is 'no'—for if we take $m = |X|$ then we have just one PD value, and this can be realized on any phylogenetic X-tree by taking appropriate branch lengths. However, Pachter and Speyer [32] recently showed that if m does not exceed $(n+1)/2$ then the tree $\mathcal{T}$ is uniquely determined by the PD scores of the m-element subsets of X. More precisely, their result states:

Theorem 6.4 *Let $\mathcal{T}$ be a phylogenetic X-tree (with $n = |X|$) and $m \geq 2$ an integer. If $n \geq 2m - 1$ then $\mathcal{T}$ is determined by the map that associates each m-element subset of X with its induced PD score.*

Moreover, even when m exceeds $(n+1)/2$ some partial information concerning $\mathcal{T}$ can be recovered from this map [24]. This paper also describes a modification of Neighbor-Joining to reconstruct trees from their induced PD values. The central idea here is to identify a cherry of the tree. The following result (the 'cherry-picking theorem' of [24]) generalizes the way that Neighbor-Joining identifies cherries in the special case $m = 2$.

Theorem 6.5 *Suppose that $\mathcal{T}$ is a phylogenetic X-tree with n leaves, and m is any integer between 2 and $n - 2$. Then any distinct pair $i, j \in X$ that minimizes the expression*

$$\left(\frac{n-2}{m-1}\right) \sum_{\substack{Y \subset X: \\ i,j \in Y, |Y|=m}} PD(Y) - \sum_{\substack{Y \subset X: \\ i \in Y, |Y|=m}} PD(Y) - \sum_{\substack{Y \subset X: \\ j \in Y, |Y|=m}} PD(Y)$$

is a cherry of $\mathcal{T}$.

Phylogenetic diversity also forms the basis of other approaches to tree reconstruction—most notably the 'balanced minimum evolution' (BME) method of Pauplin [35]. This method takes a (pairwise) distance estimate d on X as input and scores each resolved phylogenetic X-tree $\mathcal{T}$ by what d would estimate for $PD(X)$ using equation (6.2). Thus, if d is additive on $\mathcal{T}$ then this BME score is equal to the PD value of X (on $\mathcal{T}$); while if d is additive on some other resolved tree $\mathcal{T}'$, then the BME score of $\mathcal{T}$ can be shown to exceed the PD value of set X (on $\mathcal{T}'$) [11]. The balanced minimum evolution method seeks the phylogenetic tree that minimizes the associated BME score. There is a close relationship between this method and Neighbor-Joining, which can be viewed as a locally optimal method for constructing a BME tree—for details see [12], [17].

6.5.1 *Tree reconstruction from PD-values over an abelian group*

So far we have regarded the lengths of the edges of a tree as being some positive real number. However, the concept of phylogenetic diversity is well-defined when edge-weights are chosen from any abelian group $\mathcal{G}$ (briefly, an 'abelian group' is any set on which an addition can be defined which is associative and commutative, and there is a zero element and every element has an additive inverse; for details see [27]). This is both mathematically useful and potentially useful in applications. For the mathematical justification, one can ask what properties of PD depend on properties of the real numbers (such as the fact that they are ordered) and how much is just 'algebraic'. Clearly the 'Neighbor-Joining' algorithm no longer applies since the concept of minimizing or maximizing does not apply for a general abelian group. Moreover, although algebraic relations like the 3–point condition (equation (6.6)) apply in general, other results such as the representation (equation (6.2)) no longer do, as we may not be able to divide by factors such as $d(v) - 1$. Regarding tree reconstruction from pairwise PD values, the presence of elements of order 2 in a group (i.e. non-zero elements x for which $x + x = 0$) means that the classic uniqueness result no longer applies For example, consider the tree in Fig. 6.5, and the group $\mathbb{Z}_2 = \{0, 1\}$ under addition mod 2. Suppose the non-zero element (1) of this group is assigned to each edge of the tree shown in Fig. 6.5. Then we have $PD(\{x, y\}) = 0$ for any two elements x, y of the leaf set X of this tree. Moreover there exists more than one phylogenetic tree having this shape (in fact 15 such trees) so clearly PD values on pairs of elements of X are not sufficient to uniquely specify the underlying tree, in contrast to the case where the edges have real values.

It turns out, however, that if $\mathcal{G}$ has no elements of order 2 then the classic uniqueness (and existence) results for tree representations for pairwise PD values

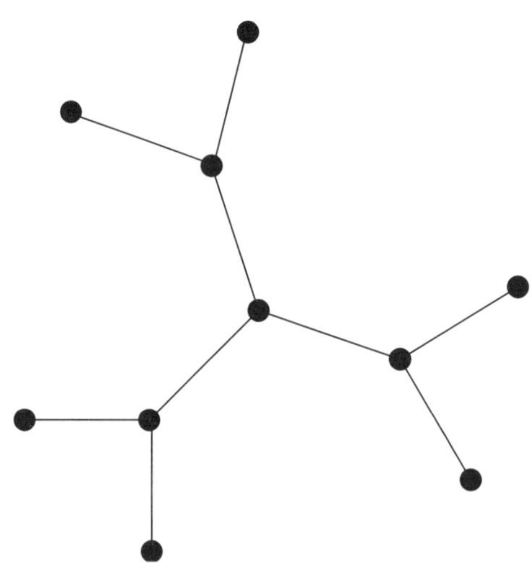

FIG. 6.5. Any leaf labelling of this tree gives $PD(\{x, y\}) = 0$ for all x, y when the element $1 \in \mathbb{Z}_2$ is assigned to each edge.

carry through to the abelian group setting. In the more general case where $\mathcal{G}$ may have elements of order 2, the uniqueness of a tree representation can be recovered, provided that one considers both pairwise and triplewise PD values [13].

More precisely the following result (from [13]) holds.

Theorem 6.6 *Let $\mathcal{T}$ be a phylogenetic X-tree, $\mathcal{G}$ an abelian group, and λ a function that assigns a non-zero element of $\mathcal{G}$ to each edge of $\mathcal{T}$. Then $\mathcal{T}$ is determined up to isomorphism (and can be reconstructed by an algorithm that runs in polynomial time in $|X|$) by the map that associates each pair and triple of elements of X with its associated $\mathcal{G}$-valued PD score.*

The existence question ('when can pairwise and triple-wise PD values be represented by a tree with edge weights drawn from an abelian group?') has also been settled—it involves the three-point condition (equation (6.6)), two four-point conditions, and a five-point condition. This last five-point condition is not required when $\mathcal{G}$ is the group of real numbers under addition, or indeed an abelian group without elements of order 2, but in general it is necessary (for details see [13]).

We end this section by outlining a situation in molecular biology where such group-based valuations arise naturally (the parity of gene orders provides another, but we will not describe this in detail here).

Consider DNA sequences of length k that have been re-coded as binary sequences (for example, by associating with each of the four bases its purine or pyrimidine class). Any two such binary sequences $(w_1, \ldots, w_k), (z_1, \ldots, z_k)$ define a $0-1$ sequence $g = (g_1, \ldots, g_k)$ of length k by setting $g_i = 0$ precisely if $w_i = z_i$, otherwise $g_i = 1$. We may regard g as an element of the abelian 2-group $\mathbb{Z}_2^k$. Now consider an evolutionary tree, where at each vertex there is some purine–pyrimidine sequence (carried by the ancestral taxon at that place in the tree). Assign to each edge the group element associated to its endpoints by the process just described. Then for any two leaves x, y the value $PD(\{x, y\})$ can be computed just from the sequences at x, y (without knowing the tree or the states assigned to other vertices)—it is simply the group element associated to the difference (or, equivalently, the sum) of the sequences at x and y. However, the value of $PD(\{x, y, z\})$ is not uniquely determined by just the sequences at x, y, and z (were this the case, then reconstructing phylogenetic trees from binary sequences would be essentially trivial). Determining $PD(\{x, y, z\})$ is equivalent to determining the sequence that was present at the *median vertex* in the tree connecting leaves x, y, z. This has a curious consequence—if one can reconstruct the ancestral sequence (of the median vertex) for any three binary sequences, then one can reconstruct the underlying tree. One might attempt to estimate this ancestral sequence as the (component-wise) median of the sequences at x, y, z but it turns out that in general the resulting PD values do not have a representation on any tree—indeed the condition for the existence of such a representation is that the splits induced by the sites of the binary sequences are compatible [13]. In practice, biological data would rarely be expected to fulfil this compatibility condition. Thus, more sophisticated approaches to estimate the ancestral sequence at a median vertex (based on models of sequence evolution, and guided by the necessary three-, four- and five-point conditions) would need to be developed before such an approach to tree reconstruction could be applied for analysing DNA sequence data.

6.6 Concluding comments

In this chapter we have investigated several applications of phylogenetic diversity: biodiversity conservation, expected patterns in biodiversity loss, and phylogenetic tree construction. This wide range of applications poses some interesting mathematical problems and provides useful approaches for managing and exploring biodiversity and tree construction.

The Noah's Ark Problem (NAP) discussed here has been applied to both conservation and genomic sequencing problems. No efficient (polynomial time) algorithm for solving the general NAP is known to the authors, and a simple exhaustive search may need to consider a large proportion of the 2^n possible subsets of taxa (this is not feasible for a problem consisting of more than a few dozen taxa). As discussed, algorithms for efficiently computing solutions to several restricted variations of the NAP exist, but some suggestions have been made in the literature that the NAP is too simplistic and needs to incorporate more realistic aspects of the problem. These extensions will further complicate the problem of finding optimal solutions.

We have also illustrated the importance of the time scale of conservation management. The magnitude assigned to the survival probabilities of the taxa determines what management time scale is being considered. For non-trivial trees, the optimal solution to the NAP is sensitive to the time scale that has been selected; selecting an inappropriate time scale may result in an inappropriate prioritization of taxa to conserve.

Investigating the expected losses in PD as taxa become extinct, is a useful approach for quantifying future expected losses in biodiversity. Here we have shown that as taxa randomly become extinct, each new extinction is expected to cause a greater loss in biodiversity, though the rate of biodiversity loss with time exhibits a different behaviour. Further work using more realistic models of extinction could provide additional insight into the loss of biodiversity. It may be particularly relevant to consider survival probabilities (the a_j values) from a skewed distribution or correlated with the distance between taxa in the phylogenetic tree (Arne Mooers, pers. comm.).

Furthermore we have considered how PD may provide a useful tool for refining tree reconstruction by using m-way comparisons of taxa. For $m = 2$ this has been well studied, and is generally referred to as 'distance-based' approaches to tree reconstruction, however many results and methods (such as Neighbor-Joining) extend naturally to larger values of m.

A final generalization is to allow the branch lengths to take values in any abelian group. The message seems to be that for groups without elements of order 2, tree reconstruction behaves just like the familiar group of real numbers (though some care is needed as concepts involving order and minimization no longer apply, so methods like Neighbor-Joining are problematic). For groups with elements of order 2, the mathematical analysis is slightly more complicated, but still tractable.

Acknowledgements

We thank Arne Mooers, Olivier Gascuel, and an anonymous referee for some helpful comments, and the *New Zealand Marsden Fund* and the *Allan Wilson Centre for Molecular Ecology and Evolution* for supporting this research.

References

[1] Altschul, S. F. and Lipman, D. J. (1990). Equal animals. *Nature*, **348** (6301), 493–494.

[2] Barker, G. M. (2002). Phylogenetic diversity: a quantitative framework for measurement of priority and achievement in biodiversity conservation. *Biological Journal of the Linnean Society*, **76**, 165–194.

[3] Bertelli, S. and Giannini, N. P. (2005). A phylogeny of extant penguins (Aves: Spenisciformes) combining morphology and mitochondrial sequences. *Cladistics*, **21**, 209–239.

[4] Bunnell, F. L. and Huggard, D. J. (1999). Biodiversity across spatial and temporal scales: problems and opportunities. *Forest Ecology and Management*, **115**, 113–126.

[5] Camm, J. D., Norman, S. K., Polasky, S., and Solow, A. R. (2006). Nature reserve site selection to maximize expected species covered. *Operations Research*, **50**(6), 946–955.

[6] Clarke, K. R. and Warwick, R. M. (1998). A taxonomic distinctness index and its statistical properties. *Journal of Applied Ecology*, **35**, 523–531.

[7] Crozier, R. H. (1992). Genetic diversity and the agony of choice. *Biological Conservation*, **61**, 11–15.

[8] Crozier, R H (1997). Preserving the information content of species: Genetic diversity, phylogeny, and conservation worth. *Annual Review of Ecology and Systematics*, **28**, 243–268.

[9] Crozier, R. H., Agapow, P., and Dunnett, L. J. (2006). Conceptual issues in phylogeny and conservation: a reply to Faith and Baker. *Evolutionary Bioinformatics Online*, **2**, 197–199.

[10] Crozier, R. H., Dunnett, L. J., and Agapow, P. M. (2005). Phylogenetic biodiversity assessment based on systematic nomenclature. *Evolutionary Bioinformatics Online*, **1**, 11–36.

[11] Desper, R. and Gascuel, O. (2004). Theoretical foundation of the balanced minimum evolution method of phylogenetic inference and its relationship to weighted least-squares tree fitting. *Molecular Biology and Evolution*, **21**(3), 587–598.

[12] Desper, R. and Gascuel, O. (2005). The minimum evolution distance-based approach to phylogenetic inference. In *Mathematics of Evolution and Phylogeny* (ed. O. Gascuel). Oxford University Press, New York.

[13] Dress, A. and Steel, M. (2006). Phylogenetic diversity over an abelian group. *Annals of Combinatorics*, In Press.

[14] Faith, D. P. (1992). Conservation evaluation and phylogenetic diversity. *Biological Conservation*, **61**, 1–10.

[15] Faith, D. P. and Baker, A. M. (2006). Phylogenetic diversity (PD) and biodiversity conservation: some bioinformatics challenges. *Evolutionary Bioinformatics Online*, **2**, 70–77.

[16] Garey, M. R. and Johnson, D. S. (1979). *Computers and Intractability.* W. H. Freemand and Company, San Francisco.

[17] Gascuel, O. and Steel, M. (2006). Neighbor-joining revealed. *Molecular Biology and Evolution*, **23**(11), 1997–2000.

[18] Gaston, K. J. (1996). Species richness: measure and measurement. In *Biodiversity: A Biology of Numbers and Difference* (ed. K. Gaston), pp. 77–113. Blackwell Science, Cambridge.

[19] Giannini, N. P. and Bertelli, S. (2004, April). Phylogeny of extant penguins based on integumentary and breeding characters. *The Auk*, **121**(2), 422–434.

[20] Haake, C., Kashiwada, A., and Su, F. E. (2005, March). The shapley value of phylogenetic trees. *IMW Working Paper #363* (363).

[21] Hartmann, K. and Steel, M. (2006). Maximizing phylogenetic diversity in biodiversity conservation: Greedy solutions to the Noah's Ark Problem. *Systematic Biology*, **55**(4), 644–651.

[22] IUCN (2004). 2004 IUCN Red list of threatened species. `http://www.iucnredlist.org`.

[23] Korte, B., Lovász, L., and Schrader, R. (1991). *Greedoids, Algorithms and Combinatorics.* Springer-Verlag Berlin.

[24] Levy, D., Yoshida, R., and Pachter, L. (2006). Neighbor joining with phylogenetic diversity estimates. *Molecular Biology and Evolution*, **23**(3), 491–498.

[25] Lewis, C. A., Lester, N. P., Bradshaw, A. D., Fitzgibbon, J. E., Fuller, K., Hakanson, L., and Richards, C. (1996). Considerations of scale in habitat conservation and restoration. *Canadian Journal of Fisheries and Aquatic Sciences*, **53**(Suppl. 1), 440–445.

[26] Lewis, L. A. and Lewis, P. O. (2005). Unearthing the molecular phylodiversity of desert soil green algae (Chlorophyta). *Systematic Biology*, **54**(6), 936–947.

[27] Maclane, S. and Birkoff, G. (1979). *Algebra* (second edn). Macmillan, New York.

[28] Minh, B. Q., Klaere, S., and von Haeseler, A. (2006). Phylogenetic diversity within seconds. *Systematic Biology*, **55**(5), 769-773.

[29] Mooers, A. Ø., Heard, S. B., and Chrostowski, E. (2005). Evolutionary heritage as a metric for conservation. In *Phylogeny and Conservation* (ed. A. Purvis, T. Brooks, and J. Gittleman), pp. 120–138. Cambridge University Press, New York.

[30] Nee, S., and May, R. M. (1997). Extinction and the loss of evolutionary history. *Science*, **278**(5338), 692–694.

[31] Norton, B. G. (1987). *Why Preserve Natural Variety?* Princeton University Press, Princeton.

[32] Pachter, L. and Speyer, D. (2004). Reconstructing trees from subtree weights. *Applied Mathematics Letters*, **17**(6), 615–621.

[33] Pardi, F. and Goldman, N. (2007). Resource aware taxon selection for maximising phylogenetic diversity. *Systematic Biology*, In Press.

[34] Pardi, F. and Goldman, N. (2005). Species choice for comparative genomics: no need for cooperation. *PLoS Genetics*, **1**(6), 71.

[35] Pauplin, Y. (2000). Direct calculation of a tree length using a distance matrix. *Journal of Molecular Evolution*, **51**, 41–47.

[36] Pavoine, S., Ollier, S., and Dufour, A. (2005). Is the originality of a species measurable? *Ecology Letters*, **8**, 579–586.

[37] Pullin, A. S. (2002). *Conservation Biology*. Cambridge University Press, New York.

[38] Redding, D. W., and Mooers, A. Ø. (2006). Incorporating evolutionary measures into conservation prioritization. *Conservation Biology*, In Press.

[39] Reist-Marti, S., Abdulai, A., and Simianer, H. (2006). Optimum allocation of conservation funds and choice of conservation programs for a set of African cattle breeds. *Genetics Selection Evolution*, **38**, 99–126.

[40] Rodrigues, A. S. L., Brooks, T. M., and Gaston, K. J. (2005). Integrating phylogenetic diversity in the selection of priority areas for conservation: does it make a difference? In *Phylogeny and Conservation* (ed. A. Purivs, J. L. Gittleman, and T. Brooks), Number 8 in Conservation Biology, Chapter 5, pp. 101–119. Cambridge University Press, New York.

[41] Sechrest, W., Brooks, T. M., da Fonseca, G. A. B., Konstant, W. R., Mittermeier, R. A., Purvis, A., Rylands, A. B., and Gittleman, J. L. (2002). Hotspots and the conservation of evolutionary history. *Proceedings of the National Academy of Sciences*, **99**(4), 2067–2071.

[42] Semple, C. and Steel, M. (2003). *Phylogenetics*. Oxford University Press, New York.

[43] Semple, C. and Steel, M. (2004). Cyclic permutations and evolutionary trees. *Advances in Applied Mathematics*, **32**(4), 669–680.

[44] Simianer, H., Marti, S. B., Gibson, J., Hanotte, O., and Rege, J. E. O. (2003). An approach to the optimal allocation of conservation funds to minimize loss of genetic diversity between livestock breeds. *Ecological Economics*, **45**, 377–392.

[45] Soutullo, A., Dodsworth, S., Heard, S. B., and Mooers, A. Ø. (2005). Distribution and correlates of carnivore phylogenetic diversity across the Americas. *Animal Conservation*, **8**(3), 249–258.

[46] Steel, M. (2005). Phylogenetic diversity and the greedy algorithm. *Systematic Biology*, **54**(4), 527–529.

[47] Steel, M. (2006). Tools to construct and study big trees: A mathematical perspective. In *Reconstructing the Tree of Life: Taxonomy and Systematics of Species Rich Taxa* (ed. T. R. Hodkinson and J. A. Parnell). CRC Press.

[48] Steel, M. A., Penny, D., and Hendy, M. D. (1988). Loss of information in genetic distance. *Nature*, **336**(6195), 118.

[49] Sumner, J. G., and Jarvis, P. D. (2005). Entanglement invariants and phylogenetic branching. *Journal of Mathematical Biology*, **51**(1), 18–36.

[50] van der Heide, C. M., van den Bergh, Jeroen C. J. M., and van Ierland, E. C. (2005). Extending Weitzman's economic ranking of biodiversity protection: combining ecological and genetic considerations. *Ecological Economics*, **55**(2), 218–223.

[51] Vane-Wright, R. I., Humphries, C. J., and Williams, P. H. (1991). What to protect? - Systematics and the agony of choice. *Biological Conservation*, **55**, 235–254.

[52] Weitzman, M. L. (1998). The Noah's Ark Problem. *Econometrica*, **66**(6), 1279–1298.

[53] Wilson, K. A., McBride, M. F., Bode, M., and Possingham, H. (2006). Prioritizing global conservation efforts. *Nature*, **440**, 337–340.

[54] Zaretskii, K. A. (1965). Constructing trees from the set of distances between pendant vertices. *Uspehi Matematiceskih Nauk*, **20**, 90–92.

IV

TREES FROM SUBTREES AND CHARACTERS

7

FRAGMENTATION OF LARGE DATA SETS IN PHYLOGENETIC ANALYSES

Michael J. Sanderson, Cécile Ané, Oliver Eulenstein, David Fernández-Baca, Junhyong Kim, Michelle M. McMahon, and Raul Piaggio-Talice

Abstract

Genome-scale data and efficient mining of sequence databases are allowing construction of very large data sets for phylogenetic inference. Sample biases and problems of homology can force these data sets to be relatively sparse, leading to fragmentation of phylogenetic information in ways that have been little explored. Here we outline several aspects of the problem of fragmentation and describe three broad classes of strategies for identifying and coping with it. The first of these treats the problem after phylogenetic analysis by attempting to extract sub-signals from the resulting collection of trees. The second attempts to provide very minimal necessary conditions for combining fragments in the first place, by identifying so-called 'groves' in the data. The third strategy is heuristic, using clustering or optimization procedures to seek strongly informative subsets of the data for separate phylogenetic analyses.

7.1 Introduction

Data sets for phylogenetic analysis of species relationships are becoming increasingly large. Genomic data ranging from whole genome sequences to EST libraries are increasing the number of loci that can be included in one analysis: many studies in the last several years have inferred trees based on 100–500 genes [12, 13, 25, 26, 36]. At the same time, easy access to GenBank and other sequence databases, which (as of March 2006) contain data on 150,000 species, or approximately 9% of all described species on Earth, coupled with development of tools to automate data mining [10, 19] has prompted increasingly broad taxonomic sampling. Phylogenies with several thousand species have now been reconstructed [19, 21, 23]. Typical 'large scale' phylogenetic analyses of the past few years have entailed data combination in some form or other: either combining information from many loci for relatively few taxa, or a few loci for many taxa. Methodologies for building trees from such large combined data sets fall into two broad categories: supermatrix (or 'superalignment') approaches that concatenate aligned sequences into one grand alignment, and supertree approaches

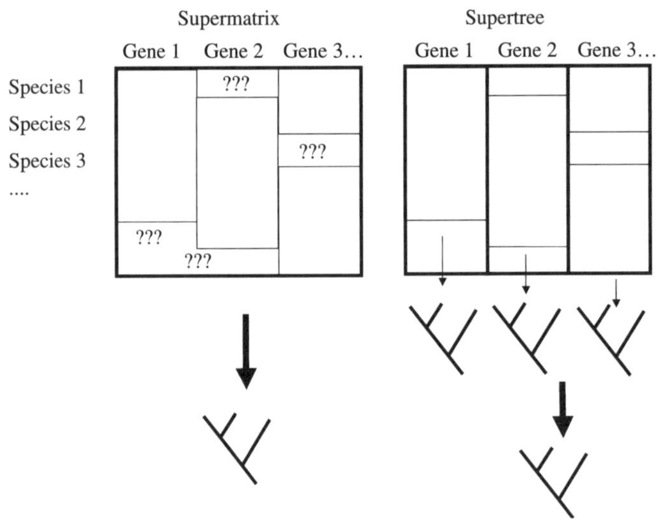

FIG. 7.1. The two main strategies for constructing phylogenomic-scale data sets: on the left is construction of supermatrix by concatenating sequence data and building a tree from this combined matrix; on right is construction of supertree by first building trees for each gene locus and then combining the trees themselves.

that algorithmically combine trees constructed from each individual alignment (Fig. 7.1) [27]. The basic observation that motivates the present paper is that there appear to be intrinsic tradeoffs that limit the density of the data assembled in these large-scale phylogenetic studies. Loosely speaking, by *density*, we mean the completeness of information available for each taxon relative to each locus (in a supermatrix analysis) or tree (for a supertree analysis). Gene loss, sequence divergence that obscures homology, and sampling biases are just three factors that make construction of large high-density data sets difficult. Here we examine data fragmentation caused by low density and outline quantitative approaches for characterizing this fragmentation and coping with it in tree inference.

Two ways to represent fragmentation in phylogenetic data are shown in Table 7.1 and Fig. 7.2. To fix ideas, let the problem be the assembly of either collections of homologous sequences ('loci') for various taxa (the supermatrix setting), or collections of trees built from those loci for various taxa (the supertree setting). In a data-availability matrix, the pattern of missing data for combinations of taxa and loci (trees) is indicated directly as a matrix. Alternatively, in a graph representation, a bipartite graph is used in which one set of nodes corresponds to taxa and the other to loci (trees). Edges in the graph indicate the presence of a sequence (or tree) and taxon. In either representation, the notion of density is clear, either as the fraction of filled cells in the matrix, or the fraction

TABLE 7.1. A character data-matrix or sequence alignment (left) and two representations of its structure. Locus 1 includes sites 1–5; locus 2 includes site 6–10; locus 3 includes sites 11–15. Gaps in alignment are denoted with dashes; missing data in alignment with '?'. Right is the data-availability matrix indicating presence or absence of sequence for these combinations of loci and taxa.

	1	2	3		123
A	ACGTT	?????	?????	A	100
B	ACGTT	?????	TCTCC	B	101
C	ACCGG	?????	TCGTC	C	101
D	ACACG	TAATA	?????	D	110
E	?????	TT-TT	?????	E	010

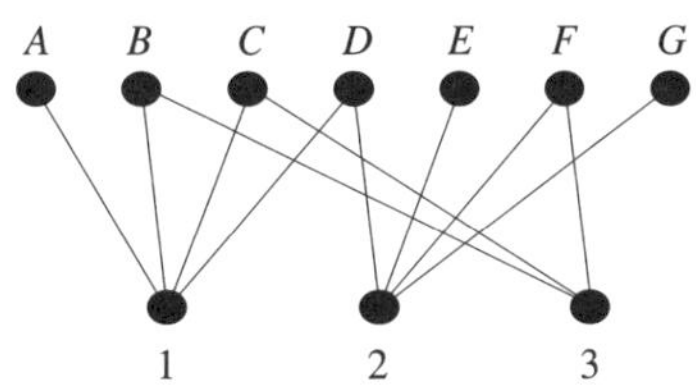

FIG. 7.2. Bipartite graph showing the same information from Table 7.1. The density in either case is 11/21.

of edges out of the maximum possible. Fragmentation refers to a pattern of low density that may entail complete lack of overlap of blocks in the data-availability matrix or strict disconnection of the bipartite graph.

It is easy to see the problems that can arise in a fragmented data set with some simple examples. Figure 7.3A shows a taxon by character data-matrix with a sizeable fraction of missing sequence data arranged in a pattern of nearly non-overlapping blocks. Each block, if analysed separately using maximum parsimony, yields one optimal tree. If the matrix is analysed as a whole, there are 55 optimal trees and the strict consensus is completely unresolved. The two trees corresponding to the blocks in this example can, however, still be recovered from the collection of 55 trees by finding its maximum agreement subtrees (MAST), which are the largest trees common to the entire collection when taxa are pruned [18]. Smaller (submaximal) agreement subtrees would reveal other signals arising from fragments whose signals are overridden by ones with more characters. Figure 7.3B shows the parallel supertree problem. Here the input is two unrooted trees that share one taxon. A supertree can be constructed by a variety of methods, such as the widely used matrix representation with parsimony (MRP: [7]).

A

```
A ACGTT?????
B ACGTT?????
C ATCCG?????
D ATCCGTAATA
E ?????TAATC
F ?????TAATC
G ?????TAATA
```

B

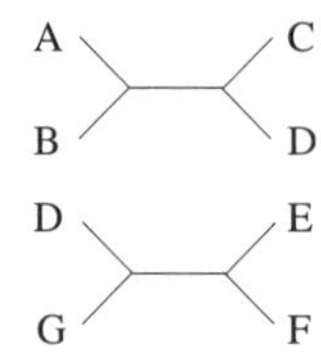

FIG. 7.3. A. Sequence alignment illustrating partially overlapping blocks of sequences. B. Collection of two unrooted input trees illustrating the same pattern of taxon overlap (Both panels indicate overlap in taxon D). Either a supermatrix analysis using parsimony or a supertree analysis using MRP methods generates a large collection of equally parsimonious trees which has an unresolved strict consensus. However, the two trees in B are returned as the maximum agreement subtrees (MAST) of this collection.

The MRP matrix for these two input trees has a structure similar to that of the matrix of Fig. 7.3A, except that instead of sites in sequences, the characters are binary and correspond to bipartitions in the input trees, missing taxa being indicated by question marks. In this very simple example, the collection of MRP supertrees is the same 55 trees found in the collection of most parsimonious trees for the supermatrix.

The main question raised by this example is whether it is better to break the data into subsets to be analysed separately, or to handle the effects of the fragmentation *after* the analysis by some method of sorting through the output trees. This question is remarkably reminiscent of the long-standing question in phylogenetics of whether and when to partition a data set into separate components (or alternatively when to combine data [11]). However, the motivation there is to avoid combining data sets that have different phylogenetic signals, arising perhaps because a different model of evolution is appropriate or perhaps because the history of the different partitions is actually different (e.g. different histories of the nuclear versus chloroplast genome). Here the question arises simply by virtue of the occurrence of missing data—or to put it another way, by the pattern of occupancy of cells in the matrix, a much more basic issue. The dichotomy between choices is a bit false, of course; there may well be methods that are intermediate.

The sparseness of large-scale phylogenetic data sets is apparent in many studies in which multiple loci are concatenated into a supermatrix. A fairly typical example is Hughes *et al.*'s [20] recent analysis of beetle phylogeny based on EST library data. They concatenated 66 loci for 20 species, but their final matrix contained 71.4% missing data. Driskell *et al.*'s [13] larger green plant and metazoan supermatrices contained 84% and 92% missing data. Other recent phylogenomic studies have somewhat denser matrices [11], but part of this reflects the authors' construction of chimeric taxa from different species, which increased the density of the matrix by effectively decreasing the number of taxa. A few studies using a small number of whole genomes (e.g. [22, 26]) have nearly complete data matrices, but, surprisingly, these matrices all have a small number of loci in them—100s out of the 10,000's found in the genome sequences themselves; which begs the question of whether lack of homology among many loci not included in these analyses is what limited the eventual size of their data matrices. In principle, as more of a genome is sampled, eventually some fraction of loci will be found for which no homologs exist in the other taxa, and these will cause fragmentation of the matrix. Low density is also a feature of supertree studies whenever there is low taxonomic overlap between input trees. This is especially evident in supertrees that assemble several shallow-level, densely-sampled phylogenies, together with deep phylogenies with sparse sampling of exemplar taxa (e.g. [35]).

In this chapter we discuss three classes of strategies for handling the fragmentation of data sets that seems to arise commonly in large-scale phylogenetic analysis. The first of these are *post-processing* strategies: ignoring the fragmentation until after phylogenetic analyses are performed, and then processing the resulting trees to tease apart the underlying signals. The other two are *pre-processing* strategies that break up the data into pieces prior to separate phylogenetic analysis. One of these pursues a strict mathematical definition of what makes a subset of the data 'ideal'. The other is more heuristic and partitions the data so as to obtain 'good' subsets according to clustering methods or optimality criteria.

7.2 Basic definitions

A *data availability matrix*, **A**, (Fig. 7.2) is a matrix of N rows (labelled by taxon names) and M columns (labelled variously: by character names, names corresponding to sets of characters—such as locus names, or by tree names in the case of a supertree analysis). Each cell of the matrix is scored 1 if data are available for that entry or 0 if not. In a supermatrix setting this matrix shows the presence or absence of sequence data for a given taxon and locus (as in Fig. 7.2). Of course, this matrix might also be defined on a finer scale, such as individual sites in a sequence, but this does not lead to a notably different set of issues. In a supertree analysis it is useful to have the columns represent the input trees, and then entries in **A** refer to whether or not a particular taxon (row) is present in that tree (column). For a given combined data set, the same data-availability matrix will be obtained whether one prefers the supermatrix or the supertree methodology. Let $m(\mathbf{A})$ be the number of entries in **A** containing a 1. The *density* of **A** is $m(\mathbf{A})/NM$. A *block* in **A** is a submatrix defined by a subset of **A**'s rows and columns entirely filled with 1s. Two columns are *non-overlapping* if no row is present that has a 1 in both columns (if the columns represent trees, for example, this means the trees share no taxa in common). Two *blocks* in a phylogenetic data matrix are non-overlapping if every pair of columns, one from each block, is non-overlapping.

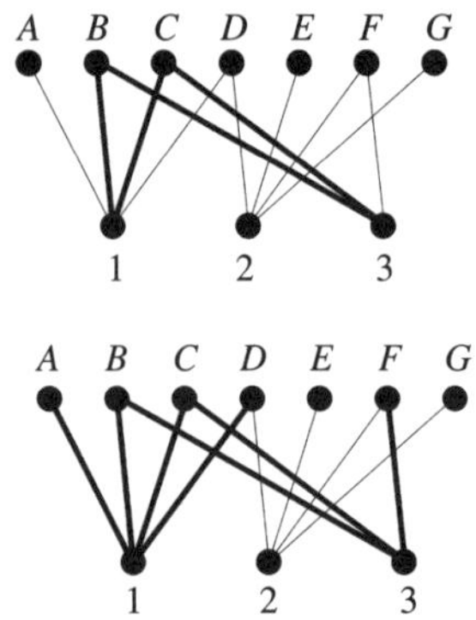

FIG. 7.4. Bicliques and quasi-bicliques. The $\mathcal{A}$ graph for the data set of Fig. 7.2 is shown below. The top graph highlights a maximal biclique comprised of taxa B and C together with loci 1 and 3 and all edges connecting them. This corresponds to a data-availability matrix for the two taxa and loci that has no missing data. The bottom graph is a quasi-biclique extension of this maximal biclique. The extension adds all taxon nodes that are connected to 50% or more of the locus nodes in the original maximal biclique. This corresponds to a data-availability matrix for taxa B, C, D, and F and loci 1 and 3 that has no more than 50% missing data (this lower bound might not hold if both node sets in the bipartite graph were extended simultaneously: see [37] for further discussion).

Alternatively (Fig. 7.2), we can construct a bipartite graph, $\mathcal{A}$, consisting of N taxon nodes and M locus (tree) nodes, in which an edge is present if data are present for the taxon and locus (tree). Similarly to the density of **A**, define the density of $\mathcal{A}$ to be $m(\mathcal{A})/NM$ where $m(\mathcal{A})$ is redefined to be the number of edges in the graph. A block in a data-availability matrix corresponds to a *biclique* in $\mathcal{A}$, that is, a subgraph in which each node of one type is connected to all nodes of the other type (Fig. 7.4).

A data set is *fragmented* if there are non-overlapping blocks in **A** or equivalently if $\mathcal{A}$ is disconnected. It may also be useful to consider a more relaxed sense of fragmentation to occur if $\mathcal{A}$ can be disconnected by the removal of only a 'few' edges. Throughout this chapter we will use the term loosely to refer to either case. Sometimes it is more convenient to discuss problems of fragmentation from the matrix perspective; sometimes from the graph perspective.

A *parent tree* of a collection of trees, is a tree that contains as subtrees all of the trees in the collection. A collection of trees is *compatible* if a parent tree exists for it.

Any set of three taxa, $\{x, y, z\}$ is a *triple*. A rooted binary tree on these taxa is a *triplet*, denoted, for example, as $xy|z$ for the case in which x and y share a more recent common ancestor than either does with z.

7.3 Strategies for handling fragmentation of data sets

7.3.1 *Strategy 1. Post-processing collections of trees*

We begin with post-processing strategies, because the phylogenetics community has considerable experience processing collections of phylogenetic trees arising from parsimony, likelihood, or Bayesian search strategies compared to pre-processing strategies discussed below. Consensus methods for summarizing information common to sets of trees are well developed [9]. However, recognition of the weaknesses of consensus has led to development of variants that specifically treat problematic taxon subsets of the data that appear to be unstable [33, 34], which can arise for many reasons including long-branch attraction [17], missing data, heterogeneous histories, hybridization, and so on.

A useful technique is identifying *maximum agreement subtrees* (MASTs: [5, 9, 18]), which are the largest subtrees common to an entire collection of input trees. We have already mentioned how, in the example of Fig. 7.3, agreement subtree algorithms can recover the signal present in two blocks of a data matrix from the collection of parsimony trees generated by the combined matrix, even when the blocks share very few taxa in common. Related to MASTs are maximum compatible trees (MCTs: [5]), a more relaxed strategy that finds collections of smaller trees that are *compatible* with all the input trees once taxa are removed; that is, some of the input trees may be refinements of the MCTs.

One of the basic limitations of a post-processing approach is the rapid combinatorial increase in the number of trees that are equally optimal—of necessity—when separate blocks of data are combined. Modifying the example of Fig. 7.3 only slightly, so that there is no overlap whatever in two blocks of four taxa each, the number of equally parsimonious binary trees skyrockets to 1155. If there are three blocks of the same size, the programme PAUP [31] cannot find all equally parsimonious trees in a reasonable time (a few hours) even using exhaustive, branch-and-bound or heuristic searches, but a simple counting argument shows that the number of solutions is $(2N-5)!!/3k$, where k is the number of blocks and $N=4k$ (for four taxon blocks). This is 24.2 million trees for $k=3$ and 2.6×10^{12} for $k=4$.

Finding the MAST is an NP-hard problem when the number of input trees is greater than two and the degree of the nodes is not bounded [2], and it is solvable in polynomial time if one of the input trees has a degree bound, but the time is exponential in that bound [16], none of which bodes well for this particular solution to the data fragmentation problem. However, heuristics may sometimes be sufficient. In the toy example above where $k=3$, a parsimony search limited to keeping even as few as 10,000 trees, lets MAST recover the three trees corresponding to the three blocks. Another approach would be to generate fewer trees but more variable ones, for example by bootstrapping the data or performing multiple stochastic search strategies from different starting positions. The clades or subtrees that occur in many replicates would then presumably correspond to well-supported relationships within the separate blocks in the original fragmented data set, with the added value that only statistically supported groups would emerge. The parallel supertree heuristic might collapse all clades on the input trees that are not well supported and then look for MASTs or MCTs.

7.3.2 *Strategy 2. Pre-processing by grove identification*

The idea of pre-processing the data is to break it into separate pieces prior to undertaking phylogenetic analysis. This kind of 'partitioning' strategy is especially appealing if nothing is lost by this dismantling, or, viewing it from the other direction, if smaller data sets are combined only if something is gained by doing so. Although opinions vary about whether the default treatment of data should be combination or partitioning [11], we focus in this section on developing in intuitive terms what could be meant by 'something is gained' by combination. See Ané *et al.* [3] for a more rigorous description. The general aim is to provide a quantitative assessment of how a data set or tree set should be partitioned for phylogenetic analysis to satisfy some very minimal requirements. We shall see, by reference to examples, that even though these are very minimal requirements, they are relevant in real phylogenomic analyses. In this section, we focus on the supertree setting, for which most of our results have been derived [3].

Sanderson *et al.* [29] speculated that in supertree analysis it is necessary for input trees to share two leaf taxa in common. This was motivated by the simple observation that two rooted input trees sharing only a single taxon in common (or having none in common), maps to a collection of parent trees that are completely incompatible (unresolved) by the strict consensus method. However, as we have seen using agreement subtrees (Fig. 7.3), the collection of parent trees can be used to recover the input trees; thus, the collection is a restriction of the set of all possible trees. However, although nothing is lost, nothing is gained in this particular case because no *new information* about phylogenetic relationships can be obtained from these two input trees.

As intuitive as the idea of 'new information' about phylogenetic relationships is, it has proven remarkably difficult to formalize [3]. Consider first the case of rooted trees that are compatible with each other such that there does exist one or more parent trees. For rooted trees, the smallest tree that yields differential subset relationships between taxa is a three-taxon tree. Therefore, information is defined in terms of subset statements on triplets. First, decompose all input trees into their rooted triplets. Then decompose all of the parent trees into their rooted triplets. If there is a triplet in *all* of the parent trees that is *not* present in any of the input trees, then this triplet reflects potential new information arising from the combination of data sets. For example, suppose one tree has the triple $\{a,d,e\}$, and another tree has $\{b,c,f\}$. An example of a new triplet that could not possibly have been present on either input tree is $a|bc$, because the *triple* $\{a,b,c\}$ is not present on either input tree. We refer to this kind of triple as a *cross-triple*, because it is composed of elements from more than one input tree. Now, if the (*cross-*)triplet $a|bc$ were present on a parent tree it might represent new information—of progress due to data combination. However, what if the other two triplets, $b|ac$ and $c|ba$ are also displayed by some of the parent trees? In that case this potential new information is not likely to be helpful: it does not

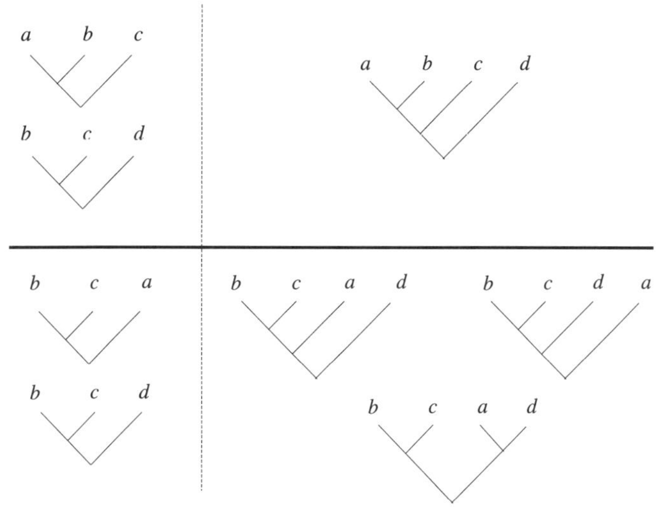

FIG. 7.5. New information and groves. On the left side of dashed line are input trees. On the right side are parent trees (supertrees). The top panel is a case of two input trees in which there exists only one parent tree that displays the input trees. The parent tree displays new information $ab|d$ and $ac|d$. The two input trees are a grove. The lower panel is a case of two input trees in which three parent trees exist. Together they display all possible triplet trees $\{ab|d, ad|b, bd|a, ac|d, ad|c, cd|a\}$ for the triples of taxa $\{a, b, d\}$ and $\{a, c, d\}$ that potentially could have provided new information—the cross-triples. Because they do not discriminate among all possible triplets, they do not provide new information and therefore these two input trees do not form a grove (after [3]).

let us easily choose among these relationships (Fig. 7.5). We refer to the case in which *only one* cross-triplet (of the three possible for that triple of taxa) is displayed by all parent trees as a *resolved* cross-triplet.

This formulation of new information is restrictive, because it begins with the assumption that the input trees are known and are compatible, when in fact the input trees are always estimates with some error and are in practice rarely compatible. Ané *et al.* [3], therefore, pursue a more general approach that assesses the potential new information in a data set irrespective of the particular method of estimating phylogenies from those data. This is dependent on the data-availability matrix, **A**, alone, which, recall, (in the supertree setting) describes the distribution of taxa among trees without requiring that the *topologies* of the trees themselves be known. Thus, they ask whether or not it is possible to *imagine* a set of input trees with taxonomic structure defined by **A** that could yield new information. Corresponding to all the triples implied by **A**, there is a much larger set of possible *triplets*. The goal is to find sets of triples for which we can assign triplets such that their parent trees agree with each other, and then to ask if any of the triplets on the parent trees are resolved cross-triplets. If no resolved cross-triplets exists for any combination of input trees, then there does not exist *any* set of input trees with the structure indicated in **A** that can generate novel phylogenetic information. This provides a strong condition for which combining trees makes no sense. [There is an important exception to this notion of combinability, however. If trees have identical label sets (as in the consensus setting), or if one tree is a subtree of another tree, there are no cross-triples whatsoever (all triples are observed triples), but it seems biologically sensible to combine information in this trivial case. See [3].]

These considerations led us to define a *grove*, loosely speaking, as a collection of trees (columns in **A** or the corresponding subgraph of $\mathcal{A}$) that are mutually informative, while sets of different groves are not. The basic idea is that a collection of columns in **A** is a grove if every partition of this collection entails some new information when combined in the sense just described. These ideas have been formulated for the case in which columns in **A** represent trees [3], but they may well apply to the supermatrix case also.

From a statistical perspective, we can view this approach in terms of identifiability. Imagine the best-case scenario in which an infinite amount of data has been applied to reconstruct each of the input trees using a statistically consistent estimator, and each of the input trees reflects a common evolutionary history (without recombination, horizontal transfer, or other processes that make the true histories different). It is still meaningful to ask whether features of the tree constructed by combining all this evidence (i.e. cross-triplets) can be identified. In fact, no triplet of the large tree becomes identifiable when combining two separate groves that was not already identifiable from one grove or the other.

Several results based on this definition of grove have been obtained. A very useful device to help both with proofs and empirical calculations on groves is the intersection graph, $\mathcal{G}$, which can be defined based on **A** or $\mathcal{A}$. Nodes in $\mathcal{G}$ correspond to loci (trees, columns in **A**), and nodes are connected by edges weighted by the number of taxa the pairs of loci have in common [30]. In the supermatrix setting this corresponds to the number of taxa having a sequence for both loci; in the supertree setting it is the number of taxa common to both trees. Let the graph $\mathcal{G}_k$ denote the graph in which any edges of weight less than k are removed. See Fig. 7.6 for an example. The $\mathcal{G}_2$ graph is especially important. The following results are proven in Ané *et al.* [3]

1. If $\mathcal{G}_2$ is connected, then it is a grove.
2. If $\mathcal{G}_2$ consists of two connected components and the two components share two taxa in common, then it is a grove. This does not automatically follow from (1) because there might be two weight-1 edges that connect the two components.

Interestingly, some graphs are groves even if all their edges have a weight of only 1 (Fig. 7.7), showing that the speculation of Sanderson *et al.* [29] was wrong, although it does appear that the structure of the intersection graph has

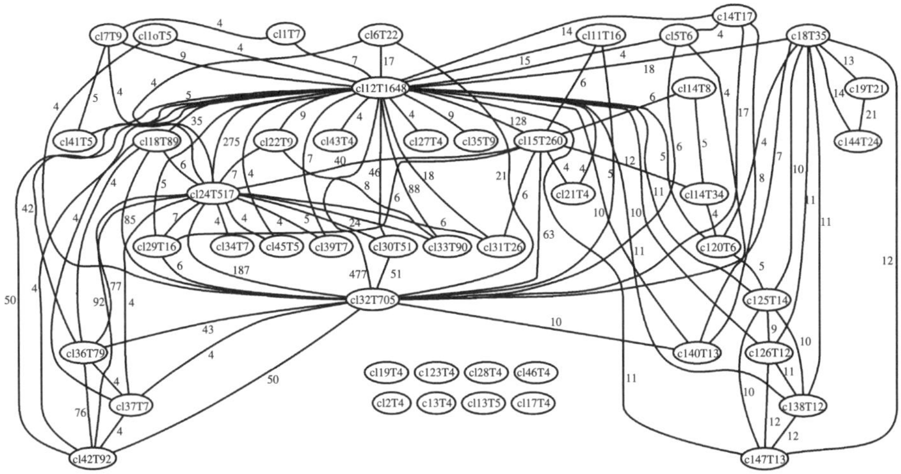

FIG. 7.6. Grove structure of 47 loci analysed in [23]. Graph shows taxonomic overlap between loci (ellipses). An edge is drawn if two loci share four or more taxa, which is our criterion for assembling loci into a supermatrix. Loci that share less than four taxa with all other loci have limited potential to contribute topological information. Eight such isolated loci were found and screened out of the analysis. Numbers next to each edge indicate total taxa shared; text inside ellipses give a reference number for the locus (cl#) followed by the number of taxa (T#) for each locus.

to be rather special in this case. Interestingly, Bininda-Emonds *et al.*'s [8] claim that a tree must share two taxa with the label set of the collection of other trees for supertree construction to be sensible is correct—as a necessary condition—but it is not a sufficient one. If the other collection of trees all overlap only by one taxa, and the one tree shares each of its two overlapping taxa with different trees, it might well not be a grove. It might seem that additional overlap would be required for new information to arise when input trees are unrooted. However, Ané *et al.* [3] also show that overlaps of one taxon between input trees—or equivalently edges of weight 1—can be sufficient, just like with rooted input trees.

Unfortunately, no general rules have been derived to determine if a graph is a grove in the general case. These rules are necessary since it is not computationally feasible in large data sets to apply the definition directly. However, it is possible to place upper and lower bounds on the minimum number of groves required to 'cover' all the data—include all the trees. This is called the grove coverage number $\mathcal{G}_r(\mathbf{A})$ (or $\mathcal{G}_r(\mathcal{A})$). In particular:

1. A lower bound on $\mathcal{G}_r(\mathbf{A})$ is the number of components in $\mathcal{G}_1$.
2. An upper bound on $\mathcal{G}_r(\mathbf{A})$ is the number of components in $\mathcal{G}_2$.

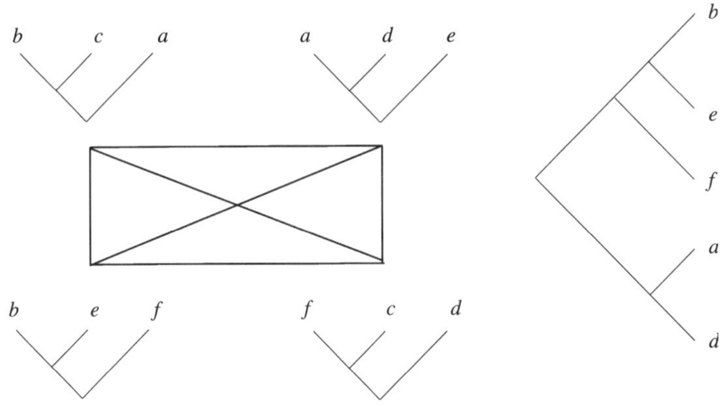

FIG. 7.7. Figure showing the case in which four trees only overlapping in one taxon is a grove. There are four input trees shown at the left along with their $\mathcal{G}_1$ overlap graph. The tree on the right is the maximum agreement subtree of five binary parent trees that each display all input trees. The five parent trees can be obtained by attaching taxon c to any of the five branches that are more closely related to b than to a (i.e. on branches in the top clade descended from the root). There are 13 new triplets displayed on the parent trees: $ad|b$, $ad|f$, $ad|c$, $be|a$, $be|d$, $bf|a$, $bf|d$, $ef|a$, $ef|d$, $cf|a$, $ce|a$, $ce|d$, $bc|d$. After [3].

3. Both of these bounds can be made tighter by consideration of additional features of the intersection graph (see Ané *et al.* [3]).

Some examples are useful at this point. Ané *et al.* [3] reanalysed the very sparse 14502 taxa $\times$ 853 genes data-availability matrix for green plants described in Driskell *et al.* [13]. The $\mathcal{G}_1$ graph has 8 components and the $\mathcal{G}_2$ graph has 32 components. Further analysis of the graph structure allowed them to narrow the bounds to between 24 and 31. McMahon and Sanderson [23] analysed data for 2236 taxa and 47 loci. The data-availability matrix was very sparse, with a density of $< 4.3\%$. Its $\mathcal{G}_1$ graph has 1 component and its $\mathcal{G}_2$ graph has 3 components. From this we can bound the number of groves between 1 and 3. The additional considerations mentioned in Ané *et al.* [3] narrow this range to between 2 and 2; in other words they allow us to determine $\mathcal{G}_r(\mathbf{A})$ exactly.

What is implied by multiple groves? At a minimum no phylogenetic analysis can expect to profit from combining data from more than one grove. Regardless of how little or much information is contained within the groves, nothing can be gained by bringing them together. No additional feature of the large tree can be identified (in the statistical sense) by combining data from more than one grove. Groves therefore set a minimum necessary condition for combinability and can be quite useful in subdividing large and sparse data sets prior to analysis. The ideas underlying groves can also be used to generate more conservative sufficient conditions for combinability by assembling data sets using the $\mathcal{G}_k$ graph with higher values of k (see below).

7.3.3 *Strategy 3. Pre-processing by clustering or optimization strategies*

Although the identification of groves can prevent wasted efforts to combine data into large analyses, there are at least two reasons to also consider subdividing data using more relaxed stringency conditions. First, with available theoretical results, only some groves can be identified unambiguously. Second, any one grove might be too inclusive in practice. It might be much more productive to require that a larger proportion of the combined analysis tree be identifiable rather than just a single triplet over and above those found on the input trees. One grove might include data that ought to be subdivided further, because further subdivision would improve the robustness of the individual phylogenetic analyses that result. After all, nothing in the definition of grove guarantees a certain level of *quality* in the resulting tree; it merely precludes mistakenly combining pieces of information that cannot under any circumstances shed new light on phylogenetic history.

Various *ad hoc* strategies can be used to subdivide (or build in the first place) the **A** matrix to pursue the goal of reliable tree construction. A widely used practice in phylogenomic analyses is to place controls on the density of the matrix or quantity of cells filled (e.g. [4, 24]). The rationale is the belief that missing data ultimately degrades the quality of phylogenetic analysis [14, 15]. Interestingly considerable simulation work [32] and now several empirical studies suggest that substantial missing data can be tolerated [13, 20]. Commonly, controls on density are imposed by placing a minimum value on the number of 1-entries in rows or columns or both. For example, Bapteste *et al.* [4] constructed a data matrix of 30 species by 123 genes in which no more than 7 missing taxa were allowed per gene (column) in the matrix. Driskell *et al.* [13] constructed a matrix in which every row (taxon) had to include 10 genes and every gene include 4 taxa (the minimum number that is informative for unrooted trees).

An extreme form of controlling missing data is to eliminate it entirely, to find blocks in **A** or bicliques in $\mathcal{A}$. For example, Moreau *et al.* [24] discarded one third of their original 149 taxa to construct a matrix of 6 genes by 102 taxa that had no missing data. Rarely have formal algorithms been used to do this in phylogenomics, but the problem is well-known elsewhere: finding *maximal* blocks in a matrix or maximal bicliques in a bipartite graph [28]. Although intractable (NP-complete), for relatively sparse graphs, exact enumeration is sometimes feasible [1]. Sanderson *et al.* [28] and Driskell *et al.* [13] investigated this approach in proteins mined from GenBank for green plants and metazoans, a sparse database, and found that the maximum size (i.e. size = NM) of the bicliques that could be constructed was surprisingly small, on the order of a few thousand sequences. These did indeed tend to produce reliable phylogenies, at least when the number of loci was large. However, one problem with this approach is that maximal bicliques can overlap, and there is no efficient algorithm to partition $\mathcal{A}$ into bicliques. Moreover, the small size of the bicliques leaves open the possibility that the collection of bicliques will not form a grove and therefore should not be combined in the first place. In Driskell *et al.* [13] a collection of bicliques of a minimal size was assembled and checked to make sure that its $\mathcal{G}_2$ graph was connected.

Obviously it should be possible to relax the notion of block or biclique in some well-defined way. Yan *et al.* [37] suggested using a-quasi-bicliques (Fig. 7.4). An a-quasi-biclique is a subgraph of $\mathcal{A}$ that 'extends' a maximal biclique of $\mathcal{A}$ by adding either taxon nodes such that each added taxon node is connected to at least a fraction a of the locus nodes in the biclique, or by adding locus nodes in a similar fashion (or both). Based on simulation studies, they concluded that phylogenies based on quasi-biclique data assemblies could often be nearly as good as those based on maximal bicliques proper.

Finally, an equally heuristic procedure could use connected components of the $\mathcal{G}_k$ graph with k set to some conservatively high value well above the values at issue for grove definition (see Fig. 7.6 for example). A high value of k would generate smaller and more numerous components, but possibly each would be more decisive because its density is higher. Simulation studies show, for example, that supertree methods tend to work better when taxon overlap is high [6]. A plot of the number of components in $\mathcal{G}_k$ versus k, which is a non-decreasing function, reveals some interesting features that might suggest ways to choose k. Figure 7.8 shows this plot for the two data sets discussed earlier. Both show a rapid increase in the number of components as k increases asymptotically to the maximum value, which is just the number of loci. Clearly, values of k greater than even some small number like 5–10 are sufficient to break up the graph into a very large number of components. This reflects the fact that it is not possible to find large collections of loci that share large numbers of taxa, Fig. 7.6 shows the $\mathcal{G}_4$ graph for the legume data set, which has 9 components, the largest of which contains 2228 taxa and formed the basis of the phylogenetic supermatrix analysis reported in [23].

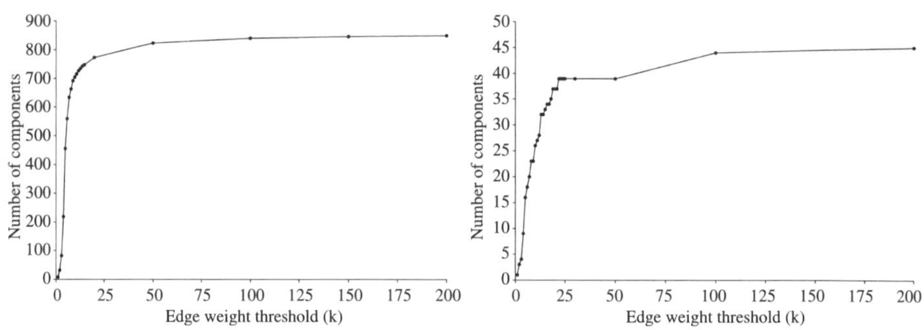

FIG. 7.8. Plot of number of connected components vs. the edge weight threshold, k, in the G_k graph (left panel) for the green plant data set of Driskell *et al.* [13] and (right panel) for the legume data set of [23].

7.4 Conclusions

A growing, but still relatively unappreciated, problem in large scale phylogenetic analyses is the fragmentation that is inevitable when many loci or trees are combined into a single analysis. Fragmentation occurs when large amounts of missing data break a data set into subsets for which a combined analysis adds little phylogenetic information that could not be obtained by analysing the subsets separately. These ideas can be formalized using the notion of grove, which provides minimal conditions for which data combination provides new information. Data subsets in separate groves may be separately informative but when combined this information is not augmented in any way. Identification of groves in large and complex data sets may save tree search algorithms from having to explore a flat likelihood or parsimony surface, i.e. much larger parts of the solution space than necessary. Even very small fragmented data sets can have a very large solution space, as shown by some simple examples. This devalues post-processing procedures that attempt to sort through large sets of solutions to tease apart the information that might be present in subsets of the data. On the other hand, computational difficulties may often preclude identification of groves *per se* in a data set, and it may sometimes be easier and more phylogenetically informative to use other kinds of heuristic procedures to partition data sets. One simple strategy, for example, is to identify the components in the taxon intersection graph defined by overlaps of $k \gg 2$ taxa. This tends to partition the data into more numerous subsets, but each subset has less missing data. Whatever the strategy used, it is unlikely that the data will cooperate to solve the problem for us, even—or especially—at a phylogenomic scale.

Acknowledgements

We thank Amy Driskell and Gordon Burleigh for insights into data analysis. This research was supported by a grant from the US National Science Foundation (NSF).

References

[1] Alexe, G., Alexe, S., Crama, Y., Foldes, S., Hammer, P. L., and Simeone, B. (2002). Consensus algorithms for the generation of all maximal bicliques. In *DIMACS Technical Report 2002-4*.

[2] Amir, A. and Keselman, D. (1994). Maximum agreement subtree in a set of evolutionary trees—metrics and efficient algorithms. In *Proceedings of the 35th Annual Symposium on Foundations of Computer Science*, pp. 758–769.

[3] Ané, C., Eulenstein, O., Piaggio-Talice, R., and Sanderson, M. J. (2006). Groves of phylogenetic trees. *Technical Report 1123. Department of Statistics, University of Wisconsin, Madison, WI.*, **http://www.stat.wisc.edu/Department/techreports/tr1123.pdf**, 1–31.

[4] Bapteste, E., Brinkmann, H., Lee, J. A., Moore, D. V., Sensen, C. W., Gordon, P., Durufle, L., Gaasterland, T., Lopez, P., Muller, M., and

Philippe, H. (2002). The analysis of 100 genes supports the grouping of three highly divergent amoebae: Dictyostelium, entamoeba, and mastigamoeba. *Proceedings of the National Academy of Sciences of the United States of America*, **99**, 1414–1419.

[5] Berry, V. and Nicolas, F. (2005). Improved parameterized complexity of the maximum agreement subtree and maximum compatible tree problems. *LIRMM Technical Report 04026*, **http://www.lirmm.fr/~vberry/Publis/parametrizedMAST-MCT.pdf**.

[6] Bininda-Emonds, O. R. P. and Sanderson, M. J. (2001). Assessment of the accuracy of matrix representation with parsimony analysis supertree construction. *Systematic Biology*, **50**, 565–579.

[7] Bininda-Emonds, O. R. P. (2004). The evolution of supertrees. *Trends in Ecology and Evolution*, **19**, 315–322.

[8] Bininda-Emonds, O. R. P., Gittleman, J., and Steel, M. (2002). The (super)tree of life: procedures, problems, and prospects. *Annual Review of Ecology and Systematics*, **33**, 265–290.

[9] Bryant, D. (2003). *A classification of consensus methods for phylogenetics.* In DIMACS Working Group Meeting on Bioconsensus. American Mathematical Society (eds. M. F. Janowitz, F.-J. Lapointe, F. R. McMorris, B. Mirkin, and F. S. Roberts).

[10] Ciccarelli, F. D., Doerks, T., von Mering, C., Creevey, C. J., Snel, B., and Bork, P. (2006). Toward automatic reconstruction of a highly resolved tree of life. *Science*, **311**, 1283–1287.

[11] De Queiroz, A., Donoghue, M. J., and Kim, J. (1995). Separate versus combined analysis of phylogenetic evidence. *Annual Review of Ecology and Systematics*, **26**, 657–681.

[12] Delsuc, F., Brinkmann, H., Chourrout, D., and Philippe, H. (2006). Tunicates and not cephalochordates are the closest living relatives of vertebrates. *Nature*, **439**, 965–968.

[13] Driskell, A. C., Ané, C., Burleigh, J. G., McMahon, M. M., O'Meara, B., and Sanderson, M. J. (2004). Prospects for building the tree of life from large sequence databases. *Science*, **306**, 1172–1174.

[14] Erdös, P. L., Steel, M. A., Szekely, L. A., and Warnow, T. J. (1999). A few logs suffice to build (almost) all trees: part (i). *Random Structures and Algorithms*, **14**, 153–184.

[15] Erdös, P. L., Steel, M. A., Szekely, L. A., and Warnow, T. J. (1999). A few logs suffice to build (almost) all trees: part ii. *Theoretical Computer Science*, **221**, 77–118.

[16] Farach, M., Przytycka, T. M., and Thorup, M. (1995). On the agreement of many trees. *Information Processing Letters*, **55**, 297–301.

[17] Felsenstein, J. (2004). *Inferring Phylogenies.* Sinauer Associates, Sunderland, MA.

[18] Finden, C. R. and Gordon, A. D. (1985). Obtaining common pruned trees. *Journal of Classification*, **2**, 255–276.

[19] Hibbett, D., Nilsson, R., Snyder, M., Fonseca, M., Costanzo, J., and Shonfeld, M. (2005). Automated phylogenetic taxonomy: An example in the homobasidiomycetes (mushroom-forming fungi). *Systematic Biology*, **54**, 660–668.

[20] Hughes, J., Longhorn, S. J., Papadopoulou, A., Theodorides, K., de Riva, A., Mejia-Chang, M., Foster, P. G., and Vogler, A. P. (2006). Dense taxonomic est sampling and its applications for molecular systematics of the coleoptera (beetles). *Molecular Biology and Evolution*, **23**, 268–278.

[21] Kllersj, M., Farris, J. S., Chase, M. W., Bremer, B., Fay, M. F., Humphries, C. J., Petersen, G., Seberg, O., and Bremer, K. (1998). Simultaneous parsimony jackknife analysis of 2538 rbcl dna sequences reveals support for major clades of green plants, land plants, seed plants and flowering plants. *Plant Systematics and Evolution*, **213**, 259–287.

[22] Lerat, E., Daubin, V., and Moran, A. (2003). From gene trees to organismal phylogeny in prokaryotes: The case of the gamma-proteobactera. *PLoS Biology*, **1**, 1–9.

[23] McMahon, M. M. and Sanderson, M. J. (2006). Phylogenetic supermatrix analysis of genbank sequences from 2228 papilionoid legumes. *Systematic Biology*. **55**, 818–836.

[24] Moreau, C. S., Bell, C. D., Vila, R., Archibald, S. B., and Pierce, N. E. (2006). Phylogeny of the ants: diversification in the age of angiosperms. *Science*, **312**, 101–104.

[25] Philippe, H., Lartillot, N., and Brinkmann, H. (2005). Multigene analyses of bilaterian animals corroborate the monophyly of ecdysozoa, lophotrochozoa, and protostomia. *Molecular Biology and Evolution*, **22**, 1246–1253.

[26] Rokas, A., Williams, B., King, N., and Carroll, S. (2003). Genome-scale approaches to resolving incongruence in molecular phylogenies. *Nature*, **425**, 798–804.

[27] Sanderson, M. J. and Driskell, A. C. (2003). The challenge of constructing large phylogenetic trees. *Trends in Plant Science*, **8**, 374–379.

[28] Sanderson, M. J., Driskell, A. C., Ree, R. H., Eulenstein, O., and Langley, S. (2003). Obtaining maximal concatenated phylogenetic data sets from large sequence databases. *Molecular Biology and Evolution*, **20**, 1036–1042.

[29] Sanderson, M. J., Purvis, A., and Henze, C. (1998). Phylogenetic supertrees: assembling the trees of life. *Trends in Ecology and Evolution*, **13**, 105–109.

[30] Schmidt, H. (2003). *Phylogenetic trees from large datasets, Ph. D. dissertation.* Ph. D. thesis, Heinrich-Heine Universitt, Dusseldorf.

[31] Swofford, D. L. (2002). *PAUP*. Phylogenetic Analysis Using Parsimony (*and Other Methods).* Sinauer Associates, Sunderland.

[32] Wiens, J. (1998). The accuracy of methods for coding and sampling higher-level taxa for phylogenetic analysis: A simulation study. *Systematic Biology*, **47**, 397–413.

[33] Wilkinson, M. (1994). Common cladistic information and its consensus representation: Reduced adams and reduced cladistic consensus trees and profiles. *Systematic Biology*, **43**, 343–368.

[34] Wilkinson, M. and Thorley, J. (2003). *Bioconsensus Vol. 61*, (eds. M. F. Janowitz, F.-J. Lapointe, F. R. McMorris, B. Mirkin, and F. S. Roberts). pp. 195–203. American Mathematical Society Providence.

[35] Wojciechowski, M. F., Sanderson, M. J., Steel, K. P., Liston., and A. (2000). *Molecular Phylogeny of the 'Temperate Herbaceous Tribes' of Papilionoid Legumes: A Supertree Approach* (eds. P. S. Herendeen and A. Bruneau.). pp. 277–298. Royal Botanic Gardens: Kew.

[36] Wolf, Y., Rogozin, I., and Koonin, E. (2004). Coelomata and not ecdysozoa: Evidence from genome-wide phylogenetic analysis. *Genome Research*, **14**, 29–36.

[37] Yan, C. H., Burleigh, J. G., and Eulenstein, O. (2005). Identifying optimal incomplete phylogenetic data sets from sequence databases. *Molecular Phylogenetics and Evolution*, **35**, 528–535.

8

IDENTIFYING AND DEFINING TREES

Stefan Grünewald and Katharina T. Huber

Abstract

Many phylogeny reconstruction methods implicitly assume that the evolution of a data set is tree-like and then go on to reconstruct a tree that best explains the data. A fundamental question therefore is: when does a data set support a tree-like evolutionary scenario and, if it does, is it unique or might there be other scenarios that are equally well supported by the data? In this chapter, we review both classical and recent results regarding this question in case the data set of interest is in terms of characters and quartets. Whenever possible, we also interpret these results from a biological point of view.

We begin our survey by presenting a standard formalization of the above question in terms of character compatibility and defining/identifying a tree. This formalization is motivated by the evolutionary idea of characters evolving without homoplasy (examples of which are SINESs, LINEs, and LTRs). Using this formalization, we then present partial and complete answers to the above question in terms of chordal graphs, closure rules, and the quartet graph. In addition, we review answers to related questions such as 'how many characters suffice to uniquely determine the evolutionary past of a taxa set if characters evolve without homoplasy'.

8.1 Introduction

Arguably, the goal of any evolutionary study is to gain insight into the evolutionary past of a set of taxa (for example, species) under consideration. In most cases, this past is assumed to be best modelled by a tree and the assumption is that the data collected will allow one to reconstruct a reasonably good approximation of that tree. From very early on, mathematicians and theoretical computer scientists have been intrigued by this assumption and have looked into the question of which premises this assumption is justified under. Early characterizations of such a tree's existence include the well-known *4-point-condition* (if the data are given in terms of distances) and a certain intersection criterion (if the data are given in terms of 2-state characters—see Section 8.3 for details). Interpreted from a biological point of view, the latter characterization means that if any two of the characters in question satisfy that criterion, then there is a tree on which they all could have evolved without homoplasy (i.e. acquired the same character state but not because of common descent [36]).

Although the concept of homoplasy has been around for some time, in recent years it has attracted a considerable amount of interest. The reasons for this are (at least) twofold. First, researchers have realized the potential of genomic data for understanding genome evolution [33] and thus evolution in general. A lack of good models to describe the former has meant that many studies so far have relied on the usage of (quantitative) characters such as rare genomic markers; examples of which include retroposons (e.g. SINEs, short interspersed elements; LINEs, long interspersed elements; and LTRs, long terminal repeats) and gene order data. These markers are known to have very low to zero amounts of homoplasy [29, 36] but can have a very large number of states. Second, there is the desire to combine phylogenetic information from different studies into an overall evolutionary picture; the most prominent example being the 'Assembling the Tree of Life' project (details can be found at *www.tolweb.org/tree*). This information may be of the form of only partially overlapping gene trees or very small evolutionary building blocks called quartets that only involve four taxa.

This chapter is aimed at reviewing recent combinatorial results concerning the following question which lies at the heart of understanding (almost) homoplasy-free evolution.

(Q) When do fundamental divisions of taxa into groups—either directly from data or from earlier phylogenetic studies—completely determine a tree on which the taxa set under consideration has evolved?

Due to space limitations, and since many of the interesting mathematical questions arise in the unrooted setting, we will only be concerned with unrooted evolutionary trees. The chapter is organized as follows: in the next section, we introduce some terminology that will allow us to formalize Question (Q). In Section 8.3, we review recent results concerning Question (Q) for fully resolved evolutionary trees within a graph theoretical framework, and in Section 8.4, we review recent results for such trees in terms of an inference rule. In the last section, we turn our attention to unresolved evolutionary trees.

Throughout this chapter, we will assume that X denotes a finite set (of, for example, taxa).

8.2 From biology to mathematics

We first formalize Question (Q) which requires the introduction of some terminology. We start with recalling concepts concerning trees and then present terminology surrounding markers to make precise what we mean by 'fundamental divisions of taxa into groups'. We conclude the section with restating (Q).

8.2.1 *Evolutionary trees and X-trees*

In many evolutionary studies, it is assumed that the evolutionary past of a set X of taxa is best modelled by a tree. Commonly, the leaves of such a tree are labelled

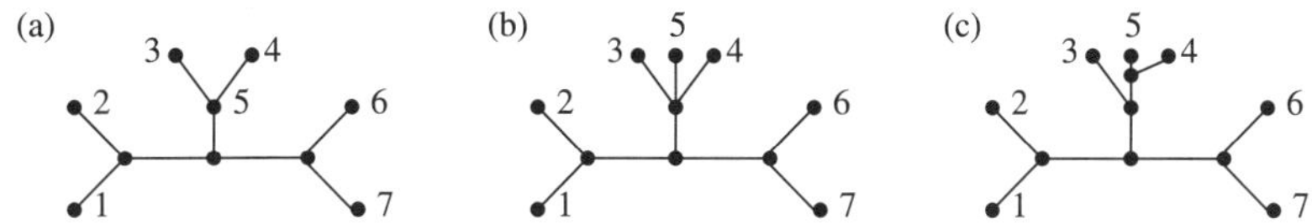

FIG. 8.1. For $X = \{1, 2, \ldots, 7\}$, a (binary) X-tree is depicted in (a). In (b), an unresolved phylogenetic tree on the same set X is pictured. Also on the same set X, a binary phylogenetic tree is presented in (c).

by the taxa under consideration and its interior vertices represent ancestral species. However, it should be noted that in some cases (e. g. viral studies involving fast evolving viruses or phylogeography studies) interior vertices may also be labelled by taxa. Due to lack of sampling, some of these vertices may be unresolved in which case they are called polytomies. These may represent simultaneous divergence (in which case the polytomy is called *hard*) or indicate uncertainty as to the order of speciation (in which case the polytomy is called *soft*).

Formally, trees used for modelling evolution are best thought of as *X-trees*, that is, pairs $\mathcal{T} = (T, \phi)$ consisting of a tree T with vertex set $V(T)$ and a labelling map $\phi : X \to V(T)$ such that every vertex v in T of degree at most two is labelled by an element in X. In case ϕ is a bijection between X and the leaf set $\mathcal{L}(\mathcal{T})$ of $\mathcal{T}$, then $\mathcal{T}$ is commonly called a *phylogenetic* (X-)tree (see Fig. 8.1 for examples). Within this framework, polytomies correspond to vertices with a high degree, i. e. vertices that are incident with four or more edges. If every interior vertex of $\mathcal{T}$ is of degree three, then $\mathcal{T}$ is said to be *binary* or *fully resolved*. Using external information, it is sometimes possible to (partially) resolve a high degree vertex of an X-tree $\mathcal{T}$ in which case we call the resulting X-tree a *refinement* of $\mathcal{T}$. Finally, to capture the idea that two X-trees with the same taxa set tell the same story but can have different representations, two X-trees $\mathcal{T}_1 = (T_1, \phi_1)$ and $\mathcal{T}_2 = (T_2, \phi_2)$ are called *isomorphic* if there is a bijection $\psi : V(T_1) \to V(T_2)$ that induces a graph isomorphism between T_1 and T_2 which is the identity on X.

8.2.2 *Characters and (partial) partitions*

A typical starting point of an evolutionary study is a collection of *(quantitative) characters* such as morphological or behavioural features, genomic markers, or DNA sequence alignment positions. Generally such characters can have two or more *(character) states* such as 'wings', 'stubs', and 'no-wings and no-stubs'.

Traditionally, a character on a taxa set X under consideration has been thought of as a map from a subset X' of X into the set of states of that character. In the ideal case, X' is X itself but because of, for example, lack of sampling we might just have $X' \subsetneq X$. To elucidate this concept, consider, for example, the set $X = \{a, b, c, d, e\}$. Then the map $\gamma : X \to \{red, green, blue\}$ with $\gamma(a) = \gamma(b) = red$, $\gamma(c) = \gamma(e) = blue$ and $\gamma(d) = green$ is a (multi-state) character and so is $\gamma' : X' = \{a, b, c, d\} \to \{red, green, blue\}$ with $\gamma'(a) = \gamma'(b) = red$, $\gamma'(c) = green$, and $\gamma'(d) = blue$.

We remark that when employing quantitative characters in a phylogenetic analysis, it is generally more important to understand which taxa share a character state rather than what the shared character state is. An alternative way to think of a character χ is therefore in terms of the partition P_χ it induces on a subset X' of a taxa set X (which may be X itself!) where all taxa sharing a character state are grouped together. We note that in case X' does not equal X then P_χ is commonly called a *partial* partition of X. Otherwise it is called a *full* partition. Although strictly speaking P_χ is a collection $\{A_1, A_2, \dots, A_m\}$ of non-empty disjoint subsets A_i of X', $1 \leq i \leq m$, whose union is X' we will write it as $A_1|A_2|\cdots|A_m$ (where the order of the A_i does not matter). In case $A_i = \{a_1, \dots, a_t\}$, $t \geq 1$, we will simplify this notation even further by replacing A_i by $a_1 \cdots a_t$. For example, for X and γ as above, the partition P_γ induced by γ is $ab|ce|d$. As suggested by this example, the number of states of a character χ always equals the number of *parts*, that is, elements of the induced partition P_χ.

To be able to introduce a key concept for partitions, consider P_γ and the partial partition $P_{\gamma'} = ab|c|d$ (induced by the above character γ'). Then, P_γ displays $P_{\gamma'}$ in the sense that removing e from the part $\{c, e\}$ of P_γ results in $P_{\gamma'}$. We formalize this by saying that a partition P *displays* a partition P' if every part in P' is contained in one part of P, and every part of P contains at most one part of P'.

We conclude this section with remarking that for the sake of clarity of exposition, we will generally view a character in terms of the partition it induces and only in very few instances in terms of a map. Consequently and also to help the reader distinguish between the two alternatives, we will denote a character by a Latin letter like P if we think of it in terms of the partition it induces and by a Greek letter like χ if we think of it in terms of a map.

8.2.3 *Homoplasy and displaying*

A natural way to synthetically produce a particularly simple character P on a finite set X is to proceed as follows. Take an X-tree $\mathcal{T}$ and arbitrarily delete n ($n \in \mathbb{N}$) edges from $\mathcal{T}$ resulting in trees $\mathcal{T}_1, \dots, \mathcal{T}_{n+1}$. Each of the trees $\mathcal{T}_1, \dots, \mathcal{T}_{n+1}$ is labelled by a proper subset of X and it is easy to see that the collection of label sets $\mathcal{L}(T_i)$, $1 \leq i \leq n+1$, is a character of X.

Obviously, there is no reason why any given X-tree $\mathcal{T}$ should display any given character P in the manner described above. For a fixed X-tree $\mathcal{T}$, we will therefore distinguish those characters P that equal a character on X obtained by deleting edges from $\mathcal{T}$ by saying that they are *displayed* by $\mathcal{T}$. Alternatively, such characters are said to be convex on $\mathcal{T}$. For later use, we will say that a set $\mathcal{P}$ of characters is *compatible* if an X-tree exists that displays $\mathcal{P}$, that is, displays every character in $\mathcal{P}$.

Although the concept of displaying seems to be purely mathematical, its importance for phylogenetic analysis lies in its close relationship with the evolutionary concept of homoplasy already mentioned in the Introduction. To see

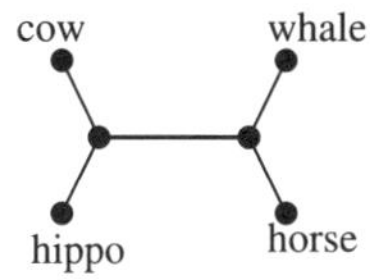

FIG. 8.2. A phylogenetic tree adapted from [30] (see also [35]) that displays the character {cow, hippo, horse}|{whale} but not {cow, horse}|{hippo, whale}.

this, assume that we are given a data set comprising of a taxa set X and a collection $\mathcal{C}$ of biological characters on X. Suppose $\mathcal{T}$ is the underlying (unknown) X-tree on which the data set has evolved. Now, if the amount of homoplasy is very low, then the elements in $\mathcal{C}$ can be readily approximated by characters on X that, over time, do not revert back to earlier character states and that do not converge on the same state by evolution in different parts of $\mathcal{T}$. In other words, the characters approximating the elements in $\mathcal{C}$ are displayed by $\mathcal{T}$ (see [40] and [41, Section 4] for more on this relationship).

To give an example, consider the tree $\mathcal{T}$ depicted in Fig. 8.2 which is adapted from [30] (see also [35]). Then the morphological character 'having legs' vs. 'having no legs' induces the character {cow, hippo, horse}|{whale} which is clearly displayed by the tree $\mathcal{T}$. Yet, the character {cow, horse}|{hippo, whale} induced by the behavioural character 'nursing offspring under water' vs. 'nursing offspring on land' is not displayed by $\mathcal{T}$. Thus, if $\mathcal{T}$ is the true tree, then the latter character cannot have evolved without homoplasy. It is therefore suggestive to interpret compatibility as the existence of a tree on which the associated characters could have evolved without homoplasy.

8.2.4 *Question (Q) restated*

Using the above introduced terminology, we are now in the position to restate our original Question (Q) as the following pair of questions:

(F1) When is a set of characters compatible?
(F2) If a given set $\mathcal{P}$ of characters on X is compatible, is the X-tree that displays every character in $\mathcal{P}$ unique (up to isomorphism)?

We say that a collection $\mathcal{P}$ of characters on X *defines* an X-tree $\mathcal{T}$ if $\mathcal{T}$ displays $\mathcal{P}$ and, up to isomorphism, $\mathcal{T}$ is the only tree with this property. Then (F2) is asking when a set of characters on X defines an X-tree. If $\mathcal{P}$ defines some X-tree, then $\mathcal{P}$ is said to be *definitive.*

A first result concerning the structure of an X-tree $\mathcal{T}$ that is defined by a set $\mathcal{P}$ of characters on X is that $\mathcal{T}$ must be fully resolved and phylogenetic. The reason is simply that $\mathcal{T}$ could otherwise be resolved to a binary phylogenetic tree $\mathcal{T}'$ that displays every character in $\mathcal{P}$, thereby violating the uniqueness of $\mathcal{T}$. The uniqueness requirement on a displaying X-tree $\mathcal{T}$ in the definition of defining can be relaxed to a possibly biologically more relevant requirement. We

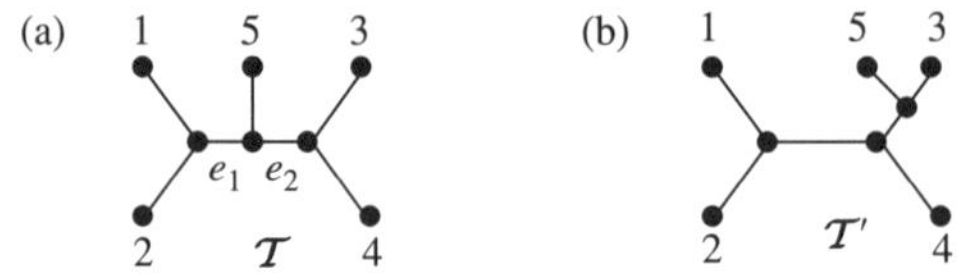

FIG. 8.3. None of the trees depicted in (a) and (b) is defined by the set $\mathcal{P}$ consisting of the characters $12|34$, $12|35$, $12|45$ plus all trivial characters on $X = \{1, 2, \ldots, 5\}$. However, they are both resolutions of a tree that is identified by $\mathcal{P}$ (see text for details). We will return to this example throughout this chapter.

say that $\mathcal{P}$ *identifies* $\mathcal{T}$ if $\mathcal{T}$ displays $\mathcal{P}$ and every X-tree that also displays $\mathcal{P}$ is a refinement of $\mathcal{T}$. Then (F2) asks when a set of characters on X identifies an X-tree.

To clarify the concepts of defining and identifying, consider for example the trees $\mathcal{T}$ and $\mathcal{T}'$ depicted in Fig. 8.3 along with the set $\mathcal{P}$ of characters $P_1 = 12|34$, $P_2 = 12|35$ and $P_3 = 12|45$ plus all *trivial* characters on $X = \{1, 2, \ldots, 5\}$ (i. e. characters of the form $x|X - \{x\}$, for all $x \in X$). Then neither $\mathcal{T}$ nor $\mathcal{T}'$ is defined by $\mathcal{P}$ as both of them display $\mathcal{P}$. However, the X-tree $\mathcal{T}''$ obtained from $\mathcal{T}$ by collapsing the interior edge of $\mathcal{T}$ that is labelled e_2 is identified by $\mathcal{P}$ since the only other X-trees that can display $\mathcal{P}$ are the three resolutions of $\mathcal{T}''$ (two of which are depicted in Fig. 8.3 and the third can be obtained from $\mathcal{T}'$ by swapping the roles of 3 and 4).

8.3 Defining trees in terms of chordal graphs

In this section, we collect together results that characterize compatible and definitive sets of characters. We will meet some of these characterizations again in Section 8.4 where we characterize identifying sets of characters.

We start our discussion by considering a special type of character set called a *split system*. These are collections of characters which are all on the same set X and all have two parts. For consistency, we will follow the common practice and call a character with two parts a *split*.

8.3.1 *Partition intersection graphs and restricted chordal completions*

In [9], Buneman showed that a split system $\mathcal{P}$ is compatible if and only if, for any two distinct splits $S_i = A_i|B_i$ in $\mathcal{P}$, $i = 1, 2$, precisely one of the following four intersections

$$A_1 \cap A_2, A_1 \cap B_2, A_2 \cap B_1, \text{ and } B_1 \cap B_2$$

is empty. Although a powerful result in many ways, it suffers from the fact that the arguments used to establish it do not lend themselves to finding an answer to the *general compatibility problem*: given a set $\mathcal{P}$ of characters, is $\mathcal{P}$ compatible? A natural question to ask in view of the relevance of character compatibility to homoplasy-free evolution pointed out above, and also the role compatibility plays in the context of recombination detection [11].

The general compatibility problem has received a considerable amount of attention in the literature from mathematics [15, 16, 17, 39, 40, 42] and computer science alike [1, 3, 7, 21, 28, 31]. For example, deciding whether a set $\mathcal{P}$ of characters is compatible or not is known to be an NP-complete problem [3, 42]. This means, we can not expect to find an efficient algorithm for deciding if an arbitrary set of characters is compatible. Having said this, the situation changes if either the size of $\mathcal{P}$ or the maximum number of parts in each partition in $\mathcal{P}$ is bounded [1, 28, 31].

It turns out that recasting Buneman's characterization of compatible split systems within a graph-theoretic framework paves the way to answering the general compatibility problem. To present this alternative way of viewing Buneman's characterization we need to introduce some terminology.

Let G be a graph that has no multiple edges and no loops. Then a sequence $P : x_0, x_1, \ldots, x_n$ of distinct but consecutively adjacent vertices is called a *path* in G and n is called the *length* of P. A path $P : x_1, x_2, \ldots, x_n$, $n \geq 3$ together with an edge between x_1 and x_n is called a *cycle* (of length n). The graph G is said to be *chordal* if every cycle in G of length at least four has a *chord*, that is an edge connecting two non-consecutive vertices.

With the definition of a chordal graph in hand, Buneman's result can be recast as follows. A collection $\mathcal{P}$ of splits is compatible precisely if the *partition intersection graph*[1] $Int(\mathcal{P})$ associated to $\mathcal{P}$—i.e., the graph whose vertex set $V(\mathcal{P})$ consists of all those pairs (P, A) with P denoting a partition in $\mathcal{P}$ and A denoting a part in P, and with an edge joining any two vertices (P, A) and (P', A') in $V(\mathcal{P})$ precisely if $A \cap A' \neq \emptyset$—is chordal. Clearly, the definition of the partition intersection graph is independent of whether or not the underlying set $\mathcal{P}$ consists solely of (a) splits or (b) full characters. Consequently, such a graph can also be associated to a set of general characters. To give an example, consider the set $\mathcal{P}$ consisting of the characters $P_1 = 12|45$, $P_2 = 34|61$ and $P_3 = 23|56$. Ignoring the dotted and dashed edges for the moment, the partition intersection graph $Int(\mathcal{P})$ associated to $\mathcal{P}$ is depicted in Fig. 8.4(a) in bold edges.

As can be seen immediately, the graph $Int(\mathcal{P})$ in Fig. 8.4(a) is not chordal as it is a cycle of length 6. However, it can be readily turned into a chordal graph by 'carefully' adding new edges to $Int(\mathcal{P})$. More precisely, only edges can be added to $Int(\mathcal{P})$ for which the first component (i.e. the character) of the resulting incident vertices are distinct. Such a graph is called a *restricted chordal completion* of $Int(\mathcal{P})$ and it should be noted that a partition intersection graph may have more than one. For example, this is the case for the partition intersection graph depicted in Fig. 8.4(a) as it has two distinct restricted chordal completions. Using again Fig. 8.4(a), they both comprise of all solid edges (as

[1] In keeping with the literature, we will use the term 'partition intersection graph'. However we remark that, in view of the remark at the end of Section 8.2.2, the name 'character intersection graph' would be more appropriate.

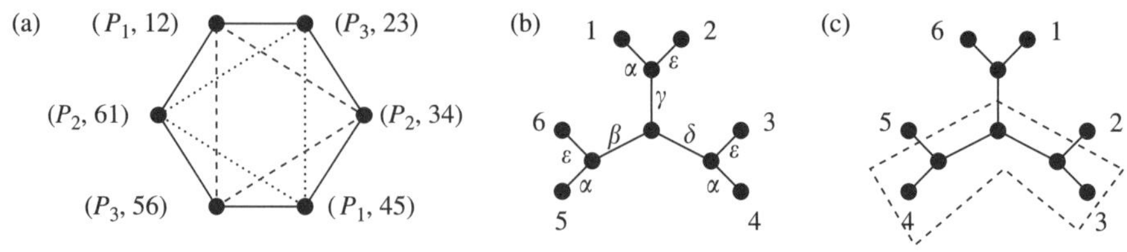

FIG. 8.4. (a) In bold edges, the partition intersection graph associated to the set $\mathcal{P}$ of characters $P_1 = 12|45$, $P_2 = 34|61$ and $P_3 = 23|56$ is presented. The edges in bold plus either all dashed or all dotted edges form a restricted chordal completion of that graph. The trees in (b) and (c) are two distinct X-trees that both display $\mathcal{P}$. The purpose of the edge labels in (b) and the dashed closed line in (c) will become clear in Sections 8.3.2 and 8.5.1, respectively, when we will return to this figure.

they are the edges of $Int(\mathcal{P})$) plus either all dashed or all dotted edges. Note that the graph containing all solid, dashed, and dotted edges is not chordal since the four vertices with P_1 or P_2 in their first component induce a four-cycle without a chord.

In general, it is unclear whether a partition intersection graph under consideration has a restricted chordal completion or not, let alone how to find one if one exists. Intrigued by this, Grünewald and Huber investigated the relationship between the *relation graph* $\mathcal{G}_{\mathcal{P}}$ associated to a set $\mathcal{P}$ of (full) characters and the partition intersection graph associated to $\mathcal{P}$ in [18]. Originally introduced in [23], the relation graph can be considered a canonical generalization of a *median network* (sometimes called a *Buneman graph*) to sets of partitions (see [24] for a recent overview on median networks). Under the assumption that $\mathcal{G}_{\mathcal{P}}$ is connected, they showed that $Int(\mathcal{P})$ does indeed have a restricted chordal completion and gave a construction how this completion can be obtained from $\mathcal{G}_{\mathcal{P}}$ (see Section 8.5.1 for a further construction for obtaining such a chordalization.

Using the idea of a restricted chordal completion of the partition intersection graph associated to a set of characters of X, Steel answered Question (F1) in [42] by showing that a set $\mathcal{P}$ of characters is compatible if and only if there exists a restricted chordal completion of $Int(\mathcal{P})$; a result already indicated in [10] and [32]. It should be noted, however, that this result does not automatically also answer Question (F2) as it only guarantees the existence of an X-tree that displays $\mathcal{P}$ but not its uniqueness (which is the concern of (F2)). For example, consider the set $\mathcal{P}$ of characters whose partition intersection graph is depicted in Fig. 8.4(a). Then, as was observed before, this graph has a restricted chordal completion. Thus, by Steel's characterization, an X-tree must exist that displays $\mathcal{P}$. However, this X-tree is not unique as is demonstrated by the two X-trees depicted in Fig. 8.4. We will return to the X-tree depicted in Fig. 8.4(b) in the next section when the edge labels will become important.

8.3.2 *Minimal restricted chordal completions and distinguishing edges*

To obtain the desired characterization of definitive sets of characters, we need two additional concepts which we introduce next.

Suppose $\mathcal{P}$ is a set of characters of a set X. Then a restricted chordal completion G of $Int(\mathcal{P})$ is called *minimal* if, for every non-empty subset F of edges in G but not in $Int(\mathcal{P})$, the graph G with the edges in F deleted is not chordal. To exemplify this concept, consider the set $\mathcal{P}$ of characters on $X = \{1, 2, 3, 4\}$ consisting of $P_1 = 12|4$, $P_2 = 23|1$, $P_3 = 2|34$, and $P_4 = 14|3$. Then three restricted chordal completions of $Int(\mathcal{P})$ are depicted in Fig. 8.5 in terms of graphs using all bold edges and one of the following:

- the dashed edge, or
- the dotted edge, or
- the dashed edge and the dotted edge.

Since from the last graph we can remove either the dotted or the dashed edge and still have a chordal graph, it is not a minimal restricted chordal completion. However, the other two restricted chordal completions of $Int(\mathcal{P})$ are clearly minimal.

To initiate the second new concept, consider the set $\mathcal{P}$ of characters $P_1 = 12|34$, $P_2 = 12|35$, and $P_3 = 12|45$ of $X = \{1, 2, 3, 4, 5\}$. Then the X-tree $\mathcal{T}$ in Fig. 8.3(a) displays P_1. The deletion of any one of the two interior edges of $\mathcal{T}$ results in two subtrees $\mathcal{T}_1$ and $\mathcal{T}_2$ so that, when ignoring the leaf labelled by 5, the leaf sets of $\mathcal{T}_1$ and $\mathcal{T}_2$ form P_1. In other words no particular interior edge in $\mathcal{T}$ is distinguished by P_1 with respect to being required for $\mathcal{T}$ to display P_1. Turning the argument around, this means that an X-tree $\mathcal{T}$ can only be defined by a set $\mathcal{P}$ of characters if every edge of $\mathcal{T}$ is, in this sense, required by some character in $\mathcal{P}$. Bearing this in mind, we say that an edge e in an X-tree $\mathcal{T}$ is *distinguished* by a character P if e is contained in every set of edges that can be deleted from $\mathcal{T}$ to display P. In addition, we say that $\mathcal{T}$ is distinguished by a set $\mathcal{P}$ of characters if every edge of $\mathcal{T}$ is distinguished by an element in $\mathcal{P}$. Note that, in Fig. 8.3(a), the edge of $\mathcal{T}$ labelled by e_1 is distinguished by $12|45$.

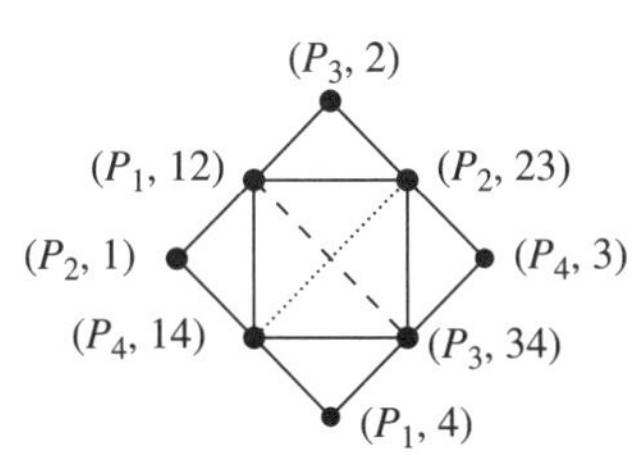

FIG. 8.5. A partition intersection graph plus its 2 minimal restricted chordal completions (see text for details).

As it turns out, the concepts of a minimal restricted chordal completion and distinguishing an edge do not suffice to characterize definitive sets of characters. However, things change when the minimal restricted chordal completion in question is unique as the following result from [39] shows.

Theorem 8.1 *Let $\mathcal{P}$ be a collection of characters of X. Then $\mathcal{P}$ defines an X-tree if and only if the following conditions hold:*

(i) there is a binary phylogenetic tree that displays $\mathcal{P}$ and is distinguished by $\mathcal{P}$; and
(ii) there is a unique minimal restricted chordal completion of $Int(\mathcal{P})$.

Moreover, if $\mathcal{T}$ is the unique X-tree displaying $\mathcal{P}$, then $\mathcal{T}$ satisfies the properties in (i).

An intriguing consequence of this result is that at most five carefully chosen characters suffice to completely determine a binary phylogenetic tree [40]. (See Section 8.4.3 for a closely related result.)

An example of a set $\mathcal{C}$ of characters on X (or more precisely the set of partitions of X induced by $\mathcal{C}$) that defines a binary phylogenetic tree $\mathcal{T}$ is provided by a set of full characters that are well-separated on $\mathcal{T}$ (c.f. [22]). Reflecting the idea that characters rarely change their state and so changes are well spread out in a tree, a character α is called *well-separated* on a phylogenetic tree $\mathcal{T}$ if for every path $a_0, a_1, \ldots, a_{n-1}, a_n$ in $\mathcal{T}$ with $n \geq 2$ and $\{a_0, a_1\}$ and $\{a_{n-1}, a_n\}$ edges in $\mathcal{T}$ on which α changes its state, the length of the sub-path $a_1, \ldots, a_{n-1}$ is at least two. To give an example, consider the phylogenetic tree $\mathcal{T}$ depicted in Fig. 8.4(b) together with the characters $\alpha, \beta, \gamma, \delta$, and ϵ whose only state changes occur on those edges of $\mathcal{T}$ that are labelled by their character name. Then each one of $\alpha, \beta, \gamma, \delta$, and ϵ is well-separated on $\mathcal{T}$.

Now if $\mathcal{C}$ is a set of full characters on X and $\mathcal{T}$ is a binary phylogenetic tree such that every character in $\mathcal{C}$ is well-separated on $\mathcal{T}$ and every edge of $\mathcal{T}$ corresponds to one character changing its state on that edge, then $\mathcal{C}$ defines $\mathcal{T}$ [22]. Interestingly, the relation graph associated to $\mathcal{C}$ as well as various other approaches (see, for example, [1, 21, 28, 31]) recover $\mathcal{T}$ in polynomial time. Returning to the previous example it follows that $\{\alpha, \beta, \gamma, \delta, \epsilon\}$ defines the tree in Fig. 8.4(b).

8.4 Defining trees in terms of closure rules

Phylogenetic trees can be thought of as a summary of information from small evolutionary building blocks called *quartets*. Being phylogenetic trees themselves, quartets have the special property that they have only four leaves and are fully resolved. In this section we will review recent results that elucidate the relationship between phylogenetic trees and sets of quartets. Motivated by the fact that any phylogenetic tree $\mathcal{T}$ gives rise to the set $\mathcal{Q}(\mathcal{T})$ of all quartets that are displayed by $\mathcal{T}$ the question we are most interested in is, when can such trees be uniquely recovered from quartet sets. To make this more precise, we start by describing a basic relationship between quartets and partial characters with two parts which are also called *partial splits*.

Suppose q is a quartet with leaf set $X = \{a, b, c, d\}$ where a and b are adjacent to the same interior vertex of q. Then deleting the interior edge of q clearly results in the split $ab|cd$ of X. Conversely, every split $ab|cd$ of X can be represented by a quartet q in which a and b are adjacent to the same interior vertex of q. Two consequences of this alternative interpretation of quartets are important. Firstly, we can extend our notation for characters to quartets. Secondly, it provides us with a way to directly extend fundamental concepts introduced for characters to quartets and thus to phylogenetic trees; important examples of which are displaying and compatibility. However, caution is required regarding the crucial concepts of defining and identifying X-trees. The reason for this is that a tree $\mathcal{T}$ can have an interior vertex v that is labelled by an element of X and both $\mathcal{T}$ and the X-tree obtained from $\mathcal{T}$ by pushing the label of v out to a leaf by adding a pendant edge to $\mathcal{T}$ display the same set of quartets. Bearing in mind that phylogenetic trees are a special type of X-tree and that for such trees the situation described above cannot occur, we adapt the definition of defining as follows: a quartet set $\mathcal{Q}$ *defines* a phylogenetic tree $\mathcal{T}$ if $\mathcal{T}$ displays $\mathcal{Q}$ and, up to isomorphism, $\mathcal{T}$ is the only phylogenetic tree with this property. If $\mathcal{Q}$ defines a phylogenetic tree, then we also call $\mathcal{Q}$ *definitive*. Similarly, we say that a quartet set $\mathcal{Q}$ *identifies* a phylogenetic tree $\mathcal{T}$ if $\mathcal{T}$ displays $\mathcal{Q}$ and every phylogenetic tree that also displays $\mathcal{Q}$ is a refinement of $\mathcal{T}$. It should be noted that the concepts of defining/identifying in terms of quartet sets and characters only differ by replacing 'X-tree' with 'phylogenetic tree'.

To elucidate these new concepts consider the phylogenetic tree $\mathcal{T}$ depicted in Fig. 8.4(b). Then $12|34$ is a quartet that is displayed by $\mathcal{T}$ since deleting the edge marked γ gives rise to the split $12|3456$ and $1, 2 \in \{1, 2\}$ and $3, 4 \in \{3, 4, 5, 6\}$. The set $\mathcal{Q} = \{12|45, 34|16, 23|56\}$ is compatible since $\mathcal{T}$ displays every quartet in $\mathcal{Q}$. However, $\mathcal{Q}$ does not define $\mathcal{T}$ since the phylogenetic tree depicted in Fig. 8.4(c) also displays every quartet in $\mathcal{Q}$. Reassuringly, every binary phylogenetic tree $\mathcal{T}$ is defined by the set $\mathcal{Q}(\mathcal{T})$ of quartets it displays [12].

We are now ready to effortlessly rephrase the questions (F1) and (F2) for the quartet framework we have been developing. Their analogues (F1') and (F2') are (F1) and (F2) with the words 'set of characters' replaced with 'quartet set' and 'X-tree' replaced with 'phylogenetic tree'.

Regarding (F2'), Theorem 8.1 almost effortlessly implies a graph-theoretical characterization of those sets of quartets (or, more generally, sets of phylogenetic trees) that define a phylogenetic tree (for details see [41, Section 6.8]). However, verifying the two conditions that make up this characterization can be very difficult for some instances, which suggests that this characterization might not lend itself as a basis for an efficient algorithm to test for defining. As it turns out, the key to efficiently checking in some cases whether a quartet set defines a phylogenetic tree is held by the notion of a quartet closure rule.

8.4.1 *Quartet closure rules*

Building on the work of Colonius and Schulze in [12] which was carried out in the context of psychology, Dekker [13] investigated rules for pasting together quartets into an overall parent tree (or supertree) that displays all the original quartets. Before we can state two of these rules which we denote by (D1) and (D2), we need to introduce some notation. Suppose $\mathcal{Q}$ is a quartet set and q is a quartet. Then we write $Q \vdash q$ if any phylogenetic tree that displays Q also displays q and call the statement $Q \vdash q$ a *quartet (closure) rule*. Dekker's rules (D1) and (D2) can then be stated as

(D1) $\{ab|cd, ab|ce\} \vdash ab|de$, and
(D2) $\{ab|cd, ac|de\} \vdash ab|ce, ab|de, bc|de$.

The rationale behind these rules is that any phylogenetic tree that displays $ab|cd$ and $ab|ce$ also displays $ab|de$, and that any phylogenetic tree that displays $ab|cd$ and $ac|de$ also displays $ab|ce$, $ab|de$, and $bc|de$.

Since for any two quartets, application of either (D1) or (D2) generates a new quartet, the question arises as to what happens if we keep applying both or one of (D1) and (D2) to the elements of a quartet set. As it turns out, for any quartet set $\mathcal{Q}$ and any one of the quartet rules (D1) or (D2) or their combination, there always exists a (unique) minimal quartet set $\mathcal{M}_{\mathcal{Q}}$ that contains $\mathcal{Q}$ and cannot be extended any further using the quartet rule(s) that one chose to apply to the elements of $\mathcal{Q}$. We will call $\mathcal{M}_{\mathcal{Q}}$ the *dyadic* closure of $\mathcal{Q}$ if both (D1) and (D2) are applied and denote it by $\mathrm{qcl}(\mathcal{Q})$. In case solely (D2) is being used, we will call $\mathcal{M}_{\mathcal{Q}}$ the *semi-dyadic* closure of $\mathcal{Q}$ and denote it by $\mathrm{qcl}_2(\mathcal{Q})$. If the type of closure for $\mathcal{Q}$ is of no relevance, we simply talk about the *quartet closure* of $\mathcal{Q}$.

Before we continue with our discussion of Dekker's rules (D1) and (D2) we pause to clarify these concepts. Consider, for example, the quartet set $\mathcal{Q} = \{12|45, 24|56, 25|34\}$. Then (D2) applied to $12|45$ and $24|56$ generates the three quartets $12|56$, $12|46$ and $14|56$. The semi-dyadic closure of $\mathcal{Q}$ consists of all 15 quartets displayed by the phylogenetic tree $\mathcal{T}$ depicted in Fig 8.4(b). It can be obtained by applying (D2) to the quartets $12|45$ and $24|56$, the quartets $12|45$ and $25|34$, and to the quartets $24|56$ and $25|34$. Note that (D1) cannot be directly applied to any two quartets in $\mathcal{Q}$. However, (D1) can be applied to $24|56$ and $12|56$ which yields $14|56$. Since we have $\mathrm{qcl}_2(\mathcal{Q}) \subseteq \mathrm{qcl}(\mathcal{Q}) \subseteq \mathcal{Q}(\mathcal{T}')$ for every phylogenetic tree $\mathcal{T}'$ that displays $\mathcal{Q}$, it follows that, for this example, $\mathrm{qcl}_2(\mathcal{Q}) = \mathrm{qcl}(\mathcal{Q})$ which, in turn, equals $\mathcal{Q}(\mathcal{T})$.

The interest in quartet closure rules for phylogenetics has recently increased considerably. One reason for this is that the dyadic closure of a quartet set can be constructed in polynomial-time. Also recent results have shed light on the problem of when a quartet closure rule reconstructs a phylogenetic tree [14, 26, 34] and the relationship between Dekker's rules (D1) and (D2) and Meacham's rules for partial splits [14, 25, 40] (which we will take up in the next section). Before we turn our attention to reviewing some of the results about definitive sets of quartets we will briefly look at Question (F1') with regards to quartet closure rules.

In general, deciding whether a quartet set $\mathcal{Q}$ is compatible or not is NP-complete [42]. Consequently, we cannot expect to find a polynomial time algorithm for deciding this problem. However, in practice, the availability of rules such as (D1) and (D2) can make it possible to determine efficiently if a quartet set is compatible since these rules often produce a conflicting pair of quartets (which implies that $\mathcal{Q}$ is not compatible), or allow one to construct a phylogenetic X-tree that displays $\mathcal{Q}$. For example, if $\mathcal{Q}$ contains k quartets and n is the number of distinct leaf labels in $\mathcal{Q}$ then Rule (D1) can be used to obtain an algorithm that can decide in $\mathcal{O}(nk2^k)$ time whether $\mathcal{Q}$ is compatible or not [41, Proposition 6.7.3]. In other words, for small sets $\mathcal{Q}$ this algorithm is not too bad. Note that for the above to hold the assumption on the size of $\mathcal{Q}$ is crucial since, as was recently established in [20], quartet rules do not suffice to detect conflicts in quartet sets. In other words, there exist quartet sets $\mathcal{Q}$ which are not compatible but every proper subset of $\mathcal{Q}$ is compatible and no quartet closure rule can be applied to a subset of $\mathcal{Q}$ to obtain further quartets.

We conclude our brief review of recent results concerning (F1') by noting that in [19] a new graph-theoretical characterization of quartet set compatibility is given which is based on so-called quartet graphs (see Section 8.6 for more).

We now turn our attention towards reviewing some of the results regarding (F2'). To put things into context, we start with a result that appeared in [6]. To be able to explain that result, we need some more terminology. Motivated by the fact that any phylogenetic tree that is defined by a set $\mathcal{Q}$ of quartets must be fully resolved (like in the case for definitive sets of characters) and $|\mathcal{Q}| - (|X| - 3) \geq 0$, Böcker and Dress studied quartet sets for which the above inequality is an equality. Loosely speaking, such quartet sets (which they called *excess-free*) contain the minimum amount of information required to possibly recover a phylogenetic tree. A consequence of their work on so-called *patchworks* [4, 5] is the following result on excess-free quartet sets which was established in more general form in [6].

Theorem 8.2 *[14] If a quartet set $\mathcal{Q}$ is compatible and contains an excess-free subset which defines a phylogenetic tree $\mathcal{T}$, then* $\mathrm{qcl}_2(\mathcal{Q}) = \mathcal{Q}(\mathcal{T})$.

An important consequence of this theorem is that it leads to a polynomial time algorithm which, for a quartet set $\mathcal{Q}$ which contains sufficient information (in the form of an excess-free subset that defines a phylogenetic tree), constructs either the unique tree that displays $\mathcal{Q}$ or returns the statement that $\mathcal{Q}$ is not compatible. However, it should be noted that the theorem does not help to decide the compatibility of quartet sets that do not contain such sufficiently informative subsets. Furthermore, as was shown in [6], the problem of deciding whether $\mathcal{Q}$ contains a definitive excess-free subset belongs to the class of NP-complete problems and therefore can not be expected to be solved efficiently.

It is natural to ask about the converse to Theorem 8.2, i.e. if $\mathcal{T}$ is a phylogenetic tree and $\mathcal{Q} \subset \mathcal{Q}(\mathcal{T})$ a quartet set so that $\mathrm{qcl}_2(\mathcal{Q}) = \mathcal{Q}(\mathcal{T})$, does $\mathcal{Q}$

contain an excess-free subset that defines $\mathcal{T}$? In general, the answer is 'no' as was recently established by Huber *et al.* in [25]. In other words, even for quartet sets whose semi-dyadic closure is the quartet set of a phylogenetic tree $\mathcal{T}$ on X we cannot expect to find $|X| - 3$ quartets that will allow us to recover $\mathcal{T}$.

We conclude this section by noting that Theorem 8.2 was recently complemented in [14] by establishing that any fully resolved phylogenetic tree $\mathcal{T}$ can be reconstructed from any sufficiently 'rich' subset of $\mathcal{Q}(\mathcal{T})$ by just repeatedly applying (D1). Such 'rich' subsets were originally introduced in [34] where it was shown that they are definitive.

8.4.2 *Split closure rules*

We start this section by returning to the observation made above that any quartet may be viewed as a split of a set on four elements into two subsets each containing two elements (and vice versa). The question we are most interested in is how the quartet closure of a quartet set $\mathcal{Q}$ compares to the so-called *split closure* of $\mathcal{Q}$. Originally formalized in [38] for sets $\mathcal{S}$ of *partial X-splits*, that is partial splits of X, the split closure of $\mathcal{S}$ relies on two rules—we will refer to them as (M1) and (M2)—which were proposed by Meacham [32]. Extending our notation of a quartet closure rule $\mathcal{Q} \vdash q$ to the setting of partial X-splits by replacing $\mathcal{Q}$ by a set of partial X-splits and q by a partial X-split, and letting $S_1 = A_1|B_1$ and $S_2 = A_2|B_2$ denote two partial X-splits, the rules (M1) and (M2) can be stated as follows:

(M1) If $A_1 \cap A_2 \neq \emptyset$ and $B_1 \cap B_2 \neq \emptyset$, then $\{S_1, S_2\} \vdash (A_1 \cap A_2)|(B_1 \cup B_2)$ and $(A_1 \cup A_2)|(B_1 \cap B_2)$, and

(M2) If none of $A_1 \cap A_2$, $A_1 \cap B_2$, $B_1 \cap B_2$ is empty but $B_1 \cap A_2 = \emptyset$, then $\{S_1, S_2\} \vdash (A_1 \cup A_2)|B_1$ and $A_2|(B_1 \cup B_2)$.

The rationale behind Meacham's rules (M1) and (M2) is similar to the one for Dekker's rules (D1) and (D2). Any phylogenetic tree $\mathcal{T}$ on X that displays two partial X-splits $S_1 = A_1|B_1$ and $S_2 = A_2|B_2$ that satisfy the pre-requisites in (M1) must also display the partial X-splits to the right of the '$\vdash$'-symbol in (M1). And if they satisfy the pre-requisite in (M2), then $\mathcal{T}$ must also display the two partial splits to the the right of '$\vdash$' in (M2).

The most striking difference between (D1) and (D2), and Meacham's rules lies in the object they generate. The former two rules enlarge a quartet set $\mathcal{Q}$ by adding new quartets to $\mathcal{Q}$ (in case the newly generated quartets are not already contained in $\mathcal{Q}$). In contrast, Meacham's rule (M2) extends the partial X-splits to which it is applied. More precisely, the two partial X-splits to the left of the '$\vdash$' symbol in (M2) can be obtained from the two partial X-splits to the right of '$\vdash$' by simply removing certain elements of X. In general, things are not as straightforward with Rule (M1) since it tends to generate new partial X-splits rather than extend given ones. For example, for the quartets $12|45$, $24|56$ (thought of as partial splits of $X = \{1, 2, \ldots, 6\}$) (M1) generates $2|456$ and $5|124$, whereas (M2) generates $12|456$ and $124|56$. Note that all four generated partial X-splits are displayed by the phylogenetic tree in Fig. 8.4(b).

Since both of Meacham's rules generate partial X-splits we may repeatedly apply either one of his rules (or their combination) to a set $\mathcal{S}$ of partial X-splits until no further partial splits can be generated. Assuming that the partial splits in $\mathcal{S}$ are compatible, the nature of (M2) implies that the resulting (unique) set $\mathcal{M}_{\mathcal{S}}$ of partial X-splits is likely to contain redundant phylogenetic information. By this we mean that if the splits S and S' are both contained in $\mathcal{M}_{\mathcal{S}}$ and S' displays S, then the phylogenetic information conveyed by S is also conveyed by S'. Consequently S can be removed from $\mathcal{M}_{\mathcal{S}}$ without losing information. The (unique) set of partial X-splits thus obtained is called the *split closure* of $\mathcal{S}$ and if only Rule (M2) was used to generate it, it is denoted by $\mathrm{spcl}(\mathcal{S})$. It should be noted that the split closure heavily depends on which rule(s) were used to generate it.

To give an example, consider the quartet set $\mathcal{Q} = \{12|45, 24|56, 25|34\}$ (thought of as a set of partial splits on $X = \{1, 2, \ldots, 6\}$). Then the split closure of $\mathcal{Q}$ obtained by exclusively applying Rule (M1) consists of $\mathcal{Q}$ and the partial splits $2|3456$, $4|1256$, and $5|1234$, whereas the split closure $\mathrm{spcl}(\mathcal{Q})$ of $\mathcal{Q}$ using (M2) consists of the three splits $12|3456$, $34|1256$ and $56|1234$ and the split closure of $\mathcal{Q}$ using both (M1) and (M2) consists of all non-trivial splits displayed by the phylogenetic tree $\mathcal{T}$ depicted in Fig. 8.4(b).

Although much could be said about these three split closures (see [14] and [20] for recent results), we will focus for the remainder of this section on the interplay between the semi-dyadic closure of $\mathcal{Q}$ and the split closure of $\mathcal{Q}$ via (M2). To keep terminology simple, in the following we will refer to the set $\mathrm{spcl}(\mathcal{Q})$ as the split closure of $\mathcal{Q}$.

One of the first things to notice about the last example is that the semi-dyadic closure of $\mathcal{Q} = \{12|45, 24|56, 25|34\}$ equals the set of all quartets displayed by the (binary) phylogenetic tree $\mathcal{T}$ depicted in Fig. 8.4(b), whereas the split closure of $\mathcal{Q}$ consists of all non-trivial splits of X that are displayed by $\mathcal{T}$. The intriguing question as to whether this is always the case suggests itself: given a binary phylogenetic tree $\mathcal{T}$ and any subset $\mathcal{Q} \subseteq \mathcal{Q}(\mathcal{T})$, is it always true that $\mathrm{qcl}_2(\mathcal{Q}) = \mathcal{Q}(\mathcal{T})$ precisely if $\mathrm{spcl}(\mathcal{Q})$ equals the set $\mathcal{S}(\mathcal{T})$ of all non-trivial splits of X displayed by $\mathcal{T}$? If true, this would allow one to choose freely between either reconstructing the corresponding binary phylogenetic tree via the split closure of a compatible quartet set $\mathcal{Q}$ or via the semi-dyadic closure of $\mathcal{Q}$. Although it is known that both closures can be computed efficiently, intuitively, the split closure of $\mathcal{Q}$ seems to be easier to find. It turns out that if $\mathcal{T}$ is a binary phylogenetic tree and $\mathrm{qcl}_2(\mathcal{Q}) = \mathcal{Q}(\mathcal{T})$ for some quartet set $\mathcal{Q} \subseteq \mathcal{Q}(\mathcal{T})$, then, indeed, $\mathrm{spcl}(\mathcal{Q}) = \mathcal{S}(\mathcal{T})$ [25]. However, somewhat surprisingly, the converse need not hold. In other words, we may have $\mathrm{spcl}(\mathcal{Q}) = \mathcal{S}(\mathcal{T})$ but $\mathrm{qcl}_2(\mathcal{Q}) \neq \mathcal{Q}(\mathcal{T})$ [25]. Loosely speaking this means that, in general, Dekker's quartet closure rules (D1) and (D2) infer less information from a quartet set $\mathcal{Q}$ for reconstructing a binary phylogenetic tree than Meacham's closure rules (M1) and (M2).

We conclude this section with remarking that Meacham's rule, in the form of the Z-closure rule, has recently been employed to construct phylogenetic supernetworks (see [27] for details).

8.4.3 *The semi-dyadic closure and homoplasy-free evolution*

As indicated above, the amount of homoplasy in some genomic characters tends to be low. Assuming that character evolution is homoplasy-free, that is, the amount of homoplasy is zero, it is therefore an interesting question to ask how many (quantitative) characters one would need to recover the underlying 'true' phylogenetic tree. Perhaps unsurprisingly, binary phylogenetic trees exist that cannot be defined by just three characters. An example of such a tree is provided by the tree depicted in Fig. 8.4(b) where each leaf is replaced by a pair of new leafs giving rise to a phylogenetic tree on 12 leaves [40]. What is surprising is that, by combining Theorem 8.1 with a certain character construction mechanism that is based on a $\mathbb{Z}_5$-edge colouring of the edge set of a binary phylogenetic tree, Semple and Steel established in [40] that at most five characters suffice. However, their arguments did not lend themselves to settling the tantalizing question of whether just four characters would be sufficient. In [26], this question was affirmatively resolved. Intriguingly, the key to settling it is held by the semi-dyadic closure of a certain carefully chosen set of quartets. The remainder of this section is devoted to explaining how this quartet set can be constructed and gives an indication on how it was finally utilized to settle the question. We will follow the approach of [26].

Suppose $\mathcal{T}$ is a binary phylogenetic tree. As a convenience for the forthcoming construction, we consider $\mathcal{T}$ to be a rooted tree by choosing any leaf r of $\mathcal{T}$ to be the root. Furthermore, we regard $\mathcal{T}$ as a rooted directed tree in which all edges are directed away from r. To simplify the explanation on how to generate the quartet set in question, assume that $\mathcal{T}$ is embedded into the plane so that we can distinguish between a left and a right child of an interior vertex of $\mathcal{T}$. For example, the rooted (and directed) phylogenetic tree on $X = \{1, 2, \ldots, 14, r\}$ depicted in Fig. 8.6 is one of many such embeddings of the unrooted version of that tree.

The definition of the quartet set in question relies heavily on a colouring of the edges of $\mathcal{T}$ so that no two incident edges of $\mathcal{T}$ have the same colour. We describe this edge colouring next. Suppose the four colours are R, L, R', and L'.

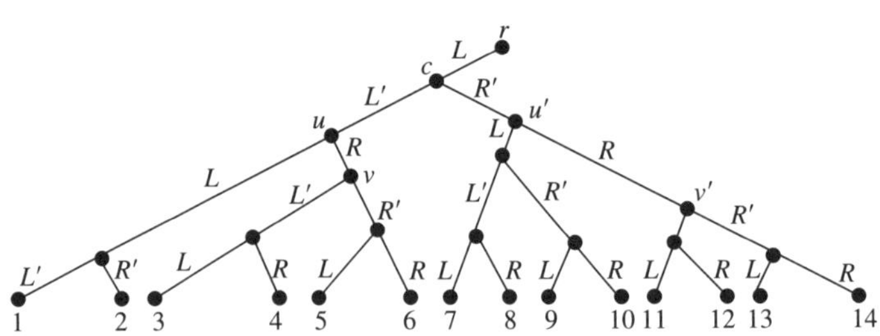

FIG. 8.6. A colouring of the edges of a binary phylogenetic tree that proved crucial for establishing that any binary phylogenetic tree can be defined by at most four characters.

Since $\mathcal{T}$ is binary, r either has a child to the left or to the right. Assume without loss of generality that r has a child c to the left (as is the case in Fig. 8.6). Then we arbitrarily colour the outgoing edge of r with either L or L'. Suppose we have coloured it L. If c is a leaf, we stop since we have coloured all edges of $\mathcal{T}$. Suppose c is not a leaf. Then, c has two children and we colour the edge incident with the left child of c by L' and the edge incident with the right child of c by R'. We continue this colouring process until we have coloured all edges of $\mathcal{T}$ always making sure that if for an interior vertex the incoming edge is coloured with the primed version of R or L, we continue with the non-primed version and vice versa. Obviously, deleting all edges coloured with the same colour results in a character of X. For example, deleting all edges coloured L' in Fig. 8.6 results in the character $\{1\}|\{3,4\}|\{2,5,6\}|\{7,8\}|\{11,12\}|\{9,10,13,14,r\}$.

Apart from giving rise to a set $\mathcal{P}$ of (at most) four characters all of which are obviously displayed by the generating tree $\mathcal{T}$, this edge colouring has a further crucial property. Namely, it allows one to capture the structure of the underlying tree $\mathcal{T}$ in terms of a quartet set $\mathcal{Q}_{\mathcal{T}}$ whose elements have the additional property that they are displayed by the characters in $\mathcal{P}$. To see how the quartets in $\mathcal{Q}_{\mathcal{T}}$ are constructed, assume that e is an interior edge of $\mathcal{T}$ coloured by R (we will consider the cases where e is coloured by L, R', or L' below) and that u is the start vertex of e and v is the end vertex of e. Then the incoming edge of u is coloured either by (i) L' or (ii) R'. In Case (i), we associate a quartet $st|xy$ to e as follows:

- s is the last vertex in the directed path that starts at v and has its first edge coloured R' and all subsequent edges coloured alternately by L and L';
- t is the last vertex of the directed path that starts at v and has edges coloured alternately by L' and L;
- x is the last vertex of the directed path that starts at u and has edges coloured alternately by L and L';
- y is the last vertex of the undirected path that starts at u, has its first two edges coloured L' and R', respectively, and all subsequent edges coloured alternately by L and L'.

For example if u and v are as in Fig. 8.6, then s is the leaf labelled 5, t is the leaf labelled 3, x is the leaf labelled 1 and, finally, y is the leaf labelled 7. In Case (ii) t, s, x are all obtained in the same way and y is the last vertex of the undirected path that starts at u and has its first edge coloured R' and all subsequent edges coloured alternately by L' and L. For example if e is the edge with start vertex u' and end vertex v' in Fig. 8.6, then s is the leaf labelled 13, t is the leaf labelled 11, x is the leaf labelled 7 and, finally, y is the leaf labelled by 1.

If the edge e is labelled by R' and starts at u and ends at v, the quartet $st|xy$ is obtained in a similar way, by following the four distinct paths whose first vertices are either u or v and whose last edges are alternately coloured using only the colours L and L'. In case e is labelled by either L or L' and again starts at u and ends at v a similar procedure is followed in which colours L and R and L' and R' are interchanged so that, in particular, the quartet $st|xy$ is obtained by following the four distinct paths whose first vertices are either u or v, and whose last edges are alternately coloured using only the colours R and R'.

This construction combined with an inductive argument on the leaf set size of $\mathcal{T}$ yields $\mathrm{qcl}_2(\mathcal{Q}_{\mathcal{T}}) = \mathcal{Q}(\mathcal{T})$ which implies the following result which appeared in slightly different form in [26].

Theorem 8.3 *Every binary phylogenetic tree can be defined by (at most) four characters.*

We mention in passing that, not surprisingly, the question of how many characters suffice to define a binary phylogenetic tree has also been looked at within a probabilistic framework. Under the assumption of a certain biologically relevant Markov model it turns out that about $\log |X|$ characters suffice in that setting (see [34] for details).

As already indicated in [41] for the five character result, a possible application of the four character result lies in the area of supertree construction which is concerned with devising methods for producing an overall parent tree for a set of input trees. A popular approach within this field is MRP (matrix representation using parsimony) [37]. However, there are concerns about MRP being biased towards large input trees due to encoding the edges of an input tree in terms of splits. A possible solution might be to employ an encoding of the input trees using a fixed number of multi-state characters (characters with two or more parts).

8.5 Identifying trees in terms of chordal graphs

So far, we have mostly been concerned with the problem of when a set $\mathcal{P}$ of characters *defines* an X-tree. In this section, we turn our attention to the biologically more relevant question of when $\mathcal{P}$ *identifies* an X-tree. The difference is that if $\mathcal{P}$ defines an X-tree $\mathcal{T}$, then $\mathcal{T}$ must necessarily be binary and phylogenetic, whereas if $\mathcal{P}$ identifies $\mathcal{T}$, then $\mathcal{T}$ need not be phylogenetic and may have unresolved vertices. The close relationship between both concepts is maybe best exemplified by the following observation. If a set $\mathcal{P}$ of characters defines an X-tree $\mathcal{T}$, then $\mathcal{T}$ is also identified by $\mathcal{P}$ and every X-tree that is identified by $\mathcal{P}$ and is binary and phylogenetic is also defined by $\mathcal{P}$.

The impression might have arisen that because of the similarity of the concepts of defining an X-tree and identifying an X-tree, a characterization of sets of characters that identify an X-tree might be the same as the one for definitive sets of characters given in Theorem 8.1 (with 'defines' replaced by 'identifies' and the word 'binary' removed). However, things are a bit more difficult. For example, consider the set $\mathcal{P}$ consisting of only the character $a|b|c$ on $X = \{a, b, c\}$. Then $\mathcal{P}$ does not identify the X-tree with edges $\{a, b\}$ and $\{b, c\}$ since the X-tree with edges $\{a, c\}$ and $\{c, b\}$ also displays $\mathcal{P}$ but neither tree is a resolution of the other. However, as required by Theorem 8.1 (adapted for identifying as outlined above) the partition intersection graph $Int(\mathcal{P})$ associated with $\mathcal{P}$ has a unique minimal restricted chordal completion (it consists of three isolated vertices and therefore is its own minimal restricted chordal completion) and the phylogenetic tree with leaf set $\{a, b, c\}$ is distinguished by $\mathcal{P}$.

8.5.1 *Restricted chordal completions revisited*

As we have seen, the concept of a minimal restricted chordal completion of the partition intersection graph associated with a set of characters is crucial for characterizing definitive sets of characters. However, we have not yet given an easy to perform construction that allows one to find such a completion (in the case that it exists!). We start this section with rectifying this as, similar to the case of definitive sets of characters, such objects lie at the heart of the sought-after characterization of identifying sets of characters.

Suppose $\mathcal{T}$ is an X-tree and $\mathcal{P}$ is a set of characters. Then it is reasonable to assume that the way the elements of the parts of a character in $\mathcal{P}$ are 'spread over' $\mathcal{T}$ will provide some information on the structure of $\mathcal{T}$. The graph theoretical tool that allows one to describe this spread is called the *subtree intersection graph* $Int(\mathcal{P}, \mathcal{T})$ associated to $\mathcal{T}$ and $\mathcal{P}$. It is formally defined in the following way. The vertices of $Int(\mathcal{P}, \mathcal{T})$ are precisely the vertices of $Int(\mathcal{P})$ and any two vertices (P, A) and (P', A') of $Int(\mathcal{P}, \mathcal{T})$ are joined by an edge if the minimal subtrees joining the vertices of $\mathcal{T}$ labelled by A and A', respectively, have a vertex in common.

To give an example, consider the set $\mathcal{P}$ of characters on $X = \{1, 2, 3, 4, 5, 6\}$ consisting of $P_1 = 12|45$, $P_2 = 34|61$ and $P_3 = 23|56$ along with the X-tree $\mathcal{T}$ depicted in Fig. 8.4(c). Then the minimal subtree of $\mathcal{T}$ that joins the vertices which are labelled by the part $\{3, 4\}$ of P_2 is circumscribed by a closed dashed line in that figure. The subtree intersection graph $Int(\mathcal{P}, \mathcal{T})$ for that example is that minimal restricted chordal completion of $Int(\mathcal{P})$ depicted in Fig. 8.4(a) that has the solid and dashed edges as its edge set.

It turns out that this agreement of $Int(\mathcal{P}, \mathcal{T})$ with a minimal restricted chordal completion of $Int(\mathcal{P})$ is not a coincidence. The reason for this is a result that appeared in Lemma 4.7.3 in [41]. It says that whenever the partition intersection graph $Int(\mathcal{P})$ associated to a set $\mathcal{P}$ of characters has a minimal restricted chordal completion G, then this chordal completion must be the one coming from a phylogenetic tree $\mathcal{T}$ (i. e. $G = Int(\mathcal{P}, \mathcal{T})$). There is good reason to believe that the converse of this result holds true too, but no reference providing a proof is known to the authors. We pause to point out an immediate consequence of this result with regards to Theorem 8.1. Suppose $\mathcal{P}$ is a set of characters that defines an X-tree $\mathcal{T}$. Then $\mathcal{T}$ must be binary and phylogenetic and, according to part (ii) of that theorem, $Int(\mathcal{P})$ has a unique minimal restricted chordal completion G. Now the general result on minimal restricted chordal completion indicated above implies $G = Int(\mathcal{P}, \mathcal{T})$.

Guided by the role restricted chordal completions play for characterizing definitive sets of characters (Theorem 8.1) it is reasonable to assume that subtree intersection graphs which are also restricted chordal completions of partition intersection graphs might help shed light on the question of when a set of characters identifies an X-tree. It turns out that this is indeed the case. For the sake

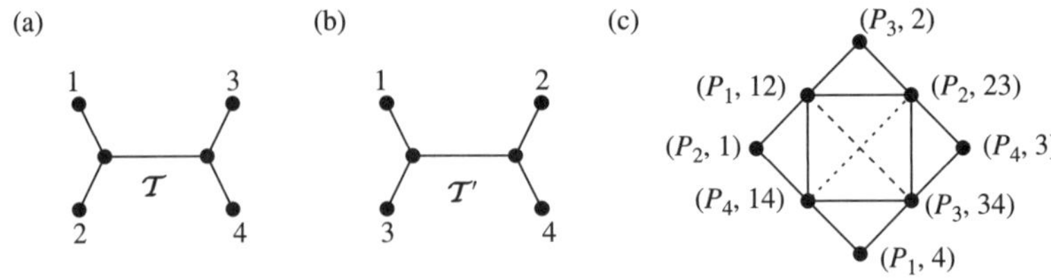

FIG. 8.7. Let $\mathcal{P}$ denote the character set consisting of $P_1 = 12|4$, $P_2 = 23|1$, $P_3 = 2|34$, and $P_4 = 14|3$ and consider the tree $\mathcal{T}$ pictured in (a). Then the subtree intersection graph $Int(\mathcal{T}, \mathcal{P})$ associated to $\mathcal{P}$ and $\mathcal{T}$ consists of all bold edges plus the dotted edge of the graph depicted in (c). Similarly, for the tree $\mathcal{T}'$ depicted in (b), $Int(\mathcal{T}', \mathcal{P})$ is the graph depicted in (c) with all bold edges plus the dashed and the dotted edges.

of clarity consider for a set $\mathcal{P}$ of characters the set $\mathcal{RCC}(\mathcal{P})$ of all restricted chordal completions G of $Int(\mathcal{P})$ for which there exists an X-tree $\mathcal{T}$ which displays $\mathcal{P}$ and $G = Int(\mathcal{P}, \mathcal{T})$. To help develop a feeling for this set, consider again the set $\mathcal{P}$ of characters $P_1 = 12|4$, $P_2 = 23|1$, $P_3 = 2|34$, and $P_4 = 14|3$ on $X = \{1, 2, 3, 4\}$. Then the edge set of $Int(\mathcal{P})$ is depicted in solid lines in Fig. 8.7(c) (which is the graph depicted in Fig. 8.5). The subtree intersection graphs associated to $\mathcal{P}$ and the X-trees $\mathcal{T}$ and $\mathcal{T}'$ depicted in Fig. 8.7(a) and (b), respectively, are $Int(\mathcal{P})$ plus the dotted edge and $Int(\mathcal{P})$ together with the dashed and the dotted edges, respectively. Hence, both graphs are elements in $\mathcal{RCC}(\mathcal{P})$. Interestingly, $Int(\mathcal{T}, \mathcal{P})$ is a proper subgraph of $Int(\mathcal{T}', \mathcal{P})$ that is, every edge in $Int(\mathcal{T}, \mathcal{P})$ is also an edge in $Int(\mathcal{T}', \mathcal{P})$ but not vice versa. As we will see later on, those subtree intersection graphs in $\mathcal{RCC}(\mathcal{P})$ that are *maximal* (i.e. they are not subgraphs of other elements in $\mathcal{RCC}(\mathcal{P})$) are crucial.

8.5.2 *Strongly distinguishing*

The purpose of this section is to provide the necessary but still missing concepts required for characterizing sets of characters that identify an X tree: *strongly distinguishing* and *inferring.*

We start with giving a definition for strongly distinguishing which is again motivated by the fact that we want to capture how the parts of the characters are 'spread' over an X-tree. Suppose $\mathcal{T} = (T, \phi)$ is an X-tree and e is an edge of T with end vertices u_1 and u_2. Then e is said to be *strongly distinguished* by a character P on X, if there exist parts A_1 and A_2 in P such that, for each $i \in \{1, 2\}$, the following hold:

(i) removing e from T results in a component so that $\phi(A_i)$ is a subset of the vertex set of that component;
(ii) $\phi^{-1}(u_i)$ is a subset of A_i;
(iii) removal of u_i from T results in components which, except for the one containing the other end vertex of e, contains an element of $\phi(A_i)$.

FIG. 8.8. Each edge in the depicted X-tree is strongly distinguished by a character in $\{12|35, 34|16, 24|56\}$.

For example, each edge in the X-tree $\mathcal{T}$ depicted in Fig. 8.8 is strongly distinguished by an element in the set $\{12|35, 34|16, 24|56\}$ of characters on $X = \{1, 2, \ldots, 6\}$. To help develop a feeling for this concept note that whenever an edge of an X-tree is strongly distinguished by a character, then it is also distinguished by it but the converse need not hold. Also note that this notion of strongly distinguishing extends the concept of strongly distinguishing introduced in [41].

Before we can state the desired characterization of identifying sets of characters, we need one more definition which is motivated by the fact that in some cases every X-tree that displays a given set of characters also displays other characters of X. Because of this, we say that a set $\mathcal{P}$ of characters *infers* a character P if every X-tree that displays $\mathcal{P}$ also displays P. For example, the split $12|345$ is inferred by the set $\{12|34, 12|35, 12|45\}$ of characters of $X = \{1, 2, 3, 4, 5\}$.

We are now in the position to present the analogous result of Theorem 8.1 for identifying sets of characters that appeared as Theorem 1.9 in [7].

Theorem 8.4 *Let $\mathcal{P}$ be a collection of characters of X. Then $\mathcal{P}$ identifies an X-tree if and only if the following conditions hold:*

(i) there is an X-tree that displays $\mathcal{P}$ and, for every edge e of this tree, there is a character of X inferred by $\mathcal{P}$ that strongly distinguishes e; and
(ii) there is a unique maximal element in $\mathcal{RCC}(\mathcal{P})$.

Moreover, if $\mathcal{P}$ identifies an X-tree $\mathcal{T}$, then $\mathcal{T}$ satisfies the properties in (i) and $Int(\mathcal{T}, \mathcal{P})$ is the unique maximal element in $\mathcal{RCC}(\mathcal{P})$.

An almost immediate consequence of Theorem 8.4 is that by replacing the words 'unique minimal chordal completion of $Int(\mathcal{P})$' in Theorem 8.1 by the words 'unique maximal element in $\mathcal{RCC}(\mathcal{P})$', we obtain a further characterization for when a set of characters defines an X-tree. Extending the concept of identifying an X-tree in terms of characters to collections of X-trees in the obvious way (see [7]), Theorem 8.4 also implies a characterization for when a collection $\mathcal{P}$ of X-trees can be amalgamated into an overall parent tree $\mathcal{T}$ so that $\mathcal{T}$ is identified by $\mathcal{P}$ (for details see Corollary 1.11 [7]).

Apart from being an interesting result in its own right, this characterization of identifying sets of characters provides important new insights into the supertree problem [2]. In the next section we will complement this new insight by a characterization for when quartet sets identify phylogenetic trees.

One of the surprising results for definitive sets of characters is that (at most) four characters suffice to define a binary phylogenetic tree. The likeness between the concepts of defining and identifying therefore raises the question of whether a similar result might also hold for identifying sets of characters. In [8], Bordewich *et al.* addressed this question. By employing a certain edge colouring for X-trees, they established that any X-tree $\mathcal{T}$ can be identified by at most $4\lceil\log_2(d-2)\rceil+4$ characters where d is the maximal degree of any vertex in $\mathcal{T}$ [8]. It should be noted that for binary X-trees $\mathcal{T}$ this result implies that, as in the case of definitive sets of characters, at most four characters are required to identify $\mathcal{T}$. Furthermore it is shown in [8] that in case of a *star tree* $\mathcal{T}$ on d leaves (a tree with precisely one interior vertex), for k characters to identify $\mathcal{T}$ we cannot have $k < \log_2 d$.

8.6 Identifying trees in terms of quartets

In this section, we will present an alternative characterization of compatible/identifying quartet sets. In addition, we present a formula for the minimal number of quartets that is necessary to identify a given phylogenetic tree. Our treatment follows [19] where detailed proofs can be found.

As we have already seen in Section 8.4 a quartet can also be interpreted as a two-by-two split of a set of size four. Moreover, there is a phylogenetic X-tree displaying a given quartet set $\mathcal{Q}$ if and only if there is an X-tree that displays $\mathcal{Q}$. This implies that quartet set compatibility can be determined by checking if the associated partition intersection graph has a restricted chordal completion. Further, a quartet set $\mathcal{Q}$ identifies a phylogenetic X-tree $\mathcal{T}$ if and only if the quartets in $\mathcal{Q}$ (thought of as partial splits) together with all trivial splits of X identify $\mathcal{T}$. Hence, Theorem 8.4 gives a characterization of identifying quartet sets in terms of partition intersection graphs.

We next present an alternative characterization for when quartet sets are compatible or identifying which provides additional insights into quartet problems.

8.6.1 *The quartet graph*

For a set $\mathcal{Q}$ of quartets on X, the *quartet graph* $G_{\mathcal{Q}}$ has the singletons of X as its vertex set and, for every quartet $q = ab|cd \in \mathcal{Q}$, there are two q-labelled edges, one joining a and b and one joining c and d. There are no further edges. Note that the quartet graph may contain parallel edges. Clearly, this edge labelling is a proper edge colouring, that is, there are no two adjacent edges with the same labelling. For the aimed for characterization we require the concept of a colour-identification sequence which we describe next.

Let G be a graph whose vertex set V contains precisely the parts of a partition of X. Then the *identification* of a subset U of V is the graph obtained from G by merging the elements of U into a single vertex (removing created loops) and

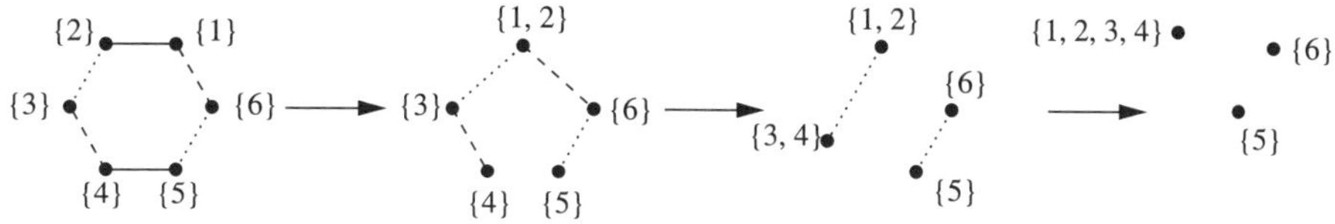

FIG. 8.9. A complete colour-identification sequence for the quartet set $\{12|45, 34|61, 23|56\}$.

retaining all other edges of G. If there is a proper edge-colouring associated with G, then those subsets U of V with the property that, for every edge label q, there is at most one q-labelled edge incident with a vertex in U have turned out to be a key object for the sought after characterization. We therefore define the *colour identification* of a vertex set U which fulfils the condition above to be the edge labelled graph obtained from G by first identifying U and then removing every edge for which the other edge with the same label has been identified. Note that the edge labelling of the resulting graph is a proper edge colouring.

Finally, we call a sequence $G_0, G_1, ..., G_k$ a *complete colour-identification sequence* of G_0 if G_i is obtained from G_{i-1} by a colour identification ($1 \le i \le k$) and the edge set of G_k is empty. For the quartet set $\mathcal{Q} = \{12|45, 34|61, 23|56\}$, a complete colour-identification sequence S_1 is depicted in Fig. 8.9 where edges with the labelling 12|45, 34|61, and 23|56 are drawn as solid, dashed, and dotted lines, respectively. We obtain S_1 by first identifying $\{1\}$ and $\{2\}$, then $\{3\}$ and $\{4\}$, and finally $\{1, 2\}$ and $\{3, 4\}$. Equipped with this procedure for shrinking a graph to a set of isolated vertices, we are now in the position to state the promised alternative characterization of compatible quartet sets.

Theorem 8.5 *Let $\mathcal{Q}$ be a collection of quartets. Then $\mathcal{Q}$ is compatible if and only if there exists a complete colour-identification sequence of $G_{\mathcal{Q}}$.*

The quartet graph can also be used to characterize quartet sets that identify a phylogenetic tree $\mathcal{T}$. To state the result we require a further generalization of 'distinguishing' edges of a tree. Let $\mathcal{Q}$ be a set of quartets on X and let $\mathcal{T}$ be a phylogenetic tree where v_1 and v_2 are two interior vertices which are connected by an edge e. For $i \in \{1, 2\}$, let $W_i^1, \ldots, W_i^{l(i)}$ be the maximal subtrees of $\mathcal{T}$ that do not contain v_j which result from deleting v_i from $\mathcal{T}$, $i \neq j$. Then $\mathcal{Q}$ *specially distinguishes* edge e if, for $i \in \{1, 2\}$, the graph with vertices $W_i^1, \ldots, W_i^{l(i)}$ and an edge between two vertices W_i^s and W_i^t if and only if there is a quartet $w_i^s w_i^t|xy \in \mathcal{Q}$ such that $w_i^s w_i^t|xy$ distinguishes e and w_i^s, w_i^t are vertices of W_i^s, W_i^t, respectively, is connected. Further, $\mathcal{Q}$ *specially distinguishes* $\mathcal{T}$ if $\mathcal{Q}$ specially distinguishes every interior edge of $\mathcal{T}$. To elucidate this definition, consider again the quartet set from the previous example together with the phylogenetic tree $\mathcal{T}$ depicted in Fig. 8.4(c). Let v_1 be the vertex incident with 2 and let v_2 be the interior vertex incident with v_1. Then W_1^1 and W_1^2 are the isolated vertices labelled 2 and 3 respectively and, W_2^1 and W_2^2 are the minimal subtrees of $\mathcal{T}$ connecting the vertices labelled 4 and 5, and 6 and 1, respectively. Then W_2^1 and W_2^2 are joined by an edge since 5 and 6 are the labels of vertices in W_2^1 and W_2^2, respectively, and $23|56 \in \mathcal{Q}$. It is easy to verify that $\mathcal{Q}$ specially distinguishes $\mathcal{T}$.

It is straightforward to check that a set of quartets which identifies a phylogenetic tree has to specially distinguish $\mathcal{T}$, but this is not sufficient. To characterize identifying quartet sets we need some condition as to which of the quartet parts are identified by a complete colour-identification sequence. To state this condition, we need a further crucial concept. Let $S = G_0, G_1, \ldots, G_k$ be a complete colour-identification sequence where, for $j \in \{1, \ldots, k\}$, G_j is obtained from G_{j-1} by identifying U_j. Then S is called *minimal* if there is no complete colour-identification sequence $G'_0, G'_1, \ldots, G'_l$ with $l < k$ such that, for $j \in \{1, \ldots, l\}$, G'_j is obtained from G'_{j-1} by identifying U'_j and U'_j is the union of the elements of a subset of $\{U_1, \ldots, U_k\}$. Minimal colour-identification sequences correspond to least resolved trees that display the given quartet set, and an example of such a sequence is the sequence S_1 constructed above.

We are now in the position to present the characterization of identifying quartet sets.

Theorem 8.6 *Let $\mathcal{Q}$ be a set of quartets on X. Then $\mathcal{Q}$ identifies a phylogenetic tree if and only if the following hold:*

(i) a phylogenetic tree $\mathcal{T}$ exists that displays $\mathcal{Q}$ and is specially distinguished by $\mathcal{Q}$, and
(ii) if $\mathcal{Q}'$ is a subset of $\mathcal{Q}$ that specially distinguishes $\mathcal{T}$ and $q = A|B \in \mathcal{Q}'$, then, whenever the last identification involving a quartet in $\mathcal{Q}'$ in a complete minimal colour-identification sequence of $G_{\mathcal{Q}}$ contains A, the choice of which part of all quartets in $\mathcal{Q}' - \{q\}$ is identified in this sequence is fixed.

A consequence of this result is that the quartet set $\mathcal{Q}$ of the previous example does not identify the tree in Fig. 8.4(c) as $\mathcal{Q}$ violates Condition (ii). This can be seen by constructing a second complete colour-identification sequence S_2 which we shall do next: first we identify $\{1\}$ and $\{6\}$, then $\{4\}$ and $\{5\}$, and finally $\{2\}$ and $\{3\}$. For both sequences S_1 and S_2, the quartet $23|56$ is the last quartet of $\mathcal{Q}$ involved in an identification and this identification contains the quartet part $\{2, 3\}$. Now consider the quartet $34|61 \in \mathcal{Q}$. In S_1, $\{3, 4\}$ is identified and in S_2, $\{6, 1\}$ is identified. Hence, the quartet part of $34|61$ that is identified is not fixed and $\mathcal{Q}$ does not identify a phylogenetic tree.

8.6.2 *Small identifying quartet sets*

As noted at the end of Section 8.5.2, $4\lceil \log_2(d-2) \rceil + 4$ characters suffice to identify an X-tree $\mathcal{T}$ where d is the maximal degree of any vertex in $\mathcal{T}$. Here we present the quartet analogue of that result which yields the smallest number of quartets necessary to identify a given phylogenetic tree $\mathcal{T}$. This number depends on the shape of $\mathcal{T}$. We denote the edge set of $\mathcal{T}$ by E and the degree of a vertex

v of $\mathcal{T}$ by $d(v)$. For every edge e connecting two vertices u and v, we define

$$q(e) = \left\lceil \frac{1}{2}(\min\{d(u), d(v)\} - 1)(\max\{d(u), d(v)\} - 2) \right\rceil .$$

Theorem 8.7 *Every quartet set that identifies a phylogenetic tree $\mathcal{T}$ contains at least*

$$q(\mathcal{T}) = \sum_{e \in E} q(e)$$

quartets. Moreover, there is a quartet set of cardinality $q(\mathcal{T})$ that identifies $\mathcal{T}$.

This result corrects Corollary 6.3.10 in [41] which states that there is a quartet set of size $n - 3$ identifying $\mathcal{T}$ for every phylogenetic X-tree $\mathcal{T}$ with $|X| = n$. Indeed it is shown in [19] that $q(\mathcal{T}) \leq \lfloor (\frac{n}{2} - 1)^2 \rfloor$ for every phylogenetic tree $\mathcal{T}$ with n leaves and that, for every $n \geq 4$, there is a tree where equality holds.

8.7 Conclusion

In this chapter we have reviewed novel results concerning the basic problem of when fundamental divisions of taxa into groups—either directly from data or from earlier phylogenetic studies—completely determine a tree on which the taxa set under consideration has evolved. We combined the standard interpretation of a biological character as a (partial) partition/map (which we also called a character) with a relatively recently introduced formalization of homoplasy-free character evolution. This led to the concept of displaying (which is at the heart of compatibility), and allows a formalization of the above recalled basic problem to the following questions:

- When is a set $\mathcal{P}$ of partial partitions compatible and,
- if $\mathcal{P}$ is compatible when does it define/identify a phylogenetic tree/X-tree?

An answer to the first question can be used to detect reticulate evolution in the form of recombination [11], hybridization, or lateral gene transfer as well as noise in the data. A positive answer to the latter question makes us confident that we have found the true tree.

We reviewed recent complete answers for these questions in terms of chordal graphs, closure rules (in the context of defining and identifying an X-tree), and quartets (in the context of identifying a phylogenetic tree). Moreover, we explained how these results can be used to shed light on the fascinating question of how many characters suffice to recover the tree asked for in the second question. In addition, we explained the relevance of the purely combinatorial concepts mentioned above for developing new and efficient supertree methods [2] and for inferring new phylogenetic relationships. The former may be useful when complex models and methods prohibit direct analysis of larger numbers of taxa and the latter for combining source trees on only partially overlapping leaf sets into an overall parent structure such as a supertree or a supernetwork [27].

We expect that future work in the area will involve the extension of the mostly deterministic results reviewed in this chapter to a probabilistic framework thereby extending work in [34]. On a more detailed level the precise relationship between the split closure and the semi-dyadic closure of a set of quartets might be of interest. Furthermore, there are several open complexity problems. While it is NP-complete to decide whether a given set of quartets or characters is compatible, the complexity of deciding whether a collection of characters or quartets is definitive or identifying is unknown.

Acknowledgements

The authors would like to thank Olivier Gascuel and Mike Steel for inviting them to write this chapter. They would also like to thank Mike Steel for his helpful comments and suggestions on an earlier version of this chapter. Finally, they would like to thank the anonymous referees for their helpful comments.

References

[1] Argawala, R. and Fernándes-Baca, D. (1994). A polynomial type algorithm for the perfect phylogeny problem when the number of characters is fixed. *SIAM Journal on Computing*, **23**(6), 1216–1224.

[2] Bininda-Emonds, O. R. P. (ed.). (2004). *Phylogenetic Supertrees. Combining Information to Reveal the Tree of Life.* Kluwer Academic Publishers, Dordrecht.

[3] Bodlaender, H., Fellows, M., and Warnow, T. (1992). Two strikes against perfect phylogeny. In *Proceedings of the 19th International Colloquium on Automata, Languages, and Programming, Lecture Notes in Computer Sciences*. Springer Verlag, Berlin, 273–283.

[4] Böcker, S. (1999). From subtrees to supertrees. Unpublished PhD thesis. Fakultät für Mathematik, Universität Bielefeld.

[5] Böcker, S. and Dress, A. (2001). Patchworks. *Advances in Mathematics*, **157**, 1–21.

[6] Böcker, S., Bryant, D., Dress, A., and Steel, M. (2000). Algorithmic aspects of tree amalgamation. *Journal of Algorithms*, **37**, 522–537.

[7] Bordewich, M., Huber, K. T., and Semple, C. (2005). Identifying phylogenetic trees. *Discrete Mathematics*, **300**(1-3), 30–43.

[8] Bordewich, M., Semple, C., and Steel, M. (2006). Identifying X-trees with few characters. *Electronic Journal of Combinatorics*, **13**(1), #R83.

[9] Buneman, P. (1971). The recovery of trees from measures of dissimilarity. In *Mathematics in the Archaeological and Historical Sciences*. pp. 387–395. Edinburgh University Press, Edinburgh.

[10] Buneman, P. (1974). A characterization of rigid circuit graphs. *Discrete Mathematics*, **9**, 205–212.

[11] Bruen, T., Philippe, H., and Bryant, D. (2006). A quick and robust statistical test to detect the presence of recombination, *Genetics*, **172**, 2665–2681.

[12] Colonius, H. and Schulze, H. H. (1981). Tree structure for proximity data. *British Journal of Mathematical and Statistical Psychology*, **34**, 167–180.

[13] Dekker, M. C. H. Reconstruction methods for derivation trees. Unpublished Masters thesis, Vrije Universiteit Amsterdam, Netherlands.

[14] Dezulian, T. and Steel, M. (2004). Phylogenetic closure operations and homoplasy-free evolution. In *Classification, Clustering, and Data Mining Applications (Proceedings of the meeting of the International Federation of Classification Societies (IFCS) 2004)* (ed. D. Banks, L. House, F.R. McMorris, P. Arabie, and W. Gaul). pp. 395–416. Springer-Verlag, Berlin.

[15] Dress, A. and Steel, M. (1992). Convex tree realizations of partitions. *Applied Mathematics Letters*, **5**(3), 3–6.

[16] Dress, A., Moulton, V., and Steel, M. (1997). Trees, taxonomy, and strongly compatible multi-state characters. *Advances in Applied Mathematics*, **19**, 1–30.

[17] Estabrook, G. F. and McMorris, F. R. (1977). When are two qualitative taxonomic characters compatible. *Journal of Mathematical Biology*, **4**, 195–200.

[18] Grünewald, S. and Huber, K. T. (2006). A novel insight into the perfect phylogeny problem. *Annals of Combinatorics*, **10**(1), 97–109.

[19] Grünewald, S., Humphries, P. J., and Semple, C. Quartet compatibility and the quartet graph. (submitted).

[20] Grünewald, S., Steel, M., and Swenson, M. S. Closure operations in phylogenetics. *Mathematical Biosciences*. in press.

[21] Gusfield, D. (1991). Efficient algorithms for inferring evolutionary trees. *Networks*, **21**, 19–28.

[22] Huber, K. T. (2004). Recovering trees from well-separated multi-state characters. *Discrete Mathematics*, **278**, 151–164.

[23] Huber, K. T. and Moulton, V. (2002). The relation graph. *Discrete Mathematics*, **244**(1-3), 153–166.

[24] Huber, K. T. and Moulton, V. (2005). Phylogenetic networks. In *Mathematics of Evolution and Phylogeny*. (ed. O. Gascuel). Oxford University Press, Oxford.

[25] Huber, K. T., Moulton, V., Semple, C., and Steel, M. (2005). Recovering a phylogenetic tree using pairwise closure operations. *Applied Mathematics Letters*, **18**(3), 361–366.

[26] Huber, K. T., Moulton, V., and Steel, M. (2005). Four characters suffice to convexly define a phylogenetic tree. *SIAM Journal on Discrete Mathematics*, **18**(4), 835–843.

[27] Huson, D. H., Dezulian, T., Klöpper, T., and Steel, M. (2004). Phylogenetic super-networks from partial trees. *IEEE/ACM Transactions on Computational Biology and Bioinformatics*, **1**(4), 151–158.

[28] Kannan, S. and Warnow, T. (1994). Inferring evolutionary histories from DNA sequences. *SIAM Journal on Computing*, **23**(3), 713–737.

[29] Kriegs, J. O., Churakov, G., Kiefmann, M., Jordan, U., Brosius, J., and Schmitz, J. (2006). Retroposed elements as archives for the evolutionary history of placental mammals. *PLoS Biology*, **4**(4) e91, 0537–0544.

[30] Lou, Z. (2000). In search of whales' sisters. *Nature*, **404**, 235–237.

[31] McMorris, F. R., Warnow, T., and Wimer, T. (1994). Triangulating vertex-coloured graphs. *SIAM Journal on Discrete Mathematics*, **7**, 296–306.

[32] Meacham, C. A. (1983). Theoretical and computational considerations of the compatibility of qualitative taxonomic characters. In *Numerical Taxonomy* (ed. J. Felsenstein). pp. 304–314, NATO ASI Series Vol. G1, Springer-Verlag, Berlin.

[33] Moret, B. M. E., Tang, J., and Warnow, T. (2005). Reconstructing phylogenies from gene-content and gene-order data. In *Mathematics of Evolution and Phylogeny* (ed. O. Gascuel). Oxford University Press.

[34] Mossel, E. and Steel, M. (2004). A phase transition for a random cluster model on phylogenetic trees. *Mathematical Biosciences*, **187**, 189–203.

[35] O'Leary, M. A. and Geisler, J. H. (1999). The position of Cetacea within Mammalia: Phylogenetic analysis of morphological data from extinct and extant taxa. *Systematic Biology*, **48**, 455–490.

[36] Rokas, A. and Holland, W. H. (2000). Rare genomic changes as a tool for phylogenetics. *TREE*, **15**, 454–458.

[37] Sanderson, M. J., Purvis, A., and Henze, C. (1998). Phylogenetic supertrees: assembling the trees of life. *Trends in Ecology and Evolution*, **13**, 105–109.

[38] Semple, C. and Steel, M. (2001). Tree reconstruction via a closure operation on partial splits. In *Computational Biology (proceedings of JOBIM 2000)*, LNCS 2066, pp. 126–134, Springer-Verlag, Berlin.

[39] Semple, C. and Steel, M. (2002). A characterization for a set of partial partitions to define an X-tree. *Discrete Mathematics*, **247**, 169–186.

[40] Semple, C. and Steel, M. (2002). Tree reconstruction from multi-state characters. *Advances in Applied Mathematics*, **28**(2), 169–184.

[41] Semple, C. and Steel, M. (2003). *Phylogenetics.* Oxford University Press, Oxford.

[42] Steel, M. (1992). The complexity of reconstructing trees from qualitative characters and subtrees. *Journal of Classification*, **9**, 91–116.

V

FROM TREES TO NETWORKS

9

SPLIT NETWORKS AND RETICULATE NETWORKS

Daniel H. Huson

Abstract

Phylogenetic networks are becoming an important tool in molecular evolution, as the role of reticulate events such as hybridization, horizontal gene transfer and recombination becomes more evident, and as the available data increases in quantity and quality. However, their usage has been hampered by a bewildering zoo of definitions and confusing terminology.

Additionally, there are two fundamental types of phylogenetic networks, namely those that aim at visualizing incompatible signals in a data set, and those that provide an explicit scenario of reticulate evolution, but this distinction is seldom appreciated. We look at split networks as a major class of the former type of networks and discuss algorithms that compute such networks from sequences, distances or trees. We then study hybridization networks, obtained from trees, and recombination networks, inferred from binary sequences, as two examples of explicit networks.

9.1 Introduction

Phylogenetic networks are becoming an important tool in molecular evolution, as the role of reticulate events such as hybridization, horizontal gene transfer, and recombination becomes more evident [7], and as the available data increases in quantity and quality. Increasingly, the problem of sampling error has been replaced by the problem of model error.

The concept of a *phylogenetic tree* is clearly defined [40] and the only real ambiguity is whether trees are rooted or unrooted (and perhaps whether the edges are weighted or not). The concept of a *phylogenetic network* is not so clear and there is much confusion in the literature [34, 21].

There appears to be three sources of confusion. Firstly, there actually are many different types of phylogenetic networks; here we list just some of them: phylogenetic trees, split networks, median networks, median-joining networks, neighbor-nets, consensus networks, reticulate networks, recombination networks, ARGs, hybridization networks, reticulgrams, haplotype networks, and the result of the netting method.

The second source of confusion is that the general term 'phylogenetic network' is often equated with some specific type of network, e.g.:

- phylogenetic network = recombination network [13],
- phylogenetic network = hybridization network [29], and
- phylogenetic network = reticulate network with multi-edges [18].

To address this problem, we suggest to define the term *phylogenetic network* to mean *any* network that represents evolutionary relationships between taxa and then to use more specific names for different types of networks.

Thirdly, a more interesting source of confusion is that there are two fundamentally different types of phylogenetic networks, namely:

- networks that provide an *explicit* picture of evolution, and
- networks that provide an *implicit* picture of evolution.

This distinction already makes sense for phylogenetic trees, as a rooted tree describes an explicit evolutionary scenario, whereas an unrooted tree does not have a direct evolutionary interpretation, but rather is a visualization of evolutionary signals. This distinction is even more relevant for phylogenetic networks, which also come in the two flavours: 'rooted' and 'unrooted'. But, more importantly, some network methods aim at displaying (incompatible) phylogenetic signals, while others aim at explicitly modeling reticulate evolution. Implicit networks are applied to 'see' what is really in a data set, whereas explicit networks are used to describe reticulate evolution.

To illustrate this distinction, in Fig. 9.1 we display two different phylogenetic networks obtained from a buttercup data set [30] that is studied in more detail below. Network (a) is an example of a 'split network' that represents all splits contained in two different gene trees. Here, each parallelogram corresponds to a pair of splits that are incompatible with each other and the network shows clearly that the two gene trees are very different. (The two underlying gene trees are based on a chloroplast JSA region and a nuclear ITS region, as discussed

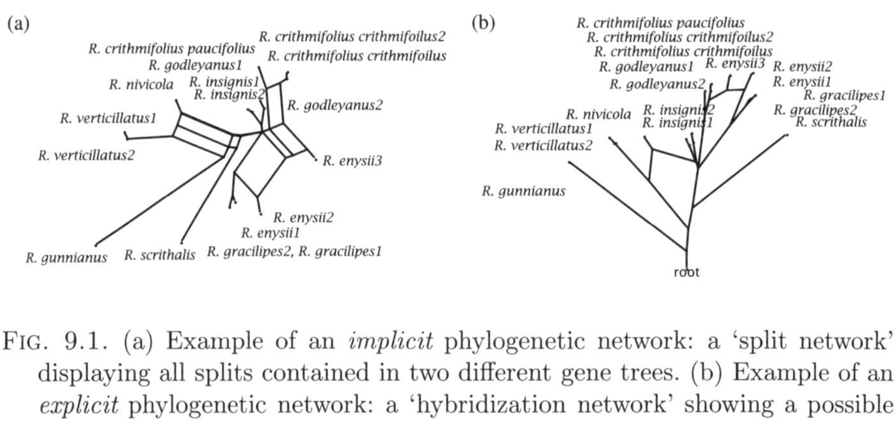

FIG. 9.1. (a) Example of an *implicit* phylogenetic network: a 'split network' displaying all splits contained in two different gene trees. (b) Example of an *explicit* phylogenetic network: a 'hybridization network' showing a possible evolutionary history involving hybridization events.

TABLE 9.1. A summary of the different approaches discussed in this chapter.

Input	Output	Method
sequences	split network	median network [2]
	recombination network	galled tree approach [12], branch and bound approach [32], split approach [24]
distances	splits	split decomposition [1], Neighbor-Net [5]
trees	splits	consensus network [16], Z-closure method [22]
	hybridization network	SPNet [35], split approach [23]
splits	split network	convex hull and equal angle algorithm [8]

below.) Network (b) is an example of a 'hybridization network' that is based on the same two trees. This network explicitly describes an evolutionary scenario, namely that the sequences evolved up the network in a tree-like fashion, but experienced two reticulate events, one producing *R. nivicola* as a hybrid of the lineages leading to *R. verticillatus* and *R. insignis*, and the other producing *R. enysii3* as a hybrid of *R. crithmifolius paucifolius* and *R. enysii3*.

In this chapter we look at split networks as a major class of *implicit networks* and then study hybridization networks and recombination networks as two examples of *explicit networks*. The discussed approaches are summarized in Table 9.1.

In Section 9.2, we introduce split networks using *consensus networks* and *super networks*, which are useful for displaying incompatible phylogenies, and form the computational basis for other types of networks. We then discuss a number of sequence and distance-based approaches that also produce splits and *split networks* in Section 9.3. In Section 9.4, we discuss how to analyse *hybridization* using *reticulate networks* based on multiple gene trees. Finally, we look at obtaining *recombination networks* from split networks in Section 9.5.

9.2 Consensus networks and super networks

In a simple model of evolution, such as the one proposed by Jukes and Cantor [26], DNA sequences evolve along a fixed tree, subject to random mutation events along the edges and speciation events at the vertices of the tree. In this section, we first discuss additional evolutionary events that are not considered in such simple models. This will lead us to the fundamental observation that: *gene trees differ*. Because of this, it may not be adequate to represent a set of gene trees by a single consensus tree, as is sometimes done. We discuss how to represent the conflicting signals using a 'consensus network' or 'super network'.

Standard models of evolution, such as the Jukes–Cantor model, are usually understood to represent the evolution of a *single* gene. These models do not consider insertions and deletions, or more complicated events. If one studies more than one gene simultaneously, additional evolutionary events must be taken into account (individual genes may be *born*, *duplicated*, or *lost*). Moreover, biological mechanisms such as *recombination*, *hybridization* or *horizontal gene transfer* may be involved.

Suppose we are given one or more genes for a set of taxa X. Consider a model in which the sequence of a gene evolves via mutations and speciation events, but in which we also allow gene *duplication* or *loss*. Note that, under this slightly more general model of evolution, the true phylogeny of a gene can differ substantially from the true species or *model* phylogeny, as exemplified in Fig. 9.2.

Let $X = \{x_1, \ldots, x_n\}$ be a set of taxa. An X-*tree* $T = (V, E)$ is a tree with vertex set V and edge (or *branch*) set E, together with a labelling of the vertices of T by elements of X, such that all taxa occur as labels and all leaves of the tree obtain at least one label [40]. An X-tree is called a *phylogenetic tree* if the leaves of T are bijectionally labelled by X. An X-tree can be *rooted* by specifying a *root* node ρ, which can be any vertex of T, or the mid-point of some edge.

An X-*split* $S = \frac{A}{B} (= \frac{B}{A})$ is a bipartitioning of X with [1]:

$$A, B \neq \emptyset,\ \ A \cap B = \emptyset \text{ and } A \cup B = X.$$

If the taxon set X is clear from the context, then we will use the terms X-split and split interchangeably. Any edge e of a tree T defines a split $\sigma_T(e) := \frac{A}{B}$,

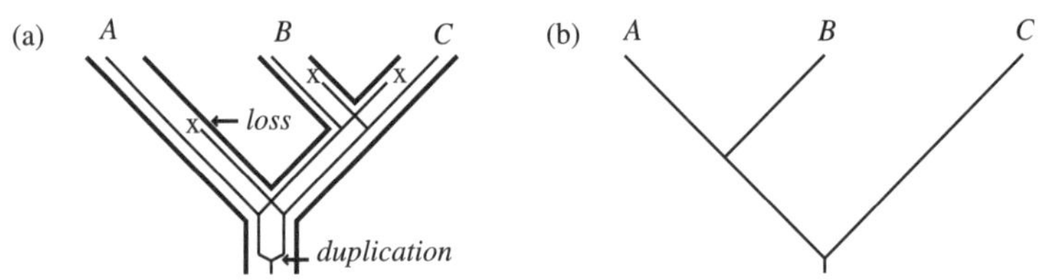

FIG. 9.2. (a) A species tree (depicted using bold parallel lines) and the history of a single gene (shown as thin lines). The gene is involved in one gene-duplication event and three subsequent gene-loss events. (b) The gene tree induced by the extant copies of the gene has a different topology (branching order) than the species tree.

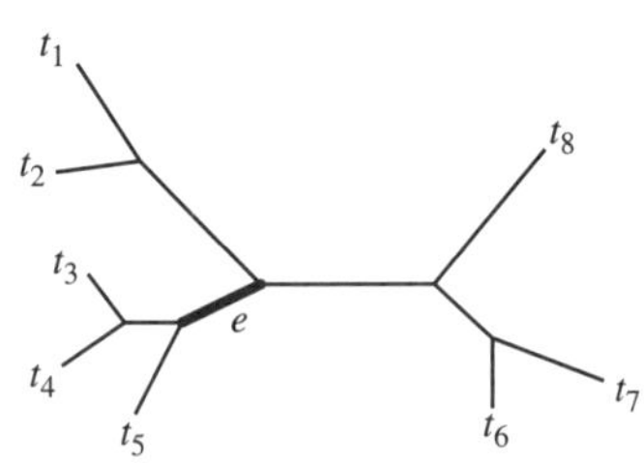

FIG. 9.3. The edge e corresponds to the split $\sigma_T(e) = \frac{A}{B}$ with $A = \{t_1, t_2, t_6, t_7, t_8\}$ and $B = \{t_3, t_4, t_5\}$.

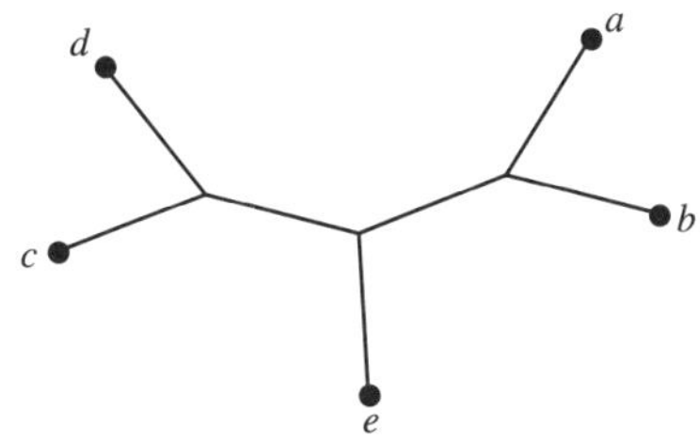

FIG. 9.4. A tree on five taxa.

where A and B are the sets of taxa contained in the two sub-trees defined by e, see Fig. 9.3. We use $\Sigma(T)$ to denote the *split encoding of* T, i.e. the set of all splits obtained from T. For example, the split encoding $\Sigma(T)$ of the tree depicted in Fig. 9.4 contains five *trivial* splits, each separating one taxon from all others:

$$\frac{\{a\}}{\{b,c,d,e\}}, \frac{\{b\}}{\{a,c,d,e\}}, \frac{\{c\}}{\{a,b,d,e\}}, \frac{\{d\}}{\{a,b,c,e\}} \text{ and } \frac{\{e\}}{\{a,b,c,d\}},$$

and two *non-trivial* splits, each separating at least two taxa from at least two other:

$$\frac{\{a,b\}}{\{c,d,e\}} \text{ and } \frac{\{a,b,e\}}{\{c,d\}}.$$

Two different X-splits $S = \frac{A}{B}$ and $S' = \frac{A'}{B'}$ are called *compatible*, if one is a refinement of the other, i.e., if one of the four following intersections is empty:

$$A \cap A', B \cap A', B \cap A' \text{ or } B \cap B'.$$

A set Σ of X-splits is called *compatible*, if every pair of splits $S, S' \in \Sigma$ is compatible, but is otherwise called *incompatible*. By definition, any trivial split is always compatible with all other splits. The *incompatibility graph* $IG(\Sigma)$ associated with Σ is a simple graph with vertex set Σ in which any two vertices $S, S' \in \Sigma$ are connected by an edge, if and only if they are incompatible. Compatibility is an important concept in phylogenetics and we have:

Lemma 9.1 *Let Σ be a set of X-splits. There exists an unique X-tree T with $\Sigma = \Sigma(T)$ if and only if Σ is compatible [6].*

Any *compatible* set of X-splits can be represented by a phylogenetic tree. What about incompatible splits sets? Consider the two trees T_1 and T_2 displayed in Fig. 9.5, for which the splits $S_p = \frac{\{a,b,c\}}{\{d,e\}} \in \Sigma(T_1)$ and $S_q = \frac{\{a,b,d\}}{\{c,e\}} \in \Sigma(T_2)$ are incompatible. The split network SN represents the incompatible set of splits $\Sigma(T_1) \cup \Sigma(T_2)$, using a cut-set of parallel edges to represent each split [8, 21].

Definition 9.2 *For a set of X-splits Σ, we define a* split network $SN = SN(\Sigma)$ *as a connected graph in which some of the nodes are labelled by taxa and all edges*

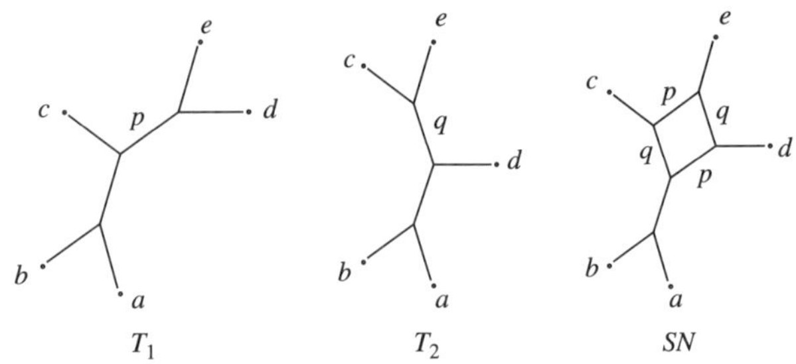

FIG. 9.5. The splits $S_p = \frac{\{a,b,c\}}{\{d,e\}} \in \Sigma(T_1)$ and $S_q = \frac{\{a,b,d\}}{\{c,e\}} \in \Sigma(T_2)$ contained in trees T_1 and T_2, respectively, are incompatible. The displayed 'split network' SN represents all splits present in T_1 or T_2, or in both. In SN, the two edges labelled p represent S_p and the two edges labelled q represent S_q.

are labelled by splits, such that:

(N1) For any split $S = \frac{A}{B} \in \Sigma$, removing all edges labelled S produces precisely two connected components, one containing all vertices with labels in A and other containing all vertices with labels in B.

(N2) The edges along any shortest path in SN all have different labels.

A collection of X-trees $\mathcal{T} = \{T_1, \ldots, T_K\}$ is often summarised using a consensus tree. Let $\Sigma_{all} = \cup_{T\in\mathcal{T}}\Sigma(T)$ be the set of all present X-splits. Let

$$\Sigma(p) = \{S \in \Sigma_{all} : |\{T \in \mathcal{T} : S \in \Sigma(T)\}| > pK\}$$

be the set of splits that occur in more than a proportion p of all trees and define $\bar{\Sigma}(p) = \{S \in \Sigma_{all} : |\{T \in \mathcal{T} : S \in \Sigma(T)\}| \geq pK\}$. Then:

- $\bar{\Sigma}(1) = \bigcap_i \Sigma(T_i)$ defines the *strict consensus*,
- $\Sigma(\frac{1}{2})$ defines the *majority consensus*, and, more generally,
- $\Sigma(\frac{1}{d+1})$ $(d \geq 2)$ defines a set of *consensus splits.*

Note that $\bar{\Sigma}(1)$ and $\Sigma(\frac{1}{2})$ are both compatible sets, the latter by the pigeonhole principle, and thus both sets can be represented by a tree. However, $\Sigma(\frac{1}{d+1})$ may be incompatible, if $d \geq 2$, and will then need to be represented by a network rather than a tree. For example, given the six trees depicted in Fig. 9.6 as input, we can obtain the consensus trees and networks shown in Fig. 9.7.

Often, a set of trees $\mathcal{T} = \{T_1, \ldots, T_K\}$ is summarized using a consensus *tree.* This may not always be appropriate, as gene trees are not necessarily just different estimations of the same true phylogeny, but may differ substantially for biological reasons.

A *consensus network* is obtained by computing the consensus splits $\Sigma(\frac{1}{d+1})$ for some fixed value $d \geq 0$. The parameter d sets the *maximum dimensionality* of the corresponding network: for $d = 1$ the network will be 1-dimensional (a

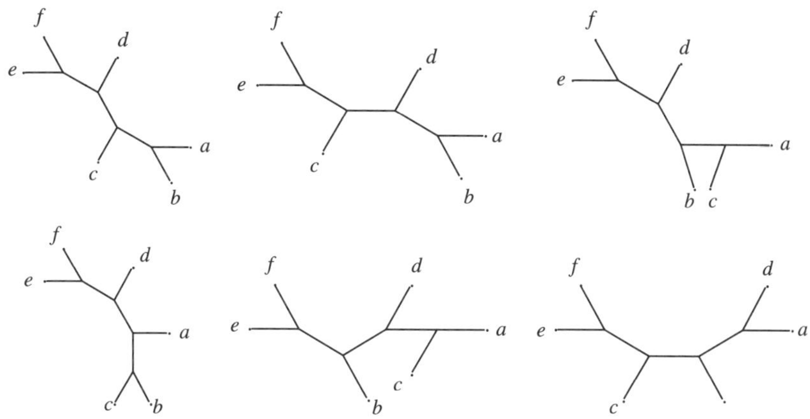

FIG. 9.6. Six different trees on $X = \{a, b, \ldots, f\}$.

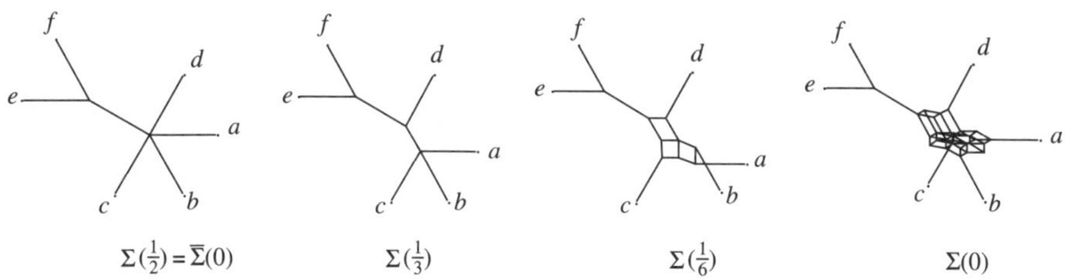

FIG. 9.7. Consensus trees and networks obtained from the six trees displayed in Fig. 9.6.

tree), for $d = 2$ the network may contain parallelograms, and in general it will contain (the complete edge skeletons of) cubes of dimension $\leq d$ [16, 15].

Consider a set of taxa $X = \{x_1, \ldots, x_n\}$ and a set of genes $G = \{g_1, \ldots, g_t\}$. It is often the case that a given gene g_i is not available for all taxa, but only for a subset $X' \subset X$. Any X'-tree inferred from such a gene g_i is called a *partial* X-tree, and any X'-split is called a *partial* X-split.

For a collection of *partial* X-trees $\mathcal{T} = \{T_1, \ldots, T_K\}$, the consensus methods above do not apply. One alternative is to compute a *super tree* T that optimally summarizes the set of input trees [3]. A second approach is to summarize the input trees in terms of a *super network* that attempts to represent as many of the input partial splits as possible.

A pair of splits $S_i = \frac{A_i}{B_i}$ and $S_j = \frac{A_j}{B_j}$ is said to be in *Z-relation* to each other, denoted by $S_i \mathsf{Z} S_j$, if $A_i \cap A_j \neq \emptyset$, $A_j \cap B_i \neq \emptyset$, $B_i \cap B_j \neq \emptyset$, but $A_i \cap B_j = \emptyset$. If $A_i \not\subseteq A_j$ or $B_j \not\subseteq B_i$, then $\{S_i, S_j\} \neq \{\frac{A_i}{B_i \cup B_j}, \frac{A_i \cup A_j}{B_j}\}$ and we say that the pair S_i, S_j is *productive*.

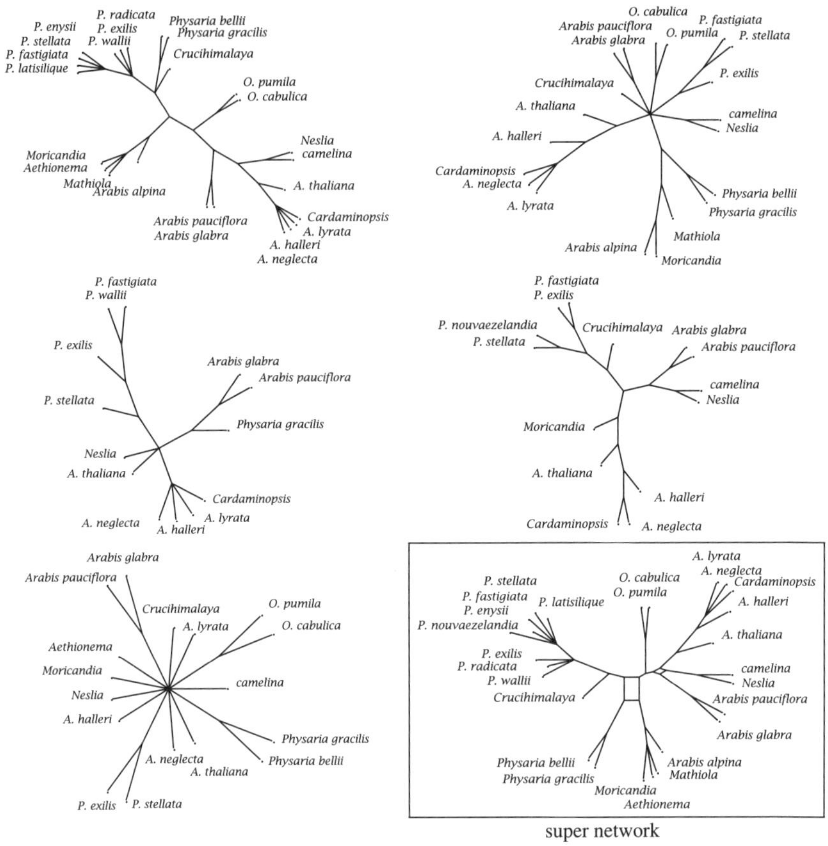

FIG. 9.8. Five partial trees, each containing between 13 and 25 species of plants [31] and the resulting super networks of 26 taxa, obtained from the input trees using the Z-closure method.

The *Z-closure method* [22] takes as input a set of partial X-trees $\mathcal{T} = \{T_1, \ldots, T_K\}$ and produces as output a set of X-splits Σ. Let $H = (S_1, \ldots, S_r)$ be an array containing all splits of the input trees. The method proceeds by repeating the following step, until no further productive pair exists: Choose a productive pair S_i and S_j, and replace the two splits in H by $S'_i = \frac{A_i}{B_i \cup B_j}$ and $S'_j = \frac{A_i \cup A_j}{B_j}$, respectively. The algorithm is fast and operates *in place*, however it is order-dependent and so should be run multiple times. In Fig. 9.8 we show an example of five partial gene trees and a summarizing super network.

Consensus networks and super networks can be used to summarize different gene trees, as discussed above, but they can also be used to summarize different tree estimations obtained by methods such as bootstrapping [10] or Bayesian sampling [37, 38].

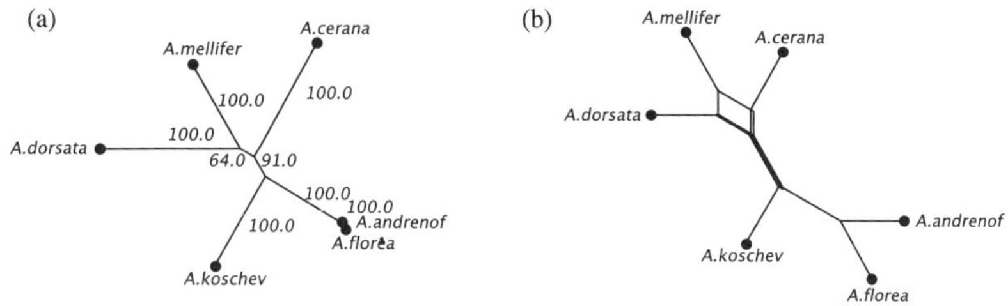

FIG. 9.9. (a) A Neighbor-Joining (NJ) tree [39] of six species of bees [46] labelled with bootstrap values obtained using 1000 bootstrap samples. (b) A split network representing all splits that occurred in any of the bootstrap replicates, with edge lengths representing the number of replicates that contain the split. The split network clearly shows that the low support of 64% of one of the central edges in the NJ tree is due to the fact that the data also contains strong support for the alternative grouping of *A.mellifer* with *A.cerana*.

One practical difference between the consensus network method and the Z-closure approach is that the former provides a parameter d with which the amount of conflict that is presented in the final split network can be controlled, which the latter method lacks.

To address this, in [25] we define the concept of the *distortion* of a split, as a measure of how much a tree needs to be modified in order to accommodate the split and extend our Z-closure to obtain a filtered super-network. The *distortion* of a (partial) X-tree T relative to a given X-split S is the parsimony score of S (interpreted as a binary character) minus one, over all trees T' that resolve T (see [25] for details). To obtain a filtered set of splits for a given set of trees, one specifies a maximal distortion per tree and a minimal number of trees on which this condition is fulfilled, and then collects all splits that meet the requirements. An example is discussed in Section 9.4 (see Fig. 9.23).

Bootstrapping is a popular way to study how robust the different branches of an inferred tree are, with respect to sampling error. In bootstrapping, one first generates many *bootstrap replicates* of input sequence alignments by randomly resampling from the original sequence alignment A. Then every branch of the originally inferred tree is labelled by the percentage of replicates that support the corresponding split. We propose to construct a *bootstrap network* [21] by collecting all splits that are present in any of the replicates and displaying them in a split network (see Fig. 9.9).

9.3 Split networks from sequences and distances

In the previous section, we saw that incompatible splits and split networks arise naturally in the context of tree consensus. In this section we discuss a number of methods that generate incompatible splits directly from aligned sequences or distances.

Consider a set of taxa $X = \{x_1, \ldots, x_n\}$ represented by an alignment A of binary sequences $a_1, \ldots, a_n$, where a_i corresponds to x_i for $i = 1, \ldots n$:

$$A = \left| \begin{array}{ll} a_1 = & a_{11}a_{12}\ldots a_{1m} \\ a_2 = & a_{21}a_{22}\ldots a_{2m} \\ & \ldots \\ a_n = & a_{n1}a_{n2}\ldots a_{nm} \end{array} \right| .$$

Every non-constant site j in such an alignment defines a split $S = \frac{A}{B}$ of X with $A = \{x_i \mid a_{ij} = 0\}$ and $B = \{x_i \mid a_{ij} = 1\}$. Vice versa, any given split $S = \frac{A}{B}$ can be represented by two distinct patterns of noughts and ones in the alignment, one obtained by choosing $a_{ij} = 1$ for all $x_i \in A$ and $= 0$ otherwise, and the other obtained by choosing $a_{ij} = 1$ for all $x_i \in B$ and $= 0$ otherwise.

Binary sequences arise in a number of ways. For example, DNA sequences are sometimes converted into the RY-alphabet, using R to represent the two purines, A and G, and Y to represent the two pyrimidines, C and T. Other sources of binary sequences include SNPs (single nucleotide polymorphisms), the presence or absence of certain restriction sites, or the presence or absence of different genes in complete genomes.

A visual representation of an alignment A of binary sequences can be obtained by constructing a split network representing all the splits defined by the columns of the alignment and then labelling each edge by the set of positions that are associated with the corresponding split (see Fig. 9.10).

If a given set of X-splits Σ is compatible, then the split network that represents Σ is a uniquely defined tree. If Σ is not compatible, then the corresponding split network is not, in general, uniquely defined. The concept of a *median network* [2] avoids this ambiguity and is defined as a split network that satisfies an additional *median closure property* which ensures that the graph is uniquely defined. In practice, the median network can be overly complicated. A simpler split network that is easier to comprehend will often exist, but at the price of being non-unique (see Fig. 9.11).

The *split decomposition* [1] and the *Neighbor-Net* method [5] each take as input a distance matrix D on X and produce as output a set of *weighted* X-splits Σ, where the sum of weights of all splits that 'separate' two taxa $x, y \in X$ is an approximation of the given distance $D(x, y)$. Both methods have the useful property that they are guaranteed to produce a tree, whenever the distance matrix fits a tree, and otherwise to produce (more or less) tree-like split networks that potentially display different and conflicting signals in a given data set.

In [1], the authors prove that the set of splits Σ computed by the split decomposition is *weakly compatible*, which means that for any three splits S_1, S_2, and S_3 in Σ and all $A_i \in S_i$ $(i = 1, 2, 3)$ and $A'_i := X \setminus A_i$, at least one of the four intersections $A_1 \cap A_2 \cap A_3$, $A_1 \cap A'_2 \cap A'_3$, $A'_1 \cap A_2 \cap A'_3$ and $A'_1 \cap A'_2 \cap A_3$ is

```
11100011111110001111001111111110000111100011101010110001111100111111110001001111100111111011111001101111100010011111111110001
11100011111100001100001111111100000011111001110001011100111110011111111100010011111110111111101111110110110111101111111010001000
11100011111100001100001111111100000011111001111001011100111110011111111100010011111110111111101111110110110111101111111010001000
11100011111100001100001011111101000011101011100010111101111110111111111100100111111001111110011111001101101101011111111000000000
011100111001000010000011111110001101100000111010101110111111001101111001011011111000111110011101101110110001101101111110000
1110011110010110100000011111110001101100000111000101110111111001101111001011111111000110110011101101110110001101101111110000
1110101110010001100011001111110000001111000111101111111001110010101010111001100111110000001100101110011011011001110011110010110
```

(a)

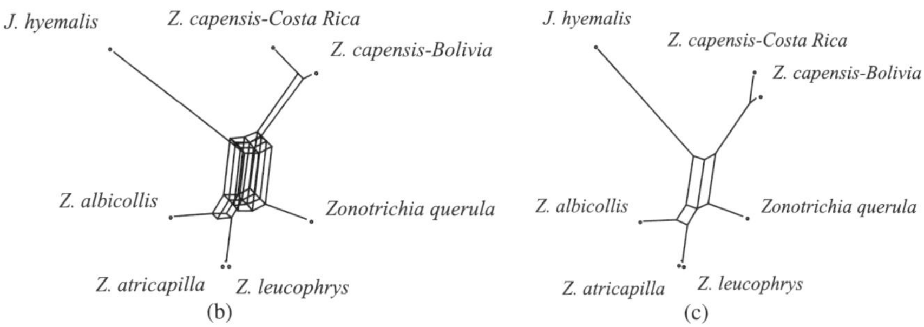

FIG. 9.10. (a) Dataset of 122 restriction sites obtained from 19 restriction endonucleases applied to mtDNA of Zonotrichia (sparrows)[47] in the following order: *Z. querula, Z. atricapilla, Z. leucophrys, Z. albicollis, Z. capensis–Bolivia, Z. capensis–Costa Rica,* and *J. hyemalis* (outgroup). (b) Split network representing all different non-constant columns of the alignment. (c) Split network representing all splits that occur in at least two different columns of the alignment.

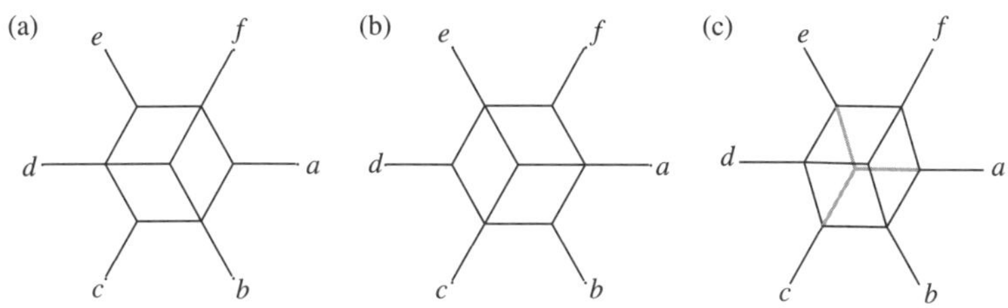

FIG. 9.11. Three different split networks all representing the same set of splits. The network shown in (c) has the *median closure property*, as discussed in [2].

empty. This is a nice generalization of compatibilty, as in practice, the resulting split networks are usually planar or only mildly non-planar.

In [5], the authors show that the set of splits Σ computed by the Neighbor-Net method is always 'cyclic', which implies that Σ can be represented by an *outer-labelled planar* split network, that is, a plane network in which all taxa appear around the perimeter of the network [8].

To illustrate the two methods, we computed the observed p-distances for the data set shown in Fig. 9.10 simply as the number positions in the alignment at

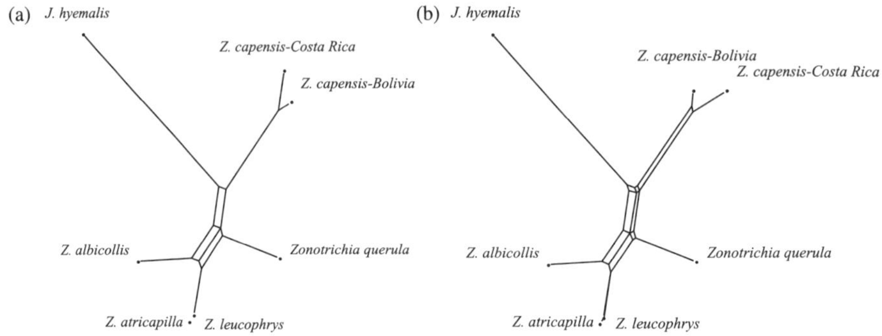

FIG. 9.12. (a) Network representing all splits obtained by applying the split decomposition method to the observed distances of the data shown in the previous figure. (b) Network representing all splits obtained by applying the Neighbor-Net method to the same distances.

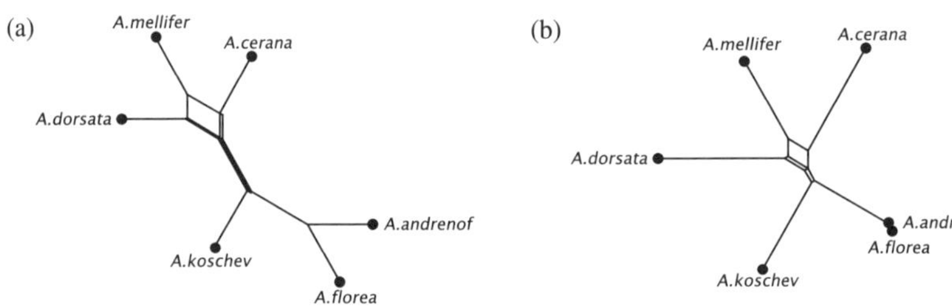

FIG. 9.13. Both the bootstrap network (a) and the split network obtained using the split decomposition method (b) clearly indicate the ambiguous grouping of *A. mellifer*.

which the state of two sequences differ. We then applied the two methods to the resulting distance matrix to obtain the two networks shown in Fig. 9.12. As recombination of mtDNA is believed to be extremely rare, the incompatibilities apparent in the figure are most likely due to multiple mutations at individual sites.

As a further illustration of such methods, we compare the bootstrap network discussed above with the network produced using the split decomposition method (see Fig. 9.13). Here, both the bootstrap analysis and split decomposition indicate that the input sequences contain two different and incompatible signals.

The split decomposition method is useful for visualizing conflicting signals in a data set. However, it is sensitive to noise and can have poor resolution

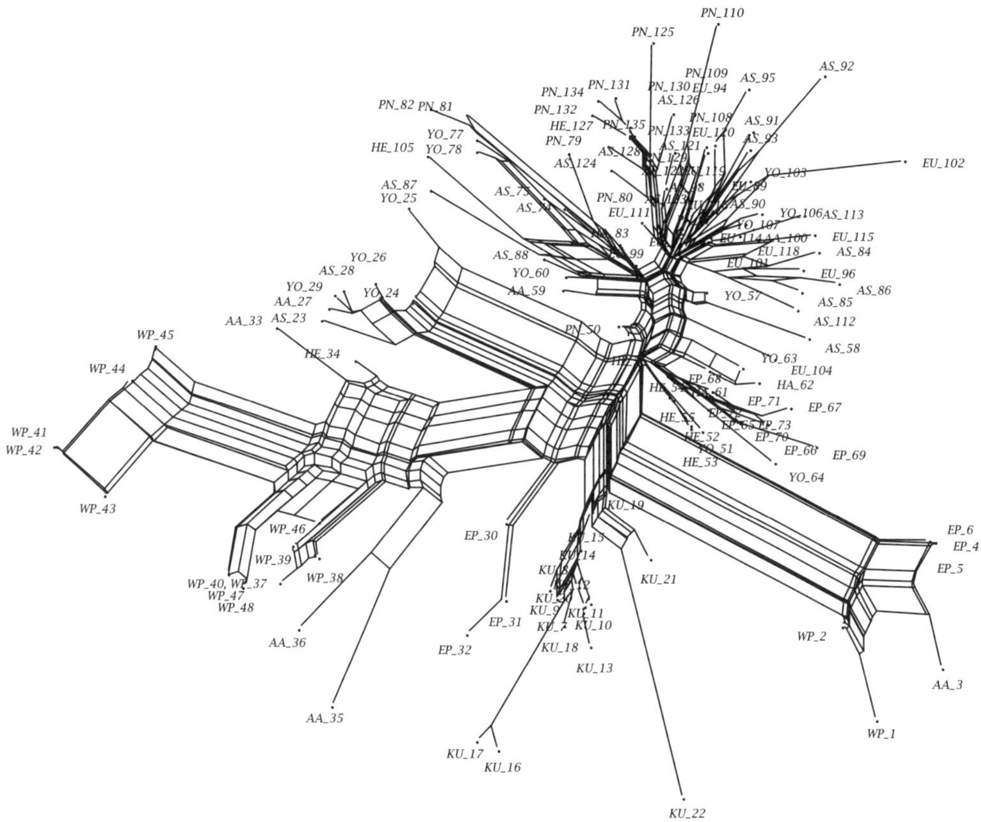

FIG. 9.14. A split network computed using the Neighbor-Net method [5], using a distance matrix computed from 133 human mtDNA sequences [44].

on large or divergent data sets. The *Neighbor-Net* method [5] is a hybrid of Neighbor-Joining and split decomposition. It is applicable to data sets containing hundreds of taxa. Figure 9.14 shows a large example based on 133 human mtDNA sequences [44].

There are currently three programmes available for computing split networks from biological data:

- SplitsTree4 [21] provides implementations of all methods described in this section, including a number of different algorithms for constructing networks from splits.
- SpectroNet [17] provides an algorithm for constructing a median network and some related methods.
- SplitsTree [20] provides an implementation of the split decomposition method.

9.4 Hybridization and reticulate networks

In this section we first discuss the concept of hybrid speciation. We then describe a simple model of evolution that incorporates gene trees and reticulation events. This is followed by an introduction to the concept of a reticulate network and a discussion of some of the approaches for inferring such networks from gene trees.

There are two main mechanisms of speciation by hybridization [29]. In *allopolyploidization*, the hybrid speciation occurs when two different lineages produce a new species that has the complete nuclear genomes of both parental species. Thus, two parents X and Y each pass on their whole diploid genomes (with $2n_1$ and $2n_2$ chromosomes respectively) to produce a polyploid offspring Z with $(2n_1 + 2n_2)$ chromosomes. Subsequently, over time it can happen that the genome is reduced to half its size and then the net result is a mosaic of genes from both ancestors. In *diploid* (or *homoploid*) hybrid speciation, each of the parents produces normal gametes (haploid) to produce a normal diploid hybrid.

Although diploid hybridization is more common, the ability of the hybrid to backcross with the parent species usually prevents a new species from arising. Although less common, allopolyploidization is believed to produce more new species. Hybridization is usually restricted to plants, frogs, and fish.

We will describe a simple model of evolution that incorporates *reticulate* events such as hybridization, and, in the next section, recombination. Consider the network shown in Fig. 9.15.

In such a *reticulate network* N, a *reticulate node* r inherits a sequence from two different ancestors P and Q. We will assume that genes are 'atomic' with respect to reticulation and thus that the evolutionary history of any given gene is a tree. Consider a gene g_1 that is inherited by r from the P ancestor. The phylogeny of g_1 is shown in Fig. 9.16. Similarly, we show the phylogeny of a gene g_2 inherited from Q in Fig. 9.17.

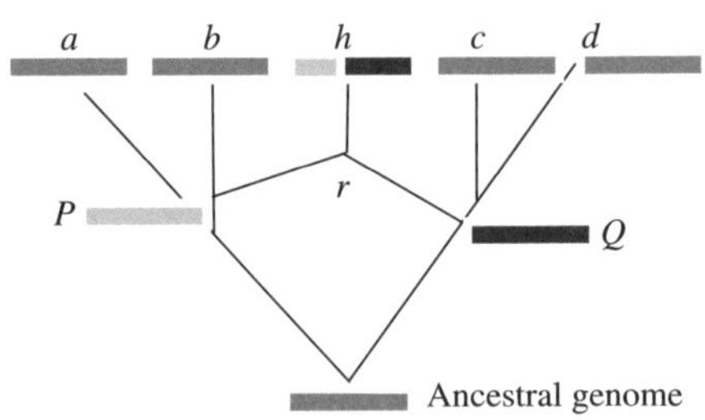

FIG. 9.15. A simple model of reticulate evolution in which a species r obtains part of its genome from one ancestor P and a complementary part from a different ancestor Q. In a hybridization scenario, one usually assumes that the two different parts are of a similar size, whereas in the context of horizontal gene transfer, one of the two contributions is much smaller than the other.

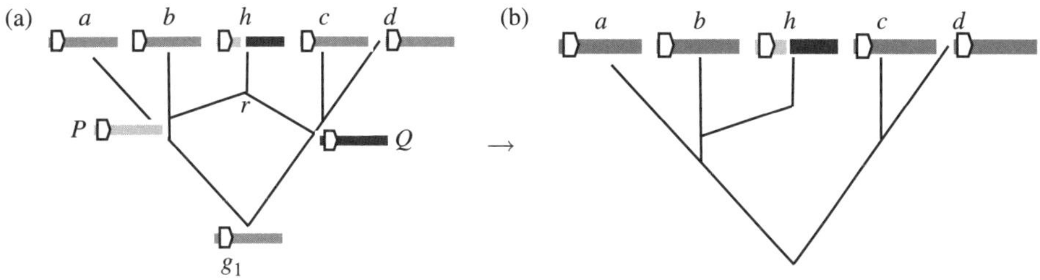

FIG. 9.16. If r inherits its copy of a gene g_1 from P as indicated in (a), then the gene tree associated with g_1 is the one displayed in (b).

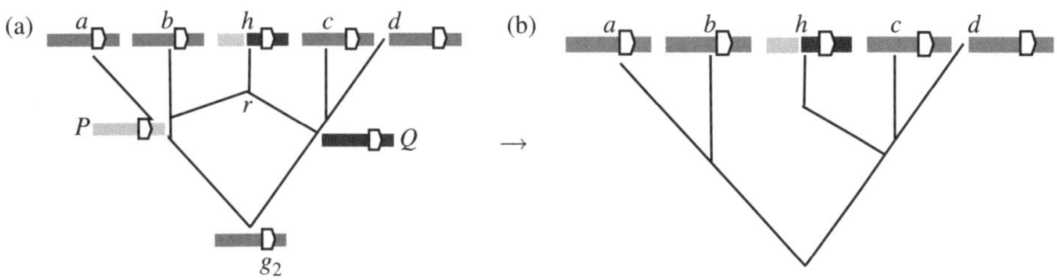

FIG. 9.17. If r inherits its copy of a gene g_2 from Q as indicated in (a), then the gene tree associated with g_2 is the one displayed in (b).

Definition 9.3 *Let X be a set of taxa. A* (rooted) reticulate network N on X *is a connected, directed acyclic graph where:*

- *there exists precisely one node of indegree* 0, *called the* root*;*
- *all other nodes are* tree nodes *of indegree* 1, *or* reticulation nodes *of indegree* 2*;*
- *every edge is either a* tree edge *incident to precisely two tree nodes, or a* reticulation edge *leading to a reticulation node; and*
- *the set of* leaves *(nodes of outdegree* 0*) labelled by X.*

Let N be a reticulate network on X with k reticulation nodes $r_1, \dots, r_k$. For any such node r_i, let p_i and q_i denote the two associated reticulation edges. We can obtain an X-tree from N by choosing and removing one reticulation edge p_i or q_i for each r_i (see Fig. 9.18), and then deleting any unlabelled leaf nodes. The set of trees $\mathcal{T} = \mathcal{T}(N)$ obtainable in this way is called the set of *induced* trees or trees that *can be sampled* from N. For any tree edge $e \in N$, let $\mathcal{T}(e) \subseteq \mathcal{T}(N)$ denote the set of all sampled trees that contain e. We define $\Sigma(e) = \{\sigma_T(e) \mid T \in \mathcal{T}(e)\}$ as the set of all splits that can be *sampled* from e.

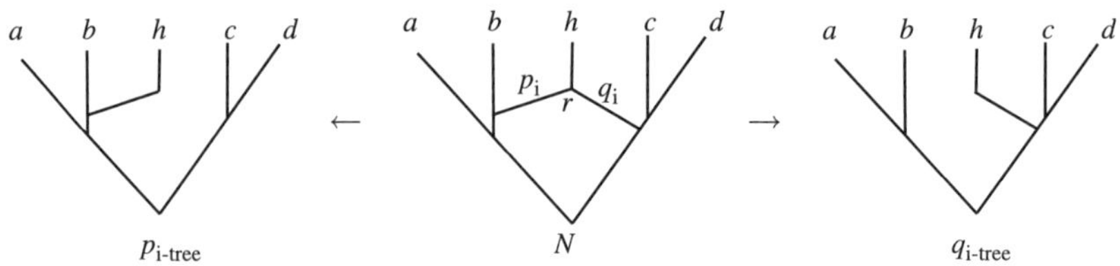

FIG. 9.18. Choosing either the p_i or q_i edge at each vertex r_i gives rise to different trees.

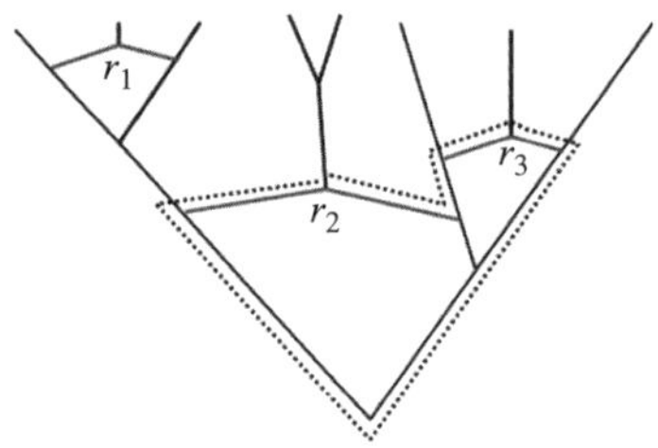

FIG. 9.19. In this reticulate network, the reticulate vertices r_2 and r_3 are contained in a common cycle (indicated by dotted lines) and are therefore not independent.

The following is easy to see:

Lemma 9.4 *The number of different trees that can be sampled from a network N with k reticulations is $|\mathcal{T}(N)| \leq 2^k$.*

Given a set of trees $\mathcal{T} = \{T_1, \ldots, T_m\}$, we would like to determine the reticulate network N from which the trees were sampled. This form of the problem is not always solvable. For example, when some of the 2^k possible trees are missing. Thus we consider the following:

Problem 1 (Most Parsimonious Network Problem). *Determine a reticulate network N such that $\mathcal{T} \subseteq \mathcal{T}(N)$ and N contains a minimum number of reticulation nodes.*

In general, this is known to be a hard problem [45, 4]. We now discuss a special case that can be solved efficiently.

Two reticulation nodes r_i, r_j in N are *independent* of each other, if they are not contained in any common undirected cycle. Consider the example shown in Fig. 9.19. There, r_1 is independent of r_2 and r_3, whereas r_2 and r_3 are not independent of each other, as the highlighted cycle shows.

A reticulation that is independent of all others is sometimes called a *gall* and a network N in which all reticulations are galls is sometimes called a *galled tree* [13] or, redundantly, a *galled-tree network* [35].

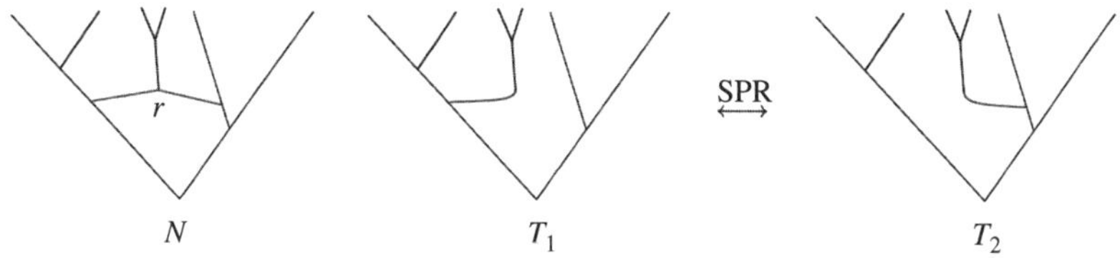

FIG. 9.20. In the reticulate network N, the subtree rooted at r attaches to the remainder of the network in two different places. The two corresponding gene trees are related by a single SPR operation between tree T_1 and tree T_2.

In [33], the author considered the situation in which the true reticulate network N contains only a single reticulation. He observed that an independent reticulation corresponds to a *sub-tree prune and regraft* (SPR) operation (see Fig. 9.20). Here is a summary of the algorithm which was employed:

- Given two bifurcating trees, compute their SPR distance.
- If the distance is 0, return a tree.
- If the distance is 1, return a network.
- In all other situations, fail.

This approach has been generalized to networks with multiple independent reticulations [35]. Unfortunately, on real data, such algorithms will usually return 'fail'. One challenge is to produce useful output in the case of real data.

In Fig. 9.21, we illustrate an important relationship between a reticulate network N and the network of all splits of all trees sampled from N [12, 23]. There exists a one-to-one correspondence between the 'netted regions' of the split network and the 'tangles' of dependent reticulations of the reticulate network.

More precisely, we prove the following result in [23]:

Theorem 9.5 (Decomposition Theorem) *Suppose N is a reticulate network. Two tree edges e, f are contained in a cycle in N, if and only if there exist two splits $S \in \Sigma(e)$ and $S' \in \Sigma(f)$ that are contained in the same connected component of the incompatibility graph $IG(\Sigma(N))$.*

The theorem inspires the following approach:

- Determine the set of all input splits.
- Determine the netted components of the split network.
- Analyse each component C separately.
- If C can be explained by a reticulate network $N(C)$, then locally replace C by $N(C)$.

Using an algorithm that allows 'overlapping' reticulations [23], this approach is implemented in the programme SplitsTree4.

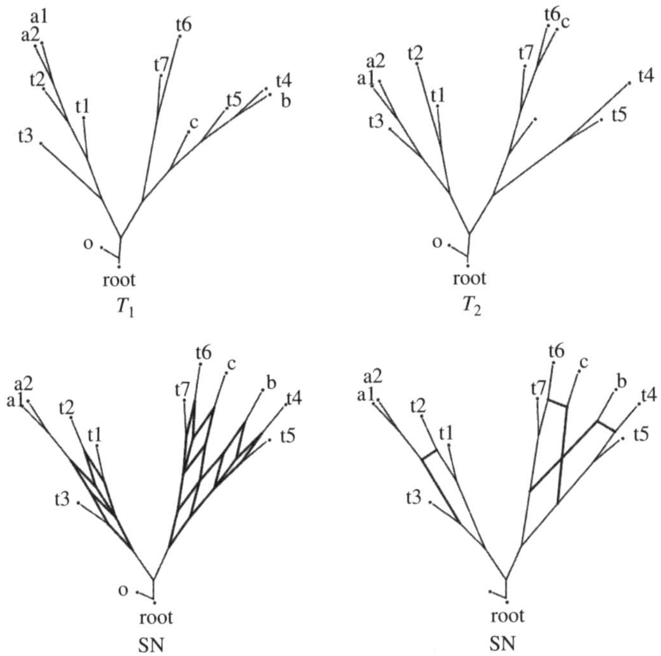

FIG. 9.21. Here we depict two trees T_1 and T_2, a split network SN and a reticulate network RN. The two trees T_1 and T_2 contain incompatible splits. The rooted split network SN displays all splits present in T_1 and T_2. Both trees can be sampled from the rooted reticulate network RN.

Consider the two trees on *Ranunculus* (buttercup) data [30], shown in Fig. 9.22. In Fig. 9.23(a) we display a split network representing all splits contained in either of the two trees. This split network suggests that *R. nivicola* may be a hybrid of the evolutionary lineages on the left- and right-hand sides.

All current algorithms for constructing reticulate networks are sensitive to false edges in the input trees and for this data set, initially no reticulation is detected. If we apply a *distortion filter* [25] to the set of splits and keep only those splits that have a distortion of at most 1 on each of the two trees, then this produces the network shown Fig. 9.23(b).

For this particular example, the distortion filter removes all confusing signals. Application of the hybridization network-construction algorithm that we have implemented in SplitsTree [23, 21] produces the network depicted in Fig. 9.24. This network clearly shows a reticulation event for *R. nivicola*, in agreement with earlier suggestions that *R. nivicola* is an allopolyploid formed between *R. insignis* and *R. verticillatus* [30]. Another clear reticulation scenario involves *Renysii3*, which has also been implicated in hybridization (Pete Lockhart, personal communication).

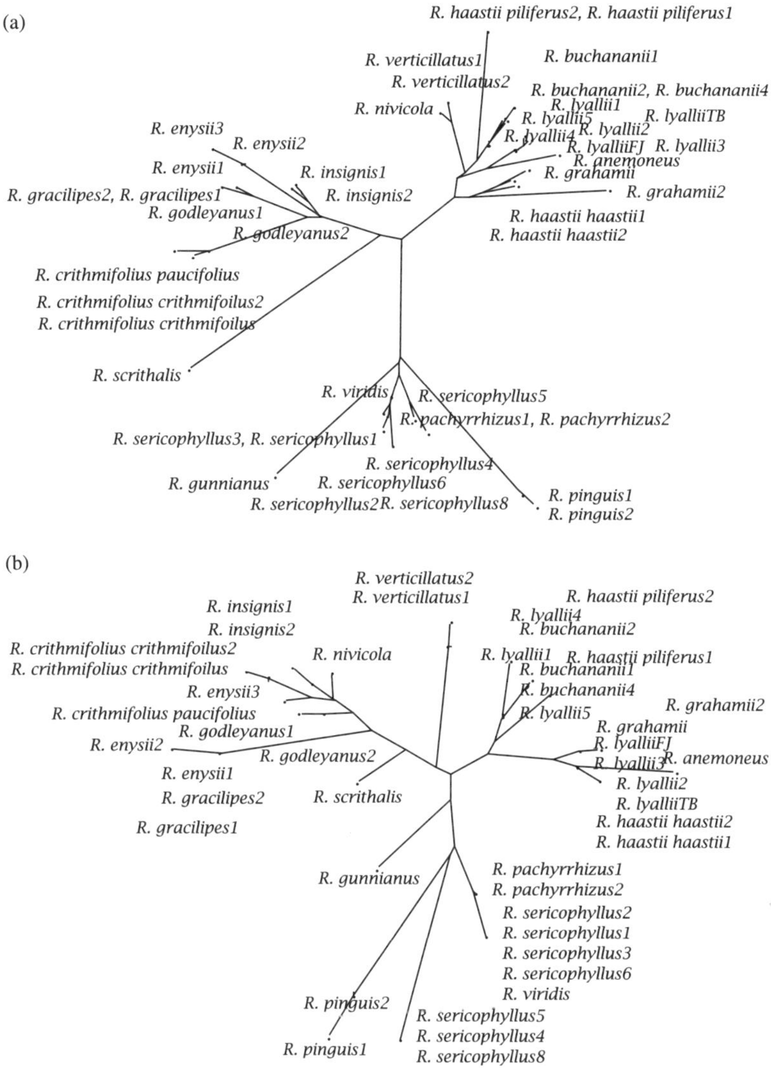

FIG. 9.22. Two phylogenetic trees for 46 buttercup species, obtained (a) using a nuclear ITS gene and (b) using a chloroplast JSA region [30].

Here is an overview of publicly available software for constructing reticulation networks from trees:

- SplitsTree4 [21] provides a method *HybridizationNetwork* that takes a list of trees or partial trees as input and produces a phylogenetic network, in which reticulate network, in which any 'unresolvable tangles' are represented by their split network, as illustrated above. (By an 'unresolvable tangle' we

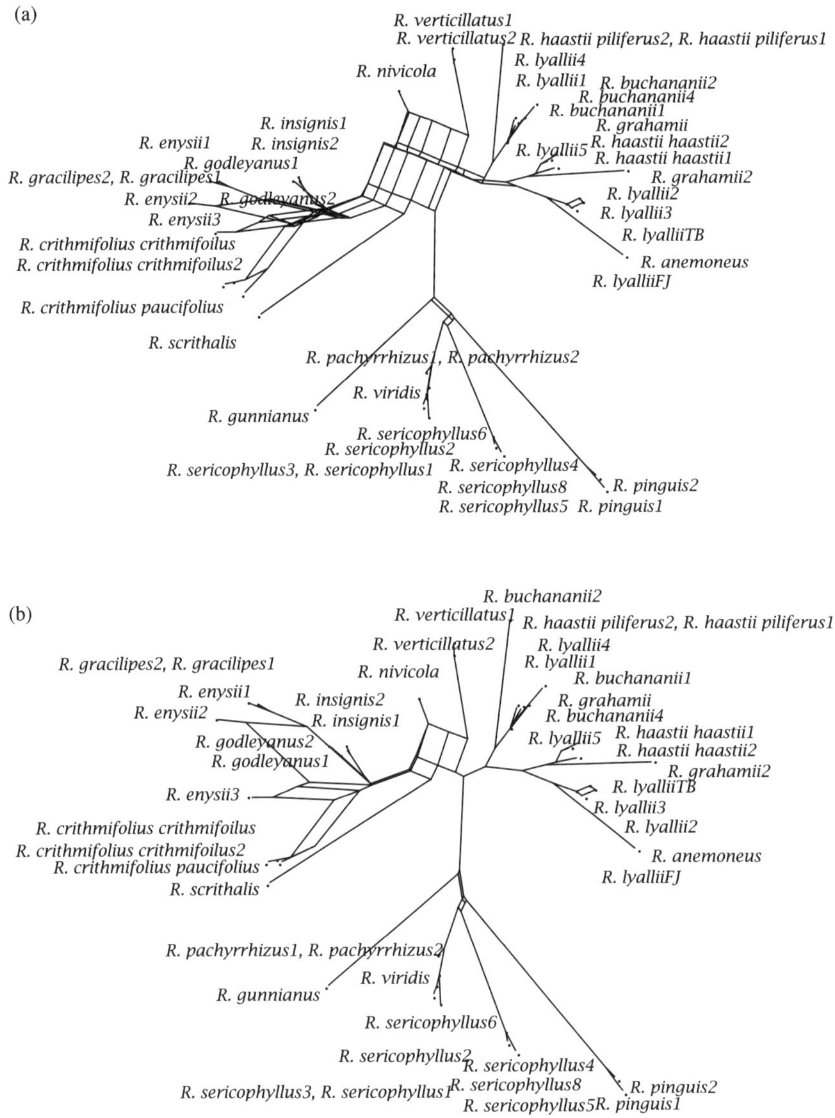

FIG. 9.23. (a) A split network displaying all splits contained in the two trees shown in Fig. 9.22. (b) The split network for those splits with distortion at most 1 on each of the two trees (see [25] for details).

mean a connected component of the incompatibility graph of the input splits that cannot be sampled from any reticulate network obtainable by the employed algorithm.)

- Reference [35] describes a programme SPNet, which is not publicly available.

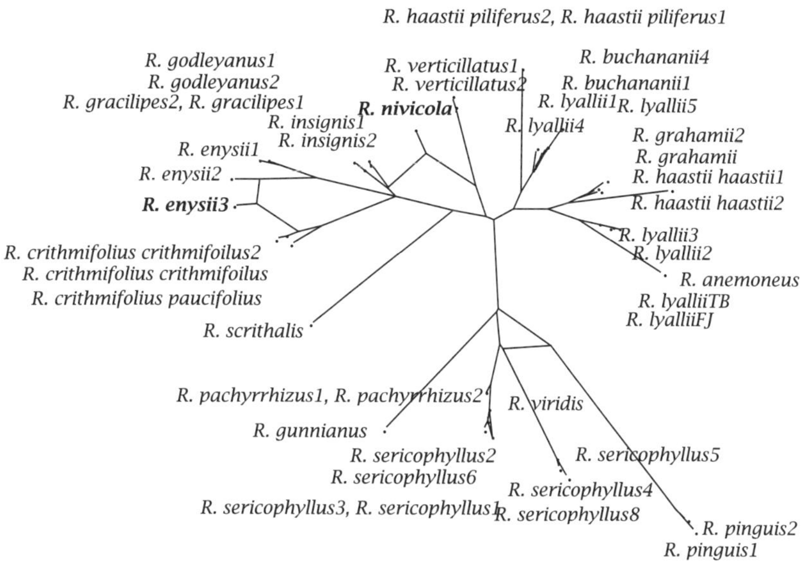

FIG. 9.24. Application of our algorithm to the filtered network gives rise to the displayed reticulate network.

9.5 Recombination networks

In this chapter, we will look at the problem of reconstructing a reticulate network from an alignment of binary sequences that have evolved under a model of mutation, speciation, and *recombination* events. This has been much studied in population genetics [19, 14, 11, 41, 42, 43] and *ancestor recombination graphs* (ARGs) are used in that context.

We will concentrate on the combinatorial aspects of the problem and thus consider *recombination networks* rather than ARGs. We make some simplifying assumptions:

- all sequences have a common ancestor, and
- any position can mutate once at most.

Given an alignment A of binary sequences of length n, a *recombination network* R [9] can be viewed as a *reticulation network* N, together with:

- a labelling of all nodes by binary sequences of length n, in such a way that the leaves of R are labelled by A,
- a corresponding labelling of each tree edge e by those positions that mutate along e, and
- a corresponding labelling of each reticulation node r indicating the crossover position for the recombination at r.

An example is shown in Fig. 9.25.

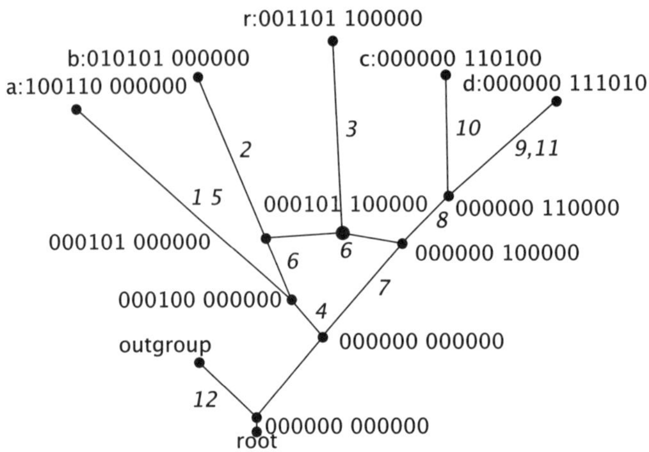

FIG. 9.25. Example of a recombination network for six sequences a, b, c, d, r, and *outgroup*, of length 12.

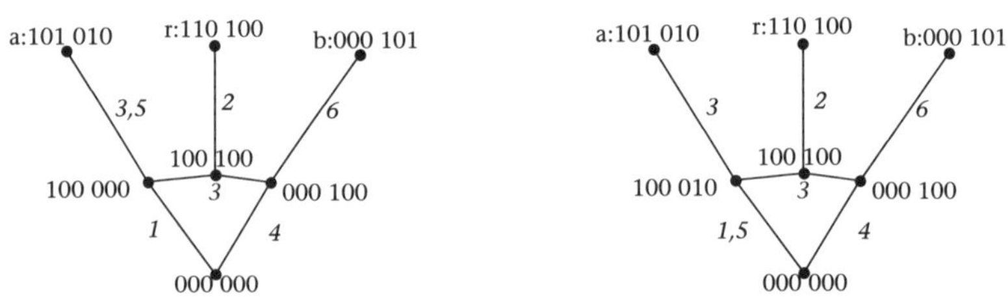

FIG. 9.26. The mutation at position 5 can be placed at two different locations, either (a) on the left-most leaf edge, or (b), inside the reticulation cycle.

Interestingly, the placement of mutations on edges is not uniquely defined. In the network depicted in Fig. 9.26, the mutation at position 5 can happen along two different edges. Faced with this choice, current algorithms [13, 24] place such ambiguous mutations outside of the reticulation cycle.

In the case of independent reticulations, Dan Gusfield and colleagues have developed an algorithm for computing a galled tree from binary sequences [13, 12]. This approach computes a galled tree as follows:

- Determine the components of the incompatibility graph.
- For each component C, do the following:
  - * Determine the restriction of the data set with respect to C, identifying with each other any taxa that are assigned identical sequences.
  - * Check whether C is bipartite and 'biconvex'.
  - * Determine whether removing one taxon produces a perfect phylogeny.
  - * If so, arrange the taxa in a gall.
  - * Return a description of the network.

An alternative splits-based approach is to first construct an underlying reticulate network using the approach described in Section 9.4 [23, 24] and then to compute an appropriate labelling of nodes and edges.

In [27], the phylogeographic structure of lineages of the fungus *Fusarium graminearum* is investigated. The papers studies 37 strains and uses the DNA sequence of six different genes to infer phylogenetic relationships between them. One result reported is that the locus *3-O-acetyltransferase* (TRI101) has undergone intragenic recombination in one of the strains (number 28721), based on the sequence of physically linked markers.

The data set for the TRI101 locus is also discussed in [36] as an example of a data set that contains a confirmed instance of recombination. The data set for this gene consists of an alignment of 28 DNA sequences of length 1336. The DNA sequences represent different strains of *F. graminearum* and are identified by numbers. The strains are partitioned into 7 lineages $(1-7)$ excluding strain 28721. In [27] the authors reported that the TRI101 sequence for 28721 arose through recombination between African lineage 2 and Asian lineage 6.

In Fig. 9.27, we show all non-constant positions of the TRI101 data set. As each character in this alignment takes on precisely two different states, we can represent this data set by a split network, as indicated in Fig. 9.28.

Application of our recombination network algorithm, as implemented in [21], computes a recombination network that correctly displays strain 28721 as resulting from a hybrid of the lineages 2 and 6. As the computed network contains a single isolated reticulation, it is a 'galled tree' and is therefore also obtainable by Gusfield's algorithm [12].

The data set shown in Fig. 9.30 is taken from restriction maps of the rDNA cistron (length $\approx 10kb$) of 12 species of mosquitoes using 8 $6bp$ recognition restriction enzymes [28]. Of 26 scored sites, 18 were polymorphic among the ingroup taxa.

This data set was analysed using a number of different tree-reconstruction methods with inconclusive results [28]. Indeed, the split network associated with this data set, shown in Fig. 9.31(a) indicates the presence of many conflicting signals. Interactive trial and error reveals that two taxa *Aedes triseriatus* and *Armigeres subalbatus* gives rise to a simpler split network, shown in Fig. 9.31(b).

A possible recombination scenario is depicted in Fig. 9.32. In this scenario, *Haemagogus equinus* arises by a single-crossover recombination, where as a second such recombination leads to *A.albopictus* and *A.flavopictus*. The main goal here is to demonstrate the general approach of using a split network to give a robust representation and then to use combinatorial algorithms in an attempt to interpret the given configuration of splits in terms of reticulations. Technically, this data set is interesting because it involves overlapping reticulations, that cannot be computed using 'galled tree' approaches. However, to establish whether recombination is indeed the true biological cause of the pattern of data observed requires a more detailed study of the biology involved, which goes beyond the scope of this chapter.

Strain	Non-constant positions of alignment
28436	gaccatcacgatgtgggtgggctcctgaacccccaactactttcagacccacctggttgtggcg
28723	....
29010	....
2903	....g......c..
28585	....g......c..
28718	....g......c..
25797	t.t.....t..c...a....................................t.a.........
29148	t.t.....t..c...a....................................t.a........a
29020	.g...g.....c........tt....a..............c.tt...tt..t.....a.ca..
26916	.g...g.....c........tt....a..............c.tt...tt..t.....a.ca..
29011	.g...g.....c........tt....a..............c.tt...tt..t.....a.ca..
29105	.g...g.....c........tt....a..............c.tt...tt..t.....a.ca..
26752	......g...gc....................t...................t...........
26754	......g..a.c....................t...................tc..........
26755	......g...gc..a.................t...................t...........
6101	.......g...c.....c....g..........a.................tt..c........
13818	.......g...c.....c....g..........a.................tt..c........
26156	.......g...c.....c....g..........a.................tt..c........
28720	.......g...c.....c....g..........a.................tt..c........
28721	....tt..c........
5883	...t.......ca......a..g.t....t......................t...c.......
6394	...t.......ca......a..g......t......................t...c.......
13383	...t.......ca......a..g......t.....g................t...c.......
28063	...t.......c.......a..g......t.g....................t...c.......
28336	...t.......ca......a..g......t......................t...c.......
28439	...........c.......a..g......t......................t...c.......
29169	...t.......c.......a..g......t.g....................t...c.......
O13393	...........c.c..a.a....t.a.gg.t...g.tcggc.c..cgtt.c.t....c.c..a.

FIG. 9.27. The 64 non-constant sites in the alignment of TRI101 sequences for different strains of *F. graminearum* and one outgroup sequence *O*13393 representing *F. lunulosporum* (from [27]).

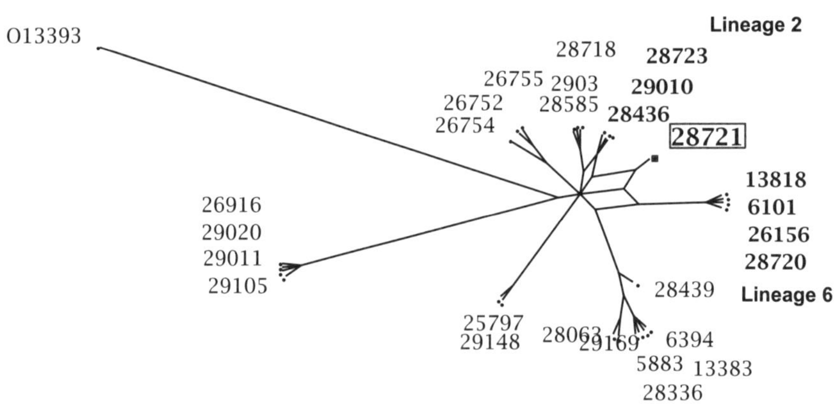

FIG. 9.28. Split network representing the 46 different splits present in the data set shown in Fig. 9.27. This network places taxon 28721 between lineage 2 and lineage 6.

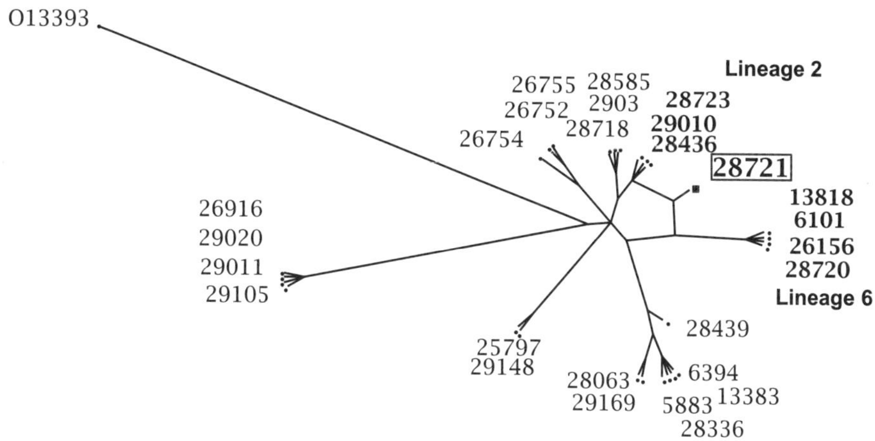

FIG. 9.29. Recombination network representing the 46 different splits present in the data set shown in Fig. 9.27. This network shows taxon 28721 arising through recombination from the lineages 2 and 6.

Species	Restriction sites
Aedes albopictus	11110101010100010101010010
Aedes aegypti	11110101000100010101000010
Aedes seatoi	11110101010100010101010000
Aedes flavopictus	11110101010100010101010010
Aedes alcasidi	11110101010100010101010000
Aedes katherinensis	11110101010100010101010000
Aedes polynesiensis	11110101000100010101010010
Aedes triseriatus	10110101000110010101000000
Aedes atropalpus	10110101000100010111000010
Aedes epactius	10110101000100010111000010
Haemagogus equinus	10110101000110010101010000
Armigeres subalbatus	10110101000100010101000000
Culex pipiens	11110111000100011101001011
Tripteroides bambusa	11110111000100010101000010
Sabethes cyaneus	11110101001100010101010000
Anopheles albimanus	11011101100101110101110100

FIG. 9.30. Restriction site data for mosquitoes [28].

Here is an overview of software for computing a recombination network from binary sequences:

- Software implementing the approach of Dan Gusfield and colleagues [13, 12] for constructing galled trees is available from: `http://www.csif.cs.ucdavis.edu/~gusfield`.

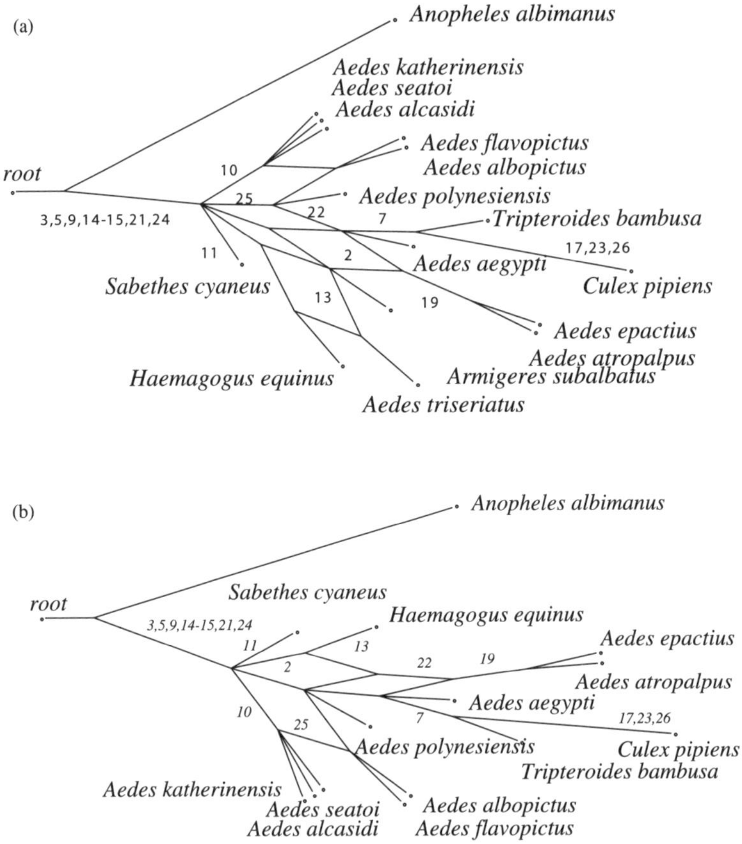

FIG. 9.31. (a) A rooted split network representing all columns of the alignment shown in Fig. 9.30. Edge labels indicate which columns are associated with a given split. (b) A slightly simpler rooted split network obtained by removing *A. triseriatus* and *A. subalbatus*.

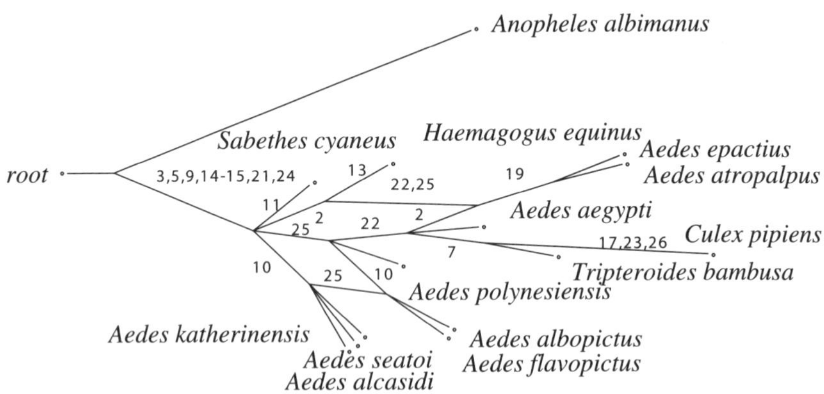

FIG. 9.32. A possible recombination scenario explaining the mosquito data set with *A. triseriatus* and *A. subalbatus* removed.

- SplitsTree4 [21] contains a method *RecombinationNetwork* for constructing galled trees and more general recombination networks from binary sequences [23, 24].
- Software is available that computes an optimal recombination network using a branch-and-bound approach (see [32]).

References

[1] Bandelt, H.-J. and Dress, A. W. M. (1992). A canonical decomposition theory for metrics on a finite set. *Advances in Mathematics*, **92**, 47–105.

[2] Bandelt, H.-J., Forster, P., Sykes, B. C., and Richards, M. B. (1995). Mitochondrial portraits of human population using median networks. *Genetics*, **141**, 743–753.

[3] Bininda-Emonds, O. (ed.). (2004). *Phylogenetic Supertrees. Combining Information to Reveal the Tree of Life.* Kluwer Academic Publishers, Dordrecht.

[4] Bordewich, M. and Semple, C. (2006). Computing the minimum number of hybridisation events for a consistent evolutionary history. To appear in: *Discrete Applied Mathematics.*

[5] Bryant, D. and Moulton, V. (2002). NeighborNet: An agglomerative method for the construction of planar phylogenetic networks. In *Proceedings of WABI, 2002 (Workshop on Algorithms in Bioinformatics)* (eds. R. Guigó and D. Gusfield), LNCS 2452, pp. 375–391. Springer-Verlag, Berlin.

[6] Buneman, P. (1971). The recovery of trees from measures of dissimilarity. In *Mathematics in the Archaeological and Historical Sciences* (eds. F. R. Hodson, D. G. Kendall, and P. Tautu), pp. 387–395. Edinburgh University Press, Edinburgh.

[7] Doolittle, W. F. (1999). Phylogenetic classification and the universal tree. *Science*, **284**, 2124–2128.

[8] Dress, A. W. M. and Huson, D. H. (2004). Constructing splits graphs. *IEEE/ACM Transactions in Computational Biology and Bioinformatics*, **1**(3), 109–115.

[9] Eddhu, S., Gusfield, D., and Langley, C. (2004). The fine structure of galls in phylogenetic networks. to appear in: *INFORMS Journal of Computing - Special Issue on Computational Biology.*

[10] Felsenstein, J. (1985). Confidence-limits on phylogenies, an approach using the bootstrap. *Evolution*, **39**(4), 783–7911.

[11] Griffiths, R. C. and Marjoram, P. (1996). Ancestral inference from samples of DNA sequences with recombination. *Journal of Computational Biology*, **3**, 479–502.

[12] Gusfield, D. and Bansal, V. (2005). A fundamental decomposition theory for phylogenetic networks and incompatible characters. In *Proceedings of the Ninth International Conference on Research in Computational Molecular Biology (RECOMB).* Volume 3500/2005. pp. 217–232. Springer-Verlag, Berlin.

[13] Gusfield, D., Eddhu, S., and Langley, C. (2003). Efficient reconstruction of phylogenetic networks with constrained recombination. In *Proceedings of the IEEE Computer Society Conference on Bioinformatics*, pp. 363–374. IEEE Computer Society, Los Alimatos.

[14] Hein, J. (1993). A heuristic method to reconstruct the history of sequences subject to recombination. *Journal of Molecular Evolution*, **36**, 396–405.

[15] Holland, B., Huber, K., Moulton, V., and Lockhart, P. J. (2004). Using consensus networks to visualize contradictory evidence for species phylogeny. *Molecular Biology and Evolution*, **21**, 1459–1461.

[16] Holland, B. and Moulton, V. (2003). Consensus networks: A method for visualizing incompatibilities in collections of trees. In *Proceedings of WABI, 2003 (Workshop on Algorithms in Bioinformatics)* (eds. G. Benson and R. Page), LNBI 2812, pp. 165–176. Springer-Verlag, Berlin.

[17] Huber, K. T., Langton, M., Penny, D., Moulton, V., and Hendy, M. (2002). Spectronet: A package for computing spectra and median networks. *Applied Bioinformatics*, **1**, 159–161.

[18] Huber, K.T. and Moulton, V. (2006). Phylogenetic networks from multi-labelled trees. *Journal of Mathematical Biology*, **52**(5), 613–632.

[19] Hudson, R. R. (1983). Properties of the neutral allele model with intergenic recombination. *Theoretical Population Biology*, **23**, 183–201.

[20] Huson, D. H. (1998). SplitsTree: A program for analyzing and visualizing evolutionary data. *Bioinformatics*, **14**(10), 68–73.

[21] Huson, D. H. and Bryant, D. (2006). Application of phylogenetic networks in evolutionary studies. *Molecular Biology and Evolution*, **23**, 254–267. Software available from `www.splitstree.org`.

[22] Huson, D. H., Dezulian, T., Kloepper, T., and Steel, M. A. (2004). Phylogenetic super-networks from partial trees. *IEEE/ACM Transactions in Computational Biology and Bioinformatics*, **1**(4), 151–158.

[23] Huson, D.H., Kloepper, T., Lockhart, P. J., and Steel, M. A. (2005). Reconstruction of reticulate networks from gene trees. In *Proceedings of the Ninth International Conference on Research in Computational Molecular Biology (RECOMB)*, LNCS 3500, pp. 233–249. Springer-Verlag, Berlin.

[24] Huson, D.H. and Kloepper, T.H. (2005). Computing recombination networks from binary sequences. *Bioinformatics*, **21**(suppl. 2), ii159–ii165. European Conferences on Computational Biology (ECCB).

[25] Huson, D. H., Steel, M. A., and Whitfield, J. (2006). Reducing distortion in phylogenetic networks. *Proceedings of WABI, 2006 (Workshop on Algorithms in Bioinformatics)* (eds. P. Bücher and B. M. E. Moret), LNBI 4175, pp. 150–161. Springer-Verlag, Berlin.

[26] Jukes, T.H. and Cantor, C.R. (1969). Evolution of protein molecules. In *Mammalian Protein Metabolism* (ed. H. N. Munro), Vol III, Chapter 24 pp. 21–132, Academic Press, New York.

[27] O'Donnell, K., Kistler, H. C., Tacke, B. K., and Casper, H. H. (2000). Gene genealogies reveal global phylogeographic structure and reproductive isolation among lineages of *fusarium graminearum*, the fungus causing wheat scab. *Proceedings of the National Academy of Sciences of the United States*, **97**(14), 7905–7910.

[28] Kumar, A., Black, W. C., and Rai, K. S. (1998). An estimate of phylogenetic relationships among culicine mosquitoes using a restriction map of the rDNA cistron. *Insect Molecular Biology*, **7**(4), 367–373.

[29] Linder, C. R. and Rieseberg, L. H. (2004). Reconstructing patterns of reticulate evolution in plants. *American Journal of Botany*, **91**(10), 1700–1708.

[30] Lockhart, P. J., McLenachan, P. A., Havell, D., Glenny, D., Huson, D. H., and Jensen, U. (2001). Phylogeny, dispersal and radiation of New Zealand alpine buttercups: molecular evidence under split decomposition. *Annals of the Missouri Botanical Garden*, **88**, 458–477.

[31] Lockhart, P. J. (2004). Unpublished data.

[32] Lyngsø, R. B., Song, Y. S., and Hein, J. (2005). Minimum recombination histories by branch and bound. In *Proceedings of WABI, 2005 (Workshop on Algorithms in Bioinformatics)*, pp. 239–250, Springer-Verlag, Berlin.

[33] Maddison, W. P. (1997). Gene trees in species trees. *Systematic Biology*, **46**(3), 523–536.

[34] Morrison, D. (2005). Networks in phylogenetic analysis: new tools for population biology. *International Journal for Parasitology*, **35**, 567–582.

[35] Nakhleh, L., Warnow, T., and Linder, C. R. (2004). Reconstructing reticulate evolution in species—theory and practice. In *Proceedings of the Eighth International Conference on Research in Computational Molecular Biology (RECOMB)* (ed. P. Bourne et al.), pp. 337–346, ACM Press, New York.

[36] Posada, D. (2002). Evaluation of methods for detecting recombination from DNA sequences. *Molecular Biology and Evolution*, **19**(5), 708–717.

[37] Rannala, B. and Yang, Z. (1996). Probability distribution of molecular evolutionary trees: A new method of phylogenetic inference. *Journal of Molecular Evolution*, **43**(3), 304–311.

[38] Ronquist, F. and Huelsenbeck, J. P. (2003). MrBayes 3: Bayesian phylogenetic inference under mixed models. *Bioinformatics*, **19**(12), 1572–4.

[39] Saitou, N. and Nei, M. (1987). The Neighbor-Joining method: a new method for reconstructing phylogenetic trees. *Molecular Biology and Evolution*, **4**, 406–425.

[40] Semple, C. and Steel, M. A. (2003). *Phylogenetics*. Oxford University Press, Oxford.

[41] Song, Y. S. and Hein, J. (2003). Parsimonious reconstruction of sequence evolution and haplotype blocks: Finding the minimum number of recombination events. In *Proceedings of WABI, 2003 (Workshop on Algorithms in Bioinformatics)*. LNBI 2812, pp. 287–302. Springer-Verlag, Berlin.

[42] Song, Y. S and Hein, J. (2004). On the minimum number of recombination events in the evolutionary history of DNA sequences. *Journal of Mathematical Biology*, **48**, 160–186.

[43] Song, Y. S and Hein, J. (2005). Constructing minimal ancestral recombination graphs. *Journal of Computational Biology*, **12**, 147–169.

[44] Vigilant, L., Stoneking, M., Harpending, H. M., Hawkes, K., and Wilson, A. (1991). African populations and the evolution of human mitochondrial DNA. *Science*, **253**(5027), 1503–1507.

[45] Wang, L., Zhang, K., and Zhang, L. (2001). Perfect phylogenetic networks with recombination. *Journal of Computational Biology*, **8**(1), 69–78.

[46] Willis, L. G., Winston, M. L., and Honda, B. M. (1992). Phylogenetic relationships in the honeybee (genus *Apis*) as determined by the sequence of the cytochrome oxidase ii region of mitochondrial DNA. *Molecular Phylogenetics and Evolution*, **1**, 169–178.

[47] Zink, R. M., Dittmann, D. L., and Roots, W. L. (1991). Mitochondrial DNA variation and the phylogeny of zonotrichia. *The Auk*, **108**(3), 578–584.

10

HYBRIDIZATION NETWORKS

Charles Semple

Abstract

Reticulate evolution is a fundamental process in the evolution of certain groups of taxa. Consequently, conflicting signals in a data set may not be the result of sampling or modelling errors, but due to the fact that reticulation has played a role in the evolutionary history of the species under consideration. Assuming that our initial data set is correct, a fundamental problem is to compute the minimum number of reticulation events that explains this set. This smallest number sets a lower bound on the number of such events and provides an indication of the extent that reticulation has had on the evolutionary history of a collection of present-day species. In this chapter, we focus our attention on this problem for when the initial set consists of two rooted binary phylogenetic trees. This may seem rather special, but there are several reasons for this. Firstly, the problem is NP-hard even when the initial set consists of two such trees. Secondly, we are interested in finding a general solution rather than one that is restricted in some way. Lastly, the problem for when the initial data set consists of binary sequences can be interpreted as a sequence of two-tree problems.

Referring to the problem of when the initial set consists of two trees, this chapter includes the problem's relationship with the rooted subtree prune and regraft distance, mathematical characterizations of the problem based on agreement forests, reduction-based algorithms for solving the problem exactly, and the problem's connection with a variant of it in which the initial data set consists of binary sequences.

10.1 Introduction

Evolutionary (phylogenetic) trees are used to represent the tree-like evolution of a collection of taxa. For many groups of taxa (for example, most mammals) this representation is appropriate. However, non-tree-like evolutionary processes such as hybridization, horizontal gene transfer, and recombination mean that some groups of taxa are not suited to this type of representation. Collectively referred to as reticulation events, these types of processes result in species being a composite of DNA regions derived from different ancestors. Frequently with bacteria, horizontal gene transfer is the transfer of a piece of DNA from one organism to another which is not its offspring. On the other hand, hybridizations combine two lineages to create a new offspring. Examples of eukaryotes whose ancestral history include hybridization are certain plant and bird species. Recombination is a type of hybridization that has been well-studied within the framework of population genetics. For informative articles on the frequency of hybridization amongst animals and the problem of distinguishing hybridization from other causes of phylogenetic incongruence, see [35] and [36], respectively.

The effect of reticulation in evolution has been recognized for quite some time. Since the 1930s, botanists suggested that the morphological variation in the New Zealand flora is due to hybridization [2]. More recently, in the context of horizontal gene transfer, Doolittle [14] wrote that 'molecular phylogeneticists will have failed to find the 'true tree', not because their methods are inadequate or because they have chosen the wrong genes, but because the history of life cannot be properly represented as a tree.' Despite this recognition, mathematical investigations into the understanding and analysis of reticulation in evolution are relatively recent.

In a separate chapter, Huson provides an overview of various ways of representing the evolutionary history of a collection of taxa that has undergone reticulate evolution. In this chapter, we focus our attention on a particular problem that is both biologically important and mathematically challenging. A fundamental problem for biologists studying the evolution of species whose past has included reticulation is the following: given a collection of rooted phylogenetic trees on sets of species that correctly represents the tree-like evolution of different parts of their genomes, what is the smallest number of reticulation events needed to explain the evolution of the species under consideration. As well as providing a lower bound on the number of such events, this smallest number also indicates the extent that reticulation has had on the evolutionary history of the collection of present-day species.

The chapter is organized as follows. In Section 10.2, we formalize the above problem and the notion of a hybridization network, the latter is central to this problem. In general, the problem is NP-hard even when the initial collection consists of two trees. However, there is an attractive and particularly useful characterization of it in this case. This characterization is described in Section 10.3, while Section 10.4 contains algorithmic applications of it. In Section 10.5, we consider the variant of the problem for when the initial collection is a set of binary sequences. The material in this section is used in the subsequent two sections. An important biological consideration of the evolutionary history of taxa is that reticulation events occur between taxa that coexist in time. We investigate this consideration in Section 10.6. Lastly, in Section 10.7, we consider some of the computational issues in computing the above smallest number.

For completeness, we end this section with some preliminaries. Unless otherwise stated, the notation and terminology in this chapter follows Semple and Steel [44].

10.1.1 *Preliminaries*

A *rooted phylogenetic X-tree* $\mathcal{T}$ is a rooted tree in which no vertex has degree 2 except possibly for the root which has degree at least 2, and whose leaf set is X. In addition, $\mathcal{T}$ is *binary* if, apart from the root which has degree 2, all interior vertices have degree 3. The set X is called the *label set* of $\mathcal{T}$ and we sometimes denote it as $\mathcal{L}(\mathcal{T})$. Examples of rooted binary phylogenetic trees are shown in Fig. 10.1 and at the top of Fig. 10.2.

For convenience, many of the examples that arise in this chapter are based on rooted caterpillar trees. A *rooted caterpillar tree* is a rooted binary phylogenetic tree that has a leaf vertex, x say, such that every other leaf vertex is attached to the path from x to the root via a pendant edge. The rooted binary phylogenetic tree shown in Fig. 10.1 is an example of a rooted caterpillar tree. Without ambiguity, we denote this rooted caterpillar tree by the n-tuple $(x_1, x_2, \ldots, x_n)$ as this is the ordering of the label set induced by the path from x_1 to the root. Note

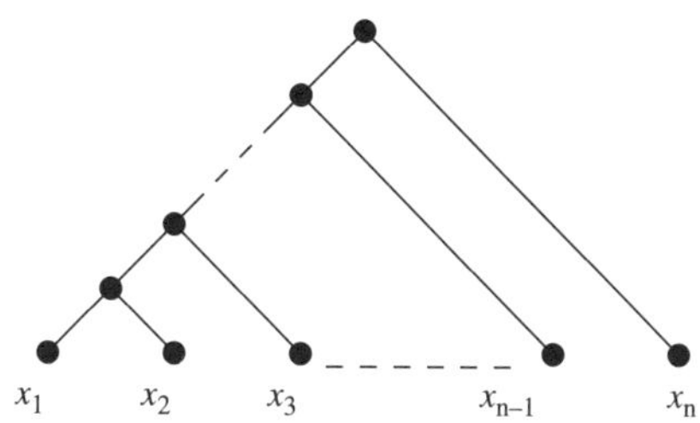

FIG. 10.1. A rooted caterpillar tree.

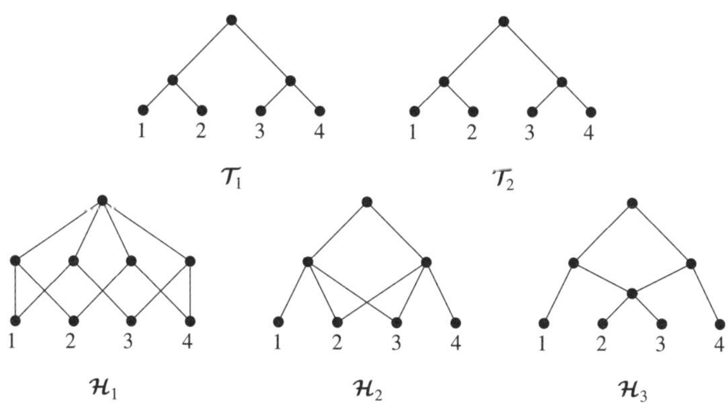

FIG. 10.2. Two rooted binary phylogenetic trees $\mathcal{T}_1$ and $\mathcal{T}_2$, and three hybridization networks $\mathcal{H}_1$, $\mathcal{H}_2$, and $\mathcal{H}_3$. Each of the hybridization networks $\mathcal{H}_1$ and $\mathcal{H}_2$ display both $\mathcal{T}_1$ and $\mathcal{T}_2$.

that the first two coordinates of this tuple could be interchanged to describe the same rooted caterpillar tree.

Let $\mathcal{T}$ be a rooted phylogenetic X-tree and let v be a vertex of $\mathcal{T}$. The subset of elements X that are descendants of v is a called a *cluster* of $\mathcal{T}$. We denote this cluster by $C_{\mathcal{T}}(v)$ or simply $C(v)$ if there is no ambiguity. We sometimes say that $C(v)$ is the cluster of $\mathcal{T}$ *corresponding to v in $\mathcal{T}$*. The set of clusters of $\mathcal{T}$ is denoted by $\mathcal{C}(\mathcal{T})$. Note here that the root of $\mathcal{T}$ gives rise to a cluster.

For a rooted phylogenetic X-tree $\mathcal{T}$, several different types of rooted subtrees will play a prominent role in this chapter. Let X' be a subset of X. The minimal rooted subtree of $\mathcal{T}$ that connects the leaves in X' is denoted by $\mathcal{T}(X')$. Furthermore, the *restriction* of $\mathcal{T}$ to X', denoted by $\mathcal{T}|X'$, is the rooted phylogenetic tree obtained from $\mathcal{T}(X')$ by suppressing any non-root vertices of degree 2. Lastly, a rooted subtree of $\mathcal{T}$ is *pendant* if it can be obtained from $\mathcal{T}$ by deleting a single edge. For example, in Fig. 10.1, the minimal rooted subtree that connects the leaves in $\{x_1, x_2, x_3\}$ is a pendant rooted subtree, but the minimal rooted subtree connecting x_2 and x_3 is not a pendant rooted subtree.

10.2 Hybridization networks

In this section, we formalize the optimization problem described in the introduction. We begin with the concept of a hybridization network which is central to this problem and this chapter. These networks are particular types of digraphs.

A *directed graph* (also known as a *digraph*) consists of a collection of vertices and a collection of directed edges called *arcs*. If an arc is directed from the vertex u to the vertex v, then it is denoted as the ordered pair (u, v). The *degree* of a vertex v is the number of arcs incident with v. To distinguish between arcs coming into v and arcs coming out of v, we refer to the number of arcs coming into v as the *indegree* of v, while the number of arcs coming out of v is referred to as the *outdegree* of v. This is denoted as $d^-(v)$ and $d^+(v)$, respectively. In evolutionary biology, directed graphs are used to represent the evolutionary history of a collection of present-day species. Vertices may represent species, individuals, or DNA sequences, while arcs represent ancestral relationships. By viewing the edges as arcs directed away from the root, rooted phylogenetic trees are examples of such digraphs.

A *directed path* in a digraph D is an alternating sequence

$$v_0, a_1, v_1, a_2, v_2, \ldots, v_{k-1}, a_k, v_k$$

of vertices and arcs in which a_i is directed from v_{i-1} to v_i for all i, and no vertex or arc appears more than once. A *directed cycle* in D is a directed path in which $v_0 = v_k$. We say that D is *acyclic* if it contains no directed cycles. An acyclic digraph D is *rooted* if the underlying graph has no parallel edges, and there is a distinguished vertex ρ with $d^-(\rho) = 0$ and the property that there is a directed path from ρ to every vertex of D.

A *hybridization network* (on X) is a rooted acyclic digraph with root ρ in which

(i) X is the set of vertices of outdegree zero,
(ii) $d^+(\rho) \geq 2$, and
(iii) for all vertices v with $d^+(v) = 1$, we have $d^-(v) \geq 2$.

The set X represents a collection of taxa and is the *label set* of $\mathcal{H}$. For convenience, it is sometimes denoted as $\mathcal{L}(\mathcal{H})$. Vertices of indegree at least two represent an exchange of genetic information between their parents. Generically, we call these vertices *hybridization vertices.* In the literature, hybridization networks have been referred to as 'hybrid phylogenies' (e.g. [6]) and 'phylogenetic networks' (e.g. [31, 40]). The latter with the additional property that hybridization vertices have indegree exactly two. Note here that vertices with indegree more than two do not represent a simultaneous exchange of genetic information between several parents but rather an uncertainty of the exact order of 'hybridization'. To illustrate the above concepts, in Fig. 10.2, $\mathcal{H}_1$, $\mathcal{H}_2$, and $\mathcal{H}_3$ are all examples of hybridization networks in which $X = \{1, 2, 3, 4\}$. Here and in all other figures, it is implicit that arcs are directed downwards. Rooted phylogenetic trees are special examples of hybridization networks in which all vertices, apart from the root, have indegree one.

Remark In the chapter written by Huson, a 'reticulate network' is simply a particular type of hybridization network. Having less restrictions on the indegree and outdegree of vertices allows for uncertainty in the exact order of speciation and hybridization. Furthermore, unlike some authors, we do not impose the condition that the outdegree of a hybridization vertex is one—this is simply for mathematical convenience and has no bearing on the results in this chapter. Lastly, we refer the reader to the figures in Huson's Chapter 9 for the biological interpretation of hybridization networks.

To quantify the number of reticulation events, the *hybridization number* of a hybridization network $\mathcal{H}$ with root ρ is

$$h(\mathcal{H}) = \sum_{v \neq \rho} (d^-(v) - 1).$$

Since $d^-(v)$ is the number of parents of v and since every vertex, apart from the root, has at least one parent, $(d^-(v) - 1)$ is the number of additional parents of v. The hybridization number of a network is at least zero. Indeed, $h(\mathcal{H}) = 0$ if and only if $\mathcal{H}$ is a rooted phylogenetic tree. In Fig. 10.2, $h(\mathcal{H}_1) = 4$, $h(\mathcal{H}_2) = 2$, and $h(\mathcal{H}_3) = 1$.

Let $\mathcal{T}$ be a rooted phylogenetic tree and let $\mathcal{H}$ be a hybridization network. We say that $\mathcal{H}$ *displays* $\mathcal{T}$ if $\mathcal{L}(\mathcal{T}) \subseteq \mathcal{L}(\mathcal{H})$ and there is a rooted subtree of $\mathcal{H}$ that is a refinement of $\mathcal{T}$. In other words, $\mathcal{T}$ can be obtained from $\mathcal{H}$ by first deleting a subset of the edges of $\mathcal{H}$ and any resulting isolated vertices, and then contracting edges. For example, in Fig. 10.2, $\mathcal{H}_1$ and $\mathcal{H}_2$ both display $\mathcal{T}_1$ and $\mathcal{T}_2$, while $\mathcal{H}_3$ displays neither $\mathcal{T}_1$ nor $\mathcal{T}_2$. We say that $\mathcal{H}$ *displays* a collection $\mathcal{P}$

of rooted phylogenetic trees if each tree in $\mathcal{P}$ is displayed by $\mathcal{H}$. Furthermore, extending the definition of the hybridization number to a collection $\mathcal{P}$ of rooted phylogenetic trees, we set

$$h(\mathcal{P}) = \min\{h(\mathcal{H}) : \mathcal{H} \text{ is a hybridization network that displays } \mathcal{P}\}.$$

If $\mathcal{P} = \{\mathcal{T}, \mathcal{T}'\}$, then we denote $h(\mathcal{P})$ by $h(\mathcal{T}, \mathcal{T}')$.

We interpret the fundamental problem for hybridization networks for when the initial collection consists of two rooted binary phylogenetic trees as the following optimization problem:

MINIMUM HYBRIDIZATION
Instance: A finite set X, and two rooted binary phylogenetic X-trees $\mathcal{T}$ and $\mathcal{T}'$.
Goal: Find a hybridization network $\mathcal{H}$ that displays $\mathcal{T}$ and $\mathcal{T}'$ with minimum hybridization number.
Measure: The value of $h(\mathcal{H})$.

In Fig. 10.2, while $\mathcal{H}_1$ displays $\mathcal{T}_1$ and $\mathcal{T}_2$, it does not minimize the hybridization number. However, it is easily checked that $\mathcal{H}_2$ has this property. Thus, in this case, $h(\mathcal{T}_1, \mathcal{T}_2) = 2$.

In its broadest sense, an instance of MINIMUM HYBRIDIZATION would consist of a collection of rooted phylogenetic trees. However, even in this simplest case when it consists of just two rooted binary phylogenetic trees, Bordewich and Semple [12] showed that MINIMUM HYBRIDIZATION is NP-hard (see Section 10.7). Nevertheless, there is an attractive characterization of this problem in the simplest case. This characterization provides valuable insight into the problem and is crucial to many of the results in this chapter. We describe this characterization and some of these results in the next section.

We end this section with several remarks. First, the input in the above problem could equally have been a set of sequences instead of a set of trees, in which case, instead of seeking a 'minimal' hybridization network, we look for a 'recombination network' that has this property. A number of authors have considered this variant of the problem and we will describe it in Section 10.5. Second, in keeping with the terminology in the chapter written by Huson and elsewhere, we use the term 'hybridization networks' as the input is unordered. In contrast, if the input is ordered in some way, as in the case of sequences, then the analogous digraphs are called 'recombination networks'. Lastly, as explicitly pointed out by Moret *et al.* [38], one needs to be careful in inferring information about hybridization events and the ancestral species involved in such events. In particular, the absence of unsampled taxa can have important ramifications in interpreting the true evolutionary history of the sampled taxa.

10.3 A characterization of Minimum Hybridization

Historically, one of the main tools that has been used to understand and model reticulate evolution is a graph-theoretic operation called 'rooted subtree prune and regraft'. Informally, this operation prunes a subtree of a rooted tree and then reattaches this subtree to another part of the tree. The use of this tool in evolutionary biology dates back to at least 1990 [23], and has been regularly used since as a way to model reticulate evolution (for example, see [6, 34, 40, 49]). The reason for this is that if two rooted binary phylogenetic X-trees are inconsistent, but this inconsistency can be explained with a single hybridization event, then one tree can be obtained from the other by a single rooted subtree prune and regraft operation. Indeed, given this, it is tempting to conjecture that the minimum number of hybridization events to explain the inconsistency of two rooted binary phylogenetic X-trees is equal to the minimum number of rooted subtree prune and regraft operations to transform one tree into the other. We will make this precise shortly, however, this is not the case. Nevertheless, these two minimum numbers are very closely related as they can both be characterized in terms of 'agreement forests'. It is one of these characterizations that is referred to at the end of Section 10.2.

10.3.1 *Rooted subtree prune and regraft operation and agreement forests*

To make the characterizations work, we regard the root of each of the two rooted binary phylogenetic X-trees $\mathcal{T}$ and $\mathcal{T}'$ in the upcoming definitions as a vertex ρ at the end of a pendant edge (called the *root edge*) adjoined to the original root. Furthermore, we regard ρ as part of the label sets of $\mathcal{T}$ and $\mathcal{T}'$, and so $\mathcal{L}(\mathcal{T}) = \mathcal{L}(\mathcal{T}') = X \cup \{\rho\}$. To illustrate, consider the two rooted binary phylogenetic trees $\mathcal{T}$ and $\mathcal{T}'$ shown at the top of Fig. 10.3. In the following, we regard $\mathcal{T}$ and $\mathcal{T}'$ as shown at the bottom of Fig. 10.3.

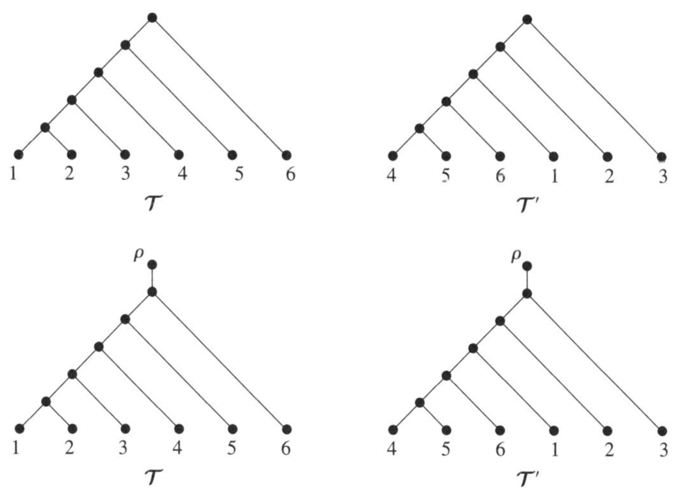

FIG. 10.3. Two rooted binary phylogenetic trees $\mathcal{T}$ and $\mathcal{T}'$ without (above) and with (below) their root labelled ρ.

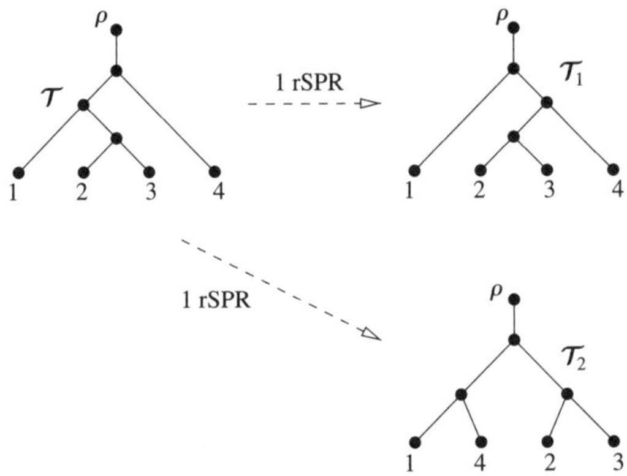

FIG. 10.4. Each of $\mathcal{T}_1$ and $\mathcal{T}_2$ are obtained from $\mathcal{T}$ by a single rooted subtree prune and regraft operation.

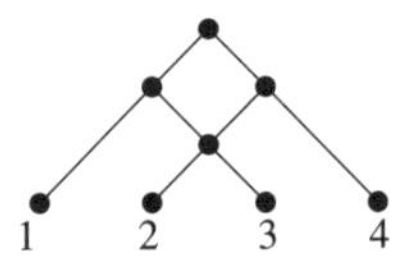

FIG. 10.5. The hybridization network resulting from the single rooted subtree prune and regraft operation that transforms $\mathcal{T}$ into $\mathcal{T}_1$ in Fig. 10.4.

Let $e = \{u, v\}$ be an edge of $\mathcal{T}$ that is not the root edge, where u is the vertex that is on the path from the root of $\mathcal{T}$ to v. Let $\mathcal{T}'$ be the rooted binary phylogenetic tree obtained from $\mathcal{T}$ by deleting e and reattaching the resulting rooted subtree via a new edge, f say, as follows. Create a new vertex u' that subdivides an edge of the component that contains ρ and adjoin f between u' and v, then suppress the degree-2 vertex u. We say that $\mathcal{T}'$ has been obtained from $\mathcal{T}$ by a *rooted subtree prune and regraft* (rSPR) operation. To illustrate, consider Fig. 10.4. Each of $\mathcal{T}_1$ and $\mathcal{T}_2$ are obtained from $\mathcal{T}$ by a single rSPR operation. Denoted by $d_{\mathrm{rSPR}}(\mathcal{T}, \mathcal{T}')$, we define the rSPR *distance* between $\mathcal{T}$ and $\mathcal{T}'$ to be the minimum number of rooted subtree prune and regraft operations that is required to transform $\mathcal{T}$ into $\mathcal{T}'$. It is well known that, for any such pair of trees, one can always obtain one tree from the other by a sequence of rSPR operations, and so this distance is well-defined. Moreover, this distance is a metric on the collection of rooted binary phylogenetic X-trees.

To explicitly highlight the connection between rooted subtree prune and regraft operations and hybridization events, consider $\mathcal{T}$ and $\mathcal{T}_1$ in Fig. 10.4. The evolutionary difference in the two trees can be explained by a single hybridization event; the corresponding hybridization vertex is the root of the pendant subtree that is pruned and regrafted in the rooted subtree prune and regraft operation shown in the figure. The resulting hybridization network is shown in Fig. 10.5.

Analogous to MINIMUM HYBRIDIZATION, we formally state the optimization problem of computing the rSPR distance between two rooted binary phylogenetic trees as follows.

MINIMUM RSPR
Instance: A finite set X, and two rooted binary phylogenetic X-trees $\mathcal{T}$ and $\mathcal{T}'$.
Goal: Find a minimum length sequence of single rSPR operations that transforms $\mathcal{T}$ into $\mathcal{T}'$.
Measure: The length of this sequence.

An *agreement forest* for $\mathcal{T}$ and $\mathcal{T}'$ is a collection $\{\mathcal{T}_\rho, \mathcal{T}_1, \mathcal{T}_2, \ldots, \mathcal{T}_k\}$ of rooted leaf-labelled trees, where $\mathcal{T}_\rho$ is a rooted tree whose label set $\mathcal{L}_\rho$ contains ρ and $\mathcal{T}_1, \mathcal{T}_2, \ldots, \mathcal{T}_k$ are rooted binary phylogenetics trees with label sets $\mathcal{L}_1, \mathcal{L}_2, \ldots, \mathcal{L}_k$, respectively, such that the following properties are satisfied:

(i) The label sets $\mathcal{L}_\rho, \mathcal{L}_1, \mathcal{L}_2, \ldots, \mathcal{L}_k$ partition $X \cup \{\rho\}$.
(ii) For each $i \in \{\rho, 1, 2, \ldots, k\}$, we have that $\mathcal{T}_i \cong \mathcal{T}|\mathcal{L}_i$ and $\mathcal{T}_i \cong \mathcal{T}'|\mathcal{L}_i$.
(iii) The trees in $\{\mathcal{T}(\mathcal{L}_i) : i \in \{\rho, 1, 2, \ldots, k\}\}$ and $\{\mathcal{T}'(\mathcal{L}_i) : i \in \{\rho, 1, 2, \ldots, k\}\}$ are vertex disjoint rooted subtrees of $\mathcal{T}$ and $\mathcal{T}'$, respectively.

It is easily seen that if $\mathcal{F}$ is an agreement forest for $\mathcal{T}$ and $\mathcal{T}'$, then, up to suppressing non-root vertices of degree two, $\mathcal{F}$ can be obtained from each of $\mathcal{T}$ and $\mathcal{T}'$ by deleting $|\mathcal{F}| - 1$ edges. An agreement forest for $\mathcal{T}$ and $\mathcal{T}'$ is a *maximum-agreement* forest if, amongst all agreement forests for $\mathcal{T}$ and $\mathcal{T}'$, it has the smallest number of components, in which case we denote this value of k by $m(\mathcal{T}, \mathcal{T}')$. For example, two agreement forests for the two trees $\mathcal{T}$ and $\mathcal{T}'$ in Fig. 10.3 are shown in Fig. 10.6. It is easily checked that the smallest number of components in any such forest is three, so $\mathcal{F}_1$ is also a maximum-agreement forest for $\mathcal{T}$ and $\mathcal{T}'$, and $m(\mathcal{T}, \mathcal{T}') = 2$.

10.3.2 *Characterizations of* MINIMUM HYBRIDIZATION *and* MINIMUM RSPR

Intuitively, the edges that are deleted to obtain an agreement forest for $\mathcal{T}$ and $\mathcal{T}'$ are those which disagree in $\mathcal{T}$ and $\mathcal{T}'$, and correspond to different paths of genetic inheritance; that is hybridization events. Thus, the fewer edges deleted,

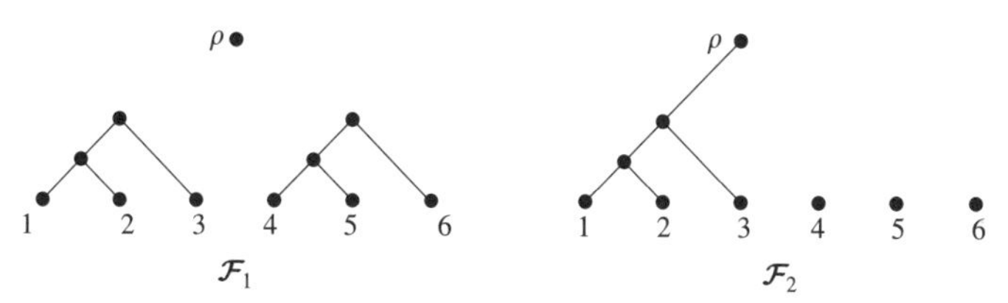

FIG. 10.6. Two possible agreement forests for $\mathcal{T}$ and $\mathcal{T}'$ in Fig. 10.3. $\mathcal{F}_1$ is a maximum-agreement forest for $\mathcal{T}$ and $\mathcal{T}'$, while $\mathcal{F}_2$ is a maximum-acyclic-agreement forest for $\mathcal{T}$ and $\mathcal{T}'$.

the smaller the number of hybridization events. Part (i) of the following theorem, due to Bordewich and Semple [11], characterizes the rSPR distance between two rooted binary phylogenetic trees in terms of agreement forests.

Theorem 10.1 *Let $\mathcal{T}$ and $\mathcal{T}'$ be two rooted binary phylogenetic X-trees. Then*

(i) $d_{\mathrm{rSPR}}(\mathcal{T}, \mathcal{T}') = m(\mathcal{T}, \mathcal{T}')$.
(ii) If $\mathcal{F}$ is an agreement forest for $\mathcal{T}$ and $\mathcal{T}'$ of size $k+1$ (i.e. $k \geq m(\mathcal{T}, \mathcal{T}')$), then there is a polynomial-time algorithm for constructing a sequence

$$\mathcal{T} = \mathcal{T}_0, \mathcal{T}_1, \mathcal{T}_2, \ldots, \mathcal{T}_k = \mathcal{T}'$$

of rooted binary phylogenetic trees such that, for all i, $\mathcal{T}_i$ is obtained from $\mathcal{T}_{i-1}$ by at most one rooted subtree prune and regraft operation (i.e. $d_{\mathrm{rSPR}}(\mathcal{T}, \mathcal{T}') \leq k$).

Remarks

1. Part (ii) of Theorem 10.1 is not explicitly stated in [11]. However, it is an immediate consequence of the inductive proof of [11, Theorem 2.1]. Although we omit the proof of this result, we will describe the algorithm in (ii) later in this section.
2. For those readers familiar with the tree rearrangement operation 'tree bisection and reconnection' (TBR), Allen and Steel [3] describe an analogous characterization for TBR in terms of agreement forests.
3. As we will soon see, agreement forests characterizations have been successfully used in gaining invaluable insights of various measures in phylogenetics. To provide intuition into why such a characterization is useful, think how much easier it is to consider deleting edges of $\mathcal{T}$ and $\mathcal{T}'$ to obtain an agreement forest as oppose to keeping track of a sequence of rSPR operations that transforms $\mathcal{T}$ into $\mathcal{T}'$.

Although it seems plausible that one could repeatedly use a single rooted subtree prune and regraft operation to represent a single hybridization event and thus the number of such events is equal to the number of such operations, the associated hybridization network that one builds in this process may contain a directed cycle. Such a cycle would mean that a vertex in this network inherits genetic information from its own descendants. As an example, consider the two rooted binary phylogenetic trees $\mathcal{T}$ and $\mathcal{T}'$ shown in Fig. 10.3. The tree $\mathcal{T}'$ can be obtained from $\mathcal{T}$ by two rSPR operations by first pruning the pendant subtree with label set $\{1, 2, 3\}$ of $\mathcal{T}$ and regrafting to obtain the tree $\mathcal{T}_1$ in Fig. 10.7(a), and then pruning the pendant subtree of $\mathcal{T}_1$ with label set $\{4, 5, 6\}$ and regrafting to obtain $\mathcal{T}'$. If one keeps each of the edges that are cut and added in this process, one obtains the 'hybridization' network shown in Fig. 10.7(b). Here e_1 is the edge that is added in the first rSPR operation and e_2 is the edge that is added in the

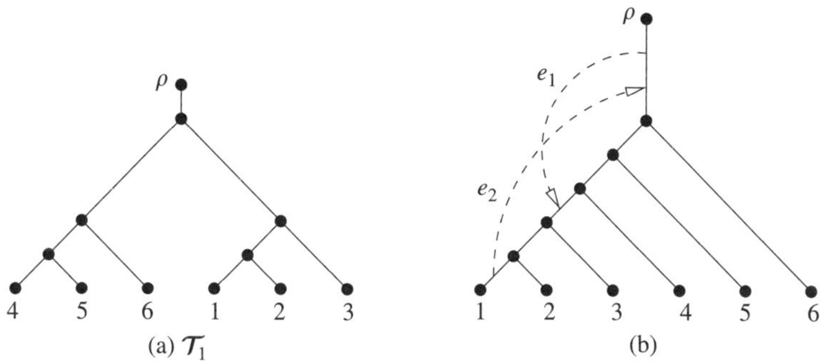

FIG. 10.7. (a) The second tree in the sequence of rSPR operations that transforms $\mathcal{T}$ into $\mathcal{T}'$, where $\mathcal{T}$ and $\mathcal{T}'$ are as shown in Fig. 10.3. (b) The network induced by the two rSPR operations that transforms $\mathcal{T}$ into $\mathcal{T}'$.

second rSPR operation. However, by viewing the (solid) edges as arcs directed away from ρ, this network contains a directed cycle. To avoid the construction of such a cycle and, in particular, rooted subtree prune and regraft operations that cause these cycles, we extend the definition of an agreement forest to an acyclic-agreement forest.

Let $\mathcal{F} = \{\mathcal{T}_\rho, \mathcal{T}_1, \mathcal{T}_2, \ldots, \mathcal{T}_k\}$ be an agreement forest for $\mathcal{T}$ and $\mathcal{T}'$. Let $G_\mathcal{F}$ be the directed graph whose vertex set is $\mathcal{F}$ and for which $(\mathcal{T}_i, \mathcal{T}_j)$ is an arc precisely if $i \neq j$ and either

(i) the root of $\mathcal{T}(\mathcal{L}_i)$ in $\mathcal{T}$ is an ancestor of the root of $\mathcal{T}(\mathcal{L}_j)$ in $\mathcal{T}$ or
(ii) the root of $\mathcal{T}'(\mathcal{L}_i)$ in $\mathcal{T}'$ is an ancestor of the root of $\mathcal{T}'(\mathcal{L}_j)$ in $\mathcal{T}'$.

Note that, as $\mathcal{F}$ is an agreement forest, the roots of $\mathcal{T}(\mathcal{L}_i)$ and $\mathcal{T}(\mathcal{L}_j)$, and the roots of $\mathcal{T}'(\mathcal{L}_i)$ and $\mathcal{T}'(\mathcal{L}_j)$ are not the same. We say that $\mathcal{F}$ is *acyclic* if $\mathcal{G}_\mathcal{F}$ has no directed cycles. If $\mathcal{F}$ is acyclic and it has the smallest number of components over all acyclic-agreement forests for $\mathcal{T}$ and $\mathcal{T}'$, then $\mathcal{F}$ is a *maximum-acyclic-agreement forest* for $\mathcal{T}$ and $\mathcal{T}'$, in which case we denote the number k by $m_a(\mathcal{T}, \mathcal{T}')$. Observe that $m_a(\mathcal{T}, \mathcal{T}') = 0$ if and only if, up to isomorphism, $\mathcal{T}$ and $\mathcal{T}'$ are identical. To illustrate these concepts, Fig. 10.8 shows the directed graph $G_{\mathcal{F}_1}$ of the agreement forest $\mathcal{F}_1$ shown in Fig. 10.6, where large open circles represent the vertices. Since this graph contains a directed cycle, $\mathcal{F}_1$ is not acyclic. However, it is easily checked that $G_{\mathcal{F}_2}$, where $\mathcal{F}_2$ is the agreement forest in Fig. 10.6, is acyclic. In fact, one can also check that this is a maximum-acyclic-agreement forest for $\mathcal{T}$ and $\mathcal{T}'$.

Analogous to Theorem 10.1, Baroni *et al.* [8] characterized the hybridization number of two rooted binary phylogenetic trees in terms of agreement forests.

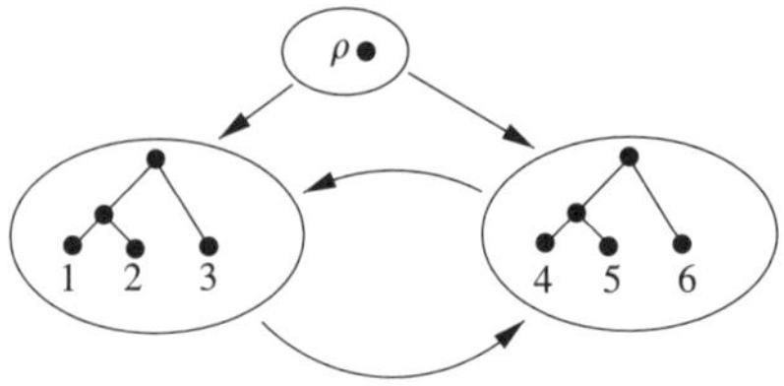

FIG. 10.8. The directed graph $G_{\mathcal{F}_1}$, where $\mathcal{F}_1$ is the agreement forest in Fig. 10.6.

Theorem 10.2 *Let $\mathcal{T}$ and $\mathcal{T}'$ be two rooted binary phylogenetic X-trees. Then*

(i) $h(\mathcal{T}, \mathcal{T}') = m_a(\mathcal{T}, \mathcal{T}')$.
(ii) If $\mathcal{F}$ is an acyclic-agreement forest for $\mathcal{T}$ and $\mathcal{T}'$ of size $k+1$ (i.e. $k \geq m_a(\mathcal{T}, \mathcal{T}')$), then there is a polynomial-time algorithm for constructing a hybridization network $\mathcal{H}$ that displays $\mathcal{T}$ and $\mathcal{T}'$ with $h(\mathcal{H}) \leq k$ (i.e. $h(\mathcal{T}, \mathcal{T}') \leq k$).

Remarks

1. Part (ii) of Theorem 10.2 is not stated in [8], but it is an immediate consequence of its inductive proof [8, Theorem 2]. Like part (ii) of Theorem 10.1, we will describe the algorithm in (ii) at the end of this section.
2. In contrast to the rSPR distance, the hybridization number is not a metric on the collection of rooted binary phylogenetic X-trees. To see this, consider $\mathcal{T}$ and $\mathcal{T}'$ in Fig. 10.3 and $\mathcal{T}_1$ in Fig. 10.7. We have already noted that $h(\mathcal{T}, \mathcal{T}') = 3$. Furthermore, it is easily checked that $h(\mathcal{T}, \mathcal{T}_1) = h(\mathcal{T}_1, \mathcal{T}') = 1$, and so the hybridization number does not satisfy the triangle inequality.
3. If one is only interested in the number of hybridization vertices (and not what each such vertex contributes to the hybridization number), then Theorem 10.2 is easily generalized to an arbitrary size collection of rooted binary phylogenetic X-trees. Here the notion of an acyclic-agreement forest for two trees is extended in the obvious way. For details, see [33].

Since every acyclic-agreement forest for two rooted binary phylogenetic X-trees $\mathcal{T}$ and $\mathcal{T}'$ is an (ordinary) agreement forest for $\mathcal{T}$ and $\mathcal{T}'$, it follows from Theorems 10.1 and 10.2 that

$$d_{\mathrm{rSPR}}(\mathcal{T}, \mathcal{T}') \leq h(\mathcal{T}, \mathcal{T}'). \qquad (10.1)$$

The fact that this inequality can be strict has been pointed out several times in the literature including [8, 24, 51]. An interesting question is just how strict? We consider this question in Section 10.3.3.

10.3.3 *Comparing $d_{\mathrm{rSPR}}(\mathcal{T}, \mathcal{T}')$ and $h(\mathcal{T}, \mathcal{T}')$*

Two natural questions arise from the inequality in (10.1).

(i) Whenever $d_{\text{rSPR}}(\mathcal{T},\mathcal{T}') = 1$, we have that $h(\mathcal{T},\mathcal{T}') = 1$, and so $d_{\text{rSPR}}(\mathcal{T},\mathcal{T}')$ provides a sharp lower bound for $h(\mathcal{T},\mathcal{T}')$. Can we find a sharp upper bound for $h(\mathcal{T},\mathcal{T}')$?

(ii) We have already seen that inequality (10.1) can be strict, so how large can the difference between $d_{\text{rSPR}}(\mathcal{T},\mathcal{T}')$ and $h(\mathcal{T},\mathcal{T}')$ be?

Consider (i). Regardless of the topology of $\mathcal{T}$ and $\mathcal{T}'$, if $X = \{x_1, x_2, \ldots, x_n\}$, then, as the forest consisting of $\mathcal{T}|\{\rho, x_1, x_2\}$ and isolated vertices $x_3, x_4, \ldots, x_n$ is an acyclic-agreement forest for $\mathcal{T}$ and $\mathcal{T}'$,

$$h(\mathcal{T},\mathcal{T}') \leq n - 2.$$

Using Theorem 10.2, Baroni *et al.* [8] showed that this upper bound is sharp. In particular, if $\mathcal{T}$ and $\mathcal{T}'$ are the two rooted caterpillars $(x_1, x_2, \ldots, x_n)$ and $(x_n, x_{n-1}, \ldots, x_1)$, then $h(\mathcal{T},\mathcal{T}') = n - 2$. In the same paper [8] and using Theorems 10.1 and 10.2, the authors also establish the following theorem.

Theorem 10.3 *For all $n \geq 4$, there are rooted binary phylogenetic trees $\mathcal{T}_1$, $\mathcal{T}_2$, and $\mathcal{T}_3$ on n leaves such that*

$$\frac{h(\mathcal{T}_1,\mathcal{T}_2)}{d_{\text{rSPR}}(\mathcal{T}_1,\mathcal{T}_2)} = \frac{1}{2}\left\lfloor\frac{n}{2}\right\rfloor$$

and

$$h(\mathcal{T}_1,\mathcal{T}_3) - d_{\text{rSPR}}(\mathcal{T}_1,\mathcal{T}_3) = n - 2\lfloor\sqrt{n}\rfloor - c,$$

where $c = 0$ if n is a square, $c = 1$ if $1 \leq n - \lfloor\sqrt{n}\rfloor^2 < \sqrt{n}$, and $c = 2$ otherwise.

Explicit examples of rooted binary phylogenetic trees that attain the equalities in Theorem 10.3 are given in [8]. For example, let $\mathcal{T}_1$ be the rooted caterpillar tree $(x_1, x_2, \ldots, x_{100})$. Let $\mathcal{T}_2$ and $\mathcal{T}_3$ be the rooted caterpillar trees on $\{x_1, x_2, \ldots, x_{100}\}$ whose orderings on their leaf sets are

$$(x_{51}, x_{52}, \ldots, x_{100}, x_1, x_2, \ldots, x_{50})$$

and

$$(x_{91}, x_{92}, \ldots, x_{100}, x_{81}, x_{82}, \ldots, x_{90}, x_{71}, \ldots, x_{19}, x_{20}, x_1, x_2, \ldots, x_{10}),$$

respectively. Then

$$\frac{h(\mathcal{T}_1,\mathcal{T}_2)}{d_{\text{rSPR}}(\mathcal{T}_1,\mathcal{T}_2)} = \frac{1}{2}\left\lfloor\frac{100}{2}\right\rfloor = 25$$

and

$$h(\mathcal{T}_1,\mathcal{T}_3) - d_{\text{rSPR}}(\mathcal{T}_1,\mathcal{T}_3) = 100 - 2\lfloor\sqrt{100}\rfloor - 0 = 80.$$

An interesting question is determine whether the ratio or difference given in Theorem 10.3 is the best possible.

The answers to (i) and (ii) in [8] both rely on Theorems 10.1 and 10.2. It seems unlikely that, without such characterizations, these results could have been attained as easily. Further applications of these theorems are given in Section 10.4.

10.3.4 *Algorithms for constructing* rSPR *sequences and hybridization networks from agreement forests*

Let $\mathcal{F}$ be an arbitrary agreement forest for two rooted binary phylogenetic X-trees $\mathcal{T}$ and $\mathcal{T}'$. The first algorithm rSPRSequence constructs a sequence of rooted binary phylogenetic trees beginning with $\mathcal{T}$ and ending with $\mathcal{T}'$ with the property that each tree in the sequence is obtained from its predecessor by a single (possibly trivial) rSPR operation. Provided $\mathcal{F}$ is acyclic, the second algorithm HybridNetwork constructs a hybridization network $\mathcal{H}$ that displays $\mathcal{T}$ and $\mathcal{T}'$ with $h(\mathcal{H}) \leq |\mathcal{F}| - 1$. Each algorithm is an immediate consequence of the inductive proofs of Theorems 10.1 and 10.2 in [11] and [8], respectively.

Algorithm: rSPRSequence($\mathcal{F}$)
Input: An agreement forest $\mathcal{F}$ of size $k+1$ of two rooted binary phylogenetic X-trees $\mathcal{T}$ and $\mathcal{T}'$.
Output: A sequence $\mathcal{T}_0, \mathcal{T}_1, \mathcal{T}_2, \ldots, \mathcal{T}_k$ of rooted binary phylogenetic X-trees with the property that $\mathcal{T}_0 = \mathcal{T}$, $\mathcal{T}_k = \mathcal{T}'$, and, for all i, either $\mathcal{T}_i$ is obtained from $\mathcal{T}_{i-1}$ by a single rSPR operation or $\mathcal{T}_i \cong \mathcal{T}_{i-1}$.

1. Set $\mathcal{T} = \mathcal{T}_0$, $\mathcal{F} = \mathcal{F}_0$, and $i = 1$.
2. Find a tree $\mathcal{S}_i$ in $\mathcal{F}_{i-1}$ such that $\mathcal{S}_i$ is a pendant subtree of $\mathcal{T}_{i-1}$.
3. In $\mathcal{T}'$, find the first subtree $\mathcal{T}'(\mathcal{L}(\mathcal{S}_j))$ corresponding to a tree $\mathcal{S}_j$ in $\mathcal{F}_{i-1}$ that is met on the path from the root of $\mathcal{T}'(\mathcal{L}(\mathcal{S}_i))$ to ρ.
4. Set $\mathcal{T}_i$ to be a tree that is obtained from $\mathcal{T}_{i-1}$ by pruning $\mathcal{S}_i$ and regrafting it so that $\mathcal{T}_i$ restricted to $\mathcal{L}(\mathcal{S}_i) \cup \mathcal{L}(\mathcal{S}_j)$ is isomorphic to $\mathcal{T}'$ restricted to $\mathcal{L}(\mathcal{S}_i) \cup \mathcal{L}(\mathcal{S}_j)$.
5. Set $\mathcal{F}_i$ to be the forest obtained from $\mathcal{F}_{i-1}$ by replacing $\mathcal{S}_i$ and $\mathcal{S}_j$ with $\mathcal{T}'$ restricted to $\mathcal{L}(\mathcal{S}_i) \cup \mathcal{L}(\mathcal{S}_j)$.

1 If $i = k$ halt; otherwise, increment i by 1 and return to Step 2.

Remarks The following comments may help the reader.

1. Step 2 is well-defined as there is always at least one tree that has this property.
2. In Step 3, the choice for $\mathcal{S}_j$ is unique because of (iii) in the definition of an agreement forest.
3. In Step 4, $\mathcal{F}_i$ is an agreement forest for $\mathcal{T}_i$ and $\mathcal{T}'$.

Before stating HybridNetwork, we need an additional concept. A simple, fast, and well-known way of deciding whether a directed graph G is acyclic is as follows. Find a vertex, v_1 say, of G that has indegree 0. If there is no such vertex, then G contains a directed cycle and so G is acyclic. Otherwise, delete v_1

(and its incident arcs) from G and find a vertex, v_2 say, of G that has indegree 0. Again, if there is no such vertex, then G is not acyclic, otherwise delete v_2 from this last digraph and continue in this way. Eventually, we either decide that G is not acyclic or obtain an ordering $v_1, v_2, \ldots, v_n$ of the vertex set of G such that, for all i, the vertex v_i has indegree 0 in the graph obtained from G by deleting the vertices $v_1, v_2, \ldots, v_{i-1}$. Such an ordering is called an *acyclic ordering* of G and it implies that G is acyclic.

Algorithm: HYBRIDNETWORK($\mathcal{F}$)
Input: An acyclic-agreement forest $\mathcal{F}$ of size $k+1$ of two rooted binary phylogenetic X-trees $\mathcal{T}$ and $\mathcal{T}'$.
Output: A hybridization network $\mathcal{H}$ that displays $\mathcal{T}$ and $\mathcal{T}'$ with $h(\mathcal{H}) \leq k$.

1. Find an acyclic ordering, $\mathcal{S}_\rho, \mathcal{S}_1, \mathcal{S}_2, \ldots, \mathcal{S}_k$ say, of $G_\mathcal{F}$.
2. Set $\mathcal{H}_0 = \mathcal{S}_\rho$ and set $i = 1$.
3. Attach $\mathcal{S}_i$ to $\mathcal{H}_{i-1}$ via two new arcs. Each arc joins the root of $\mathcal{S}_i$ to some (possibly distinct) arc of $\mathcal{H}_{i-1}$ and is directed towards the root of $\mathcal{S}_i$. These arcs are added so that the resulting network displays both $\mathcal{T}$ restricted to $\mathcal{L}(\mathcal{H}_{i-1}) \cup \mathcal{L}(\mathcal{S}_i)$ and $\mathcal{T}'$ restricted to $\mathcal{L}(\mathcal{H}_{i-1}) \cup \mathcal{L}(\mathcal{S}_i)$.
   Set $\mathcal{H}_i$ to be the resulting network and return $\mathcal{H}_i$ if $i = k$.
4. Increment i by 1 and return to Step 3.

Remark In Step 3 of the algorithm, it may be possible that only one new edge is required. This implies that $\mathcal{F}$ is not maximum and that a new acyclic-agreement forest for $\mathcal{T}$ and $\mathcal{T}'$ can be obtained by attaching one component $\mathcal{S}$ of $\mathcal{F}$ to another via an edge directed towards the root of $\mathcal{S}$.

10.4 Algorithmic applications of agreement forests

For two rooted binary phylogenetic trees $\mathcal{T}$ and $\mathcal{T}'$, agreement forests are a particularly useful tool for analysing the individual values $d_{\mathrm{rSPR}}(\mathcal{T}, \mathcal{T}')$ and $h(\mathcal{T}, \mathcal{T}')$. In this section, we consider ways that agreement forests can be used for this analysis and the resulting algorithmic implications, while in Section 10.7 we see that this tool provides invaluable leverage in understanding the computation complexity of finding these values.

As we formally state in Section 10.7, both MINIMUM RSPR and MINIMUM HYBRIDIZATION are NP-hard problems. Nevertheless, they are both susceptible to approaches that effectively reduce the size of the problem instance. Interestingly, these approaches are different and it appears that they are unique to the particular problem. For MINIMUM RSPR, we reduce the size of the problem instance while preserving the rooted subtree prune and regraft distance, while, for MINIMUM HYBRIDIZATION, we use a divide-and-conquer type approach, that is, we break the problem into a number of smaller problems. To avoid some repetition, the proofs of the first four results in this section rely on either Theorem 10.1 or Theorem 10.2.

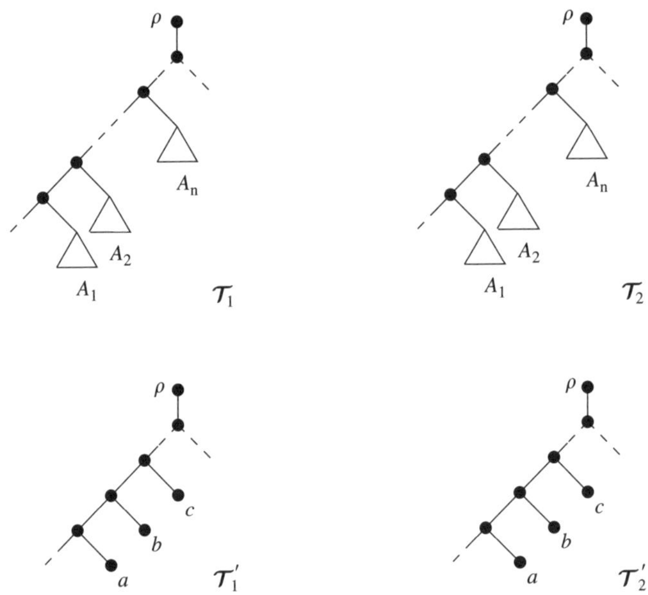

FIG. 10.9. Applying Rule 2 to two rooted binary phylogenetic trees $\mathcal{T}_1$ and $\mathcal{T}_2$, we obtain $\mathcal{T}_1'$ and $\mathcal{T}_2'$, respectively.

10.4.1 *Reduction rules*

For MINIMUM RSPR, consider the following two reduction rules:

1. Replace a pendant subtree that occurs identically in both trees by a single leaf with a new label.
2. Replace a chain of at least three pendant subtrees that occur identically and with the same orientation relative to the root in both trees by three new leaves with new labels correctly orientated to preserve the direction of the chain.

Rule 2 is illustrated in Fig. 10.9, where $A_1, A_2, \ldots, A_n$ is the chain of pendant subtrees common to both $\mathcal{T}_1$ and $\mathcal{T}_2$, and a, b, and c are the three new leaf labels orientated appropriately.

The following theorem is due to Bordewich and Semple [11].

Theorem 10.4 *Let $\mathcal{T}_1$ and $\mathcal{T}_2$ be two rooted binary phylogenetic X-trees, and let $\mathcal{T}_1'$ and $\mathcal{T}_2'$ be the two rooted binary phylogenetic X'-trees obtained from $\mathcal{T}_1$ and $\mathcal{T}_2$, respectively, by applying either Rule 1 or Rule 2. Then*

$$d_{\mathrm{rSPR}}(\mathcal{T}_1, \mathcal{T}_2) = d_{\mathrm{rSPR}}(\mathcal{T}_1', \mathcal{T}_2').$$

The proof of Theorem 10.4 relies on Theorem 10.1 and is the basis of showing that MINIMUM RSPR is fixed-parameter tractable in $d_{\mathrm{rSPR}}(\mathcal{T}_1, \mathcal{T}_2)$. Intuitively, this simply means that if the rSPR distance is small, it may be possible to efficiently compute this distance even if X is large. The reason for this is that, for small rSPR distance, one would expect the problem instance to be significantly reduced by repeatedly applying Rules 1 and 2. Note that, by Theorem 10.4, such repeated applications preserve the rSPR distance. For further details, see Section 10.7.

For MINIMUM HYBRIDIZATION, we have the following theorem due to Baroni *et al.* [7], which provides a divide-and-conquer type approach to the problem.

Theorem 10.5 *Let $\mathcal{T}$ and $\mathcal{T}'$ be two rooted binary phylogenetic X-trees, and suppose that $A \subset X$ is a cluster of both $\mathcal{T}$ and $\mathcal{T}'$. Then*

$$h(\mathcal{T}, \mathcal{T}') = h(\mathcal{T}|A, \mathcal{T}'|A) + h(\mathcal{T}_a, \mathcal{T}_a'),$$

where $\mathcal{T}_a$ and $\mathcal{T}_a'$ are obtained from $\mathcal{T}$ and $\mathcal{T}'$, respectively, by replacing the pendant subtrees $\mathcal{T}(A)$ and $\mathcal{T}'(A)$ with a single new leaf labelled a. Furthermore, if $\mathcal{H}_a$ is a hybridization network that displays $\mathcal{T}_a$ and $\mathcal{T}_a'$ with $h(\mathcal{H}_a) = h(\mathcal{T}_a, \mathcal{T}_a')$ and $\mathcal{H}_A$ is a hybridization network that displays $\mathcal{T}|A$ and $\mathcal{T}'|A$ with $h(\mathcal{H}_A) = h(\mathcal{T}|A, \mathcal{T}'|A)$, then the hybridization network obtained from $\mathcal{H}_a$ by identifying the root of $\mathcal{H}_A$ with a displays $\mathcal{T}$ and $\mathcal{T}'$, and has hybridization number $h(\mathcal{T}, \mathcal{T}')$.

We will discuss the obvious divide-and-conquer algorithm resulting from Theorem 10.5 and highlight its usefulness by applying the algorithm to a biological data set in Section 10.4.2.

Recalling that if, up to isomorphism, two rooted binary phylogenetic trees are identical, then their hybridization number is 0, we get the following corollary as an immediate consequence of Theorem 10.5.

Corollary 10.6 *Let $\mathcal{T}_1$ and $\mathcal{T}_2$ be two rooted binary phylogenetic X-trees, and let $\mathcal{T}_1'$ and $\mathcal{T}_2'$ be the two rooted binary phylogenetic X'-trees obtained from $\mathcal{T}_1$ and $\mathcal{T}_2$, respectively, by applying Rule 1. Then*

$$h(\mathcal{T}_1, \mathcal{T}_2) = h(\mathcal{T}_1', \mathcal{T}_2').$$

Curiously, despite Corollary 10.6, Rule 2 does not preserve the hybridization number of two rooted binary phylogenetic trees. We illustrate with a simple example. The argument used in the example is indicative of the arguments based on agreement forests. Let $\mathcal{T}_1$ and $\mathcal{T}_2$ be the rooted caterpillar trees

$$(b_1, b_2, b_3, b_4, b_5, b_6, a_1, a_2, a_3, a_4)$$

and

$$(b_1, a_1, a_2, a_3, a_4, b_2, b_3, b_4, b_5, b_6),$$

respectively. Let $\mathcal{T}_1'$ and $\mathcal{T}_2'$ be the rooted caterpillar trees obtained from $\mathcal{T}_1$ and $\mathcal{T}_2$, respectively, by applying Rule 2 to the chain of pendant subtrees corresponding to the labels a_1, a_2, a_3, a_4. Let a, b, and c denote the resulting new leaves.

Thus $\mathcal{T}_1'$ and $\mathcal{T}_2'$ are the rooted caterpillar trees

$$(b_1, b_2, b_3, b_4, b_5, b_6, a, b, c)$$

and

$$(b_1, a, b, c, b_2, b_3, b_4, b_5, b_6),$$

respectively. First observe that the agreement forest $\mathcal{F}$ of $\mathcal{T}_1$ and $\mathcal{T}_2$ for which the partition of $X \cup \{\rho\}$ induced by the label sets of its trees is

$$\{\{b_1, b_2, b_3, b_4, b_5, b_6, \rho\}, \{a_1\}, \{a_2\}, \{a_3\}, \{a_4\}\}$$

acyclic. Thus the number of components of a maximum-acyclic-agreement forest of $\mathcal{T}_1$ and $\mathcal{T}_2$ is at most 5. We next show that this number is exactly 5 and that $\mathcal{F}$ is the unique maximum-acyclic agreement forest for $\mathcal{T}_1$ and $\mathcal{T}_2$. Let $\mathcal{F}'$ be a maximum-acyclic-agreement forest for $\mathcal{T}_1$ and $\mathcal{T}_2$. If $b_j \in \mathcal{L}_\rho$ for some j, then, by the maximality of $\mathcal{F}'$, $\{a_1\}, \{a_2\}, \{a_3\}, \{a_4\}$ are label sets of $\mathcal{F}'$ and so, as $\mathcal{F}'$ is maximum, $\mathcal{F}' = \mathcal{F}$. Furthermore, if $a_i \in \mathcal{L}_\rho$ for some i, then $\{b_2\}, \{b_3\}, \{b_4\}, \{b_5\}, \{b_6\}$ are label sets of $\mathcal{F}'$ and so $|\mathcal{F}'| \geq 6$; a contradiction to maximality. Thus $\{\rho\}$ is a label set of $\mathcal{F}'$, in particular $\mathcal{L}_\rho \cap X$ is empty. But, because of the necessity of being acyclic, $\mathcal{L}_\rho \cap X$ is non-empty in any maximum-acyclic-agreement forest for $\mathcal{T}_1$ and $\mathcal{T}_2$ [8]. This last contradiction shows that $\mathcal{F}$ is the unique maximum-acyclic-agreement forest for $\mathcal{T}_1$ and $\mathcal{T}_2$. Using similar arguments, the unique maximum-acyclic-agreement forest for $\mathcal{T}_1'$ and $\mathcal{T}_2'$ is the forest for which the partition of $X \cup \{\rho\}$ induced by the label sets of its trees is

$$\{\{b_1, b_2, b_3, b_4, b_5, b_6, \rho\}, \{a\}, \{b\}, \{c\}\}.$$

But then $h(\mathcal{T}_1, \mathcal{T}_2) = 4$, while $h(\mathcal{T}_1', \mathcal{T}_2') = 3$. Thus Rule 2 does not preserve the hybridization number of two trees. The main point of the argument above is that, unlike the situation for (ordinary) agreement forests, there is no maximum-acyclic-agreement forest that contains a tree whose label set contains the set $\{a_1, a_2, a_3, a_4\}$, the union of the label sets of the chain of pendant subtrees that are replaced by the three new leaves.

In comparison to the last paragraph, the rSPR distance only satisfies a weaker version of Theorem 10.5. In particular, we have the following result [11].

Proposition 10.7 *Let $\mathcal{T}$ and $\mathcal{T}'$ be two rooted binary phylogenetic X-trees, and suppose that $A \subset X$ is a cluster of both $\mathcal{T}$ and $\mathcal{T}'$. Then*

$$d_{\mathrm{rSPR}}(\mathcal{T}, \mathcal{T}') \leq d_{\mathrm{rSPR}}(\mathcal{T}|A, \mathcal{T}'|A) + d_{\mathrm{rSPR}}(\mathcal{T}_a, \mathcal{T}_a') \leq d_{\mathrm{rSPR}}(\mathcal{T}, \mathcal{T}') + 1,$$

where $\mathcal{T}_a$ and $\mathcal{T}_a'$ are obtained from $\mathcal{T}$ and $\mathcal{T}'$, respectively, by replacing the pendant subtrees $\mathcal{T}(A)$ and $\mathcal{T}'(A)$ with a single new leaf labelled a. Moreover, these bounds are sharp.

To see that the first bound in Proposition 10.7 is sharp, simply choose $\mathcal{T}$ and $\mathcal{T}'$ so that $d_{\mathrm{rSPR}}(\mathcal{T}, \mathcal{T}') = 1$, and choose A to be the cluster of the pendant subtree that is pruned. For the sharpness of the second bound, choose $\mathcal{T}$ and $\mathcal{T}'$ to be the rooted caterpillar trees $(1, 2, 3, 4, 5, 6, 7, 8)$ and $(4, 5, 6, 1, 2, 3, 8, 7)$, and

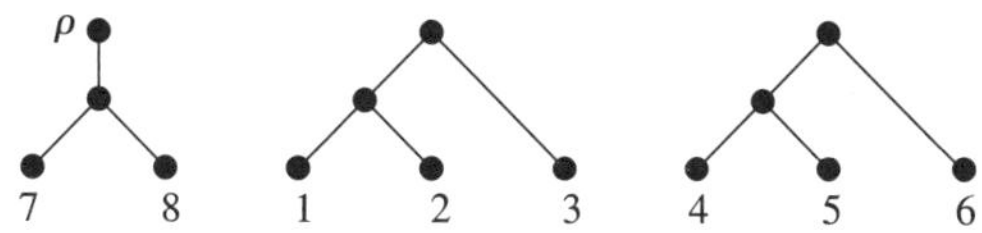

FIG. 10.10. Illustrating strict inequality in Proposition 10.7.

choose A to be the common cluster $\{1,2,3,4,5,6\}$. Then $d_{\mathrm{rSPR}}(\mathcal{T}_a, \mathcal{T}'_a) = 1$ and, as we have seen previously, $d_{\mathrm{rSPR}}(\mathcal{T}|A, \mathcal{T}'|A) = 2$, so

$$d_{\mathrm{rSPR}}(\mathcal{T}|A, \mathcal{T}'|A) + d_{\mathrm{rSPR}}(\mathcal{T}_a, \mathcal{T}'_a) = 3.$$

But the forest shown in Fig. 10.10 is an agreement forest for $\mathcal{T}$ and $\mathcal{T}'$, and therefore $d_{\mathrm{rSPR}}(\mathcal{T}, \mathcal{T}') \leq 2$.

In Sections 10.4.2 and 10.4.3, we describe two applications of Theorem 10.5.

10.4.2 *A simple divide-and-conquer algorithm for* MINIMUM HYBRIDIZATION

Theorem 10.5 and Corollary 10.6 provides us with the following simple divide-and-conquer approach to MINIMUM HYBRIDIZATION that is somewhat better than the naive approach of exhaustively searching for edges in $\mathcal{T}$ (or $\mathcal{T}'$) whose deletion results in an acyclic-agreement forest. This exact algorithm initially applies Rule 1 to $\mathcal{T}$ and $\mathcal{T}'$ as much as possible, and then locates the smallest pendant subtrees, $\mathcal{W}$ and $\mathcal{W}'$ say, in the resulting trees whose leaf sets are equal. Intuitively, these pendant subtrees localize conflicting signals in the evolutionary history of these parts of $\mathcal{T}$ and $\mathcal{T}'$. The algorithm finds a maximum-acyclic-agreement forest for these pendant subtrees $\mathcal{W}$ and $\mathcal{W}'$, and then repeats this process for the rooted binary phylogenetic trees obtained from $\mathcal{T}$ and $\mathcal{T}'$ by replacing the pendant subtrees with a single new vertex. Summing the hybridization number $h(\mathcal{W}, \mathcal{W}')$ at each iteration gives $h(\mathcal{T}, \mathcal{T}')$.

Algorithm: HYBRIDNUMBER($\{\mathcal{T}, \mathcal{T}'\}$)
Input: Two rooted binary phylogenetic X-trees $\mathcal{T}$ and $\mathcal{T}'$.
Output: The value of $h(\mathcal{T}, \mathcal{T}')$.

1. Set $\mathcal{T}_0 = \mathcal{T}$ and $\mathcal{T}'_0 = \mathcal{T}'$, and set $i = 1$.
2. Repeatedly apply Rule 1 to $\mathcal{T}_{i-1}$ and $\mathcal{T}'_{i-1}$ until the rule can no longer be applied, and set $\mathcal{S}_{i-1}$ and $\mathcal{S}'_{i-1}$ to be the resulting rooted binary phylogenetic trees, respectively. If each of $\mathcal{S}_{i-1}$ and $\mathcal{S}'_{i-1}$ consist of a single vertex, then go to Step 7.
3. Find a minimal cluster W_{i-1} in $\mathcal{C}(\mathcal{S}_{i-1}) \cap \mathcal{C}(\mathcal{S}'_{i-1})$ of size at least two.
4. Find a maximum-acyclic-agreement forest $\mathcal{F}_{i-1}$ for $\mathcal{S}_{i-1}|W_{i-1}$ and $\mathcal{S}'_{i-1}|W_{i-1}$.
5. Set $\mathcal{T}_i$ and $\mathcal{T}'_i$ to be the rooted binary phylogenetic trees obtained from $\mathcal{S}_{i-1}$ and $\mathcal{S}'_{i-1}$, respectively, by replacing $\mathcal{S}_{i-1}|W_{i-1}$ and $\mathcal{S}'_{i-1}|W_{i-1}$ with a single new vertex w_{i-1}.

6. Increment i by 1 and return to Step 2.
7. Output the sum $|\mathcal{F}_0| - 1 + |\mathcal{F}_1| - 1 + \cdots + |\mathcal{F}_{i-1}| - 1$.

Remarks

1. A naive approach to Step 4 is to exhaustively delete edges from one of the trees, $\mathcal{T}$ say, and then see if the resulting forest is an acyclic-agreement forest for $\mathcal{T}$ and $\mathcal{T}'$.
2. Observe that, if one ignores the task of finding a maximum-acyclic-agreement forest in Step 4, then HybridNumber provides a fast lower bound for $h(\mathcal{T}, \mathcal{T}')$. In particular, the number of iterations of the algorithm.

Clearly, Step 4 is the computationally most expensive part of the algorithm. However, although there is no theoretical foundations for the complexity of this algorithm, it will work well in practice provided it breaks the problem into a number of isolated parts for which the associated hybridization number is relatively small. To see whether this proviso is realistic or not, Bordewich *et al.* [10] have carried out an experimental analysis of HybridNumber on a particular grass (*Poaceae*) data set that has previously been considered by Schmidt [43]. Because of earlier findings of Ellstrand *et al.* [16], this data set is appropriate for such an analysis as it is more likely that the conflicting signals in the data is due to hybridization rather than other factors. Without going into the details, the analysis involves the running of the algorithm on pairs of trees with up to 40 taxa. The results highlight the usefulness of the reduction rules that underlie HybridNumber. We describe one particularly successful example next.

The grass data set consists of sequence data for six loci. The two phylogenetic trees shown in Fig. 10.11 are the result of applying the fastDNAml programme [41] to two of the sequences—a nuclear sequence (internal transcribed spacer) and a chloroplast sequence (phytochrome B). For convenience, as this example is simply illustrating how the algorithm works and nothing more, we have replaced the species names with numbers. Taking these two trees as the input to HybridNumber, the algorithm initially finds all common subtrees and replaces each such subtree by a single leaf with a new label. The resulting trees are shown in Fig. 10.12 where, for clarity, each common subtree has been replaced by a single leaf whose label is a concatenation of the subtree labels. The next step is to search for a minimal cluster of size at least two that is common to both trees in Fig. 10.12.

One such cluster, as shown by the inside square brackets in Fig. 10.12, is $\{1, 20, 15, 19, 4, 3, 5, 29, 12, 16, 9\}$ and the corresponding subtrees are shown at the top of Fig. 10.13. This essentially completes the first iteration of the algorithm. At the completion of two further iterations, we obtain the two further pairs of subtrees (as indicated by the middle and outside square brackets shown in Fig. 10.12) and these are shown in Fig. 10.13. Again, the trees on the left come from the nuclear sequence, while the trees on the right come from the chloroplast sequence. At this stage the original inputted trees have been reduced to two trees

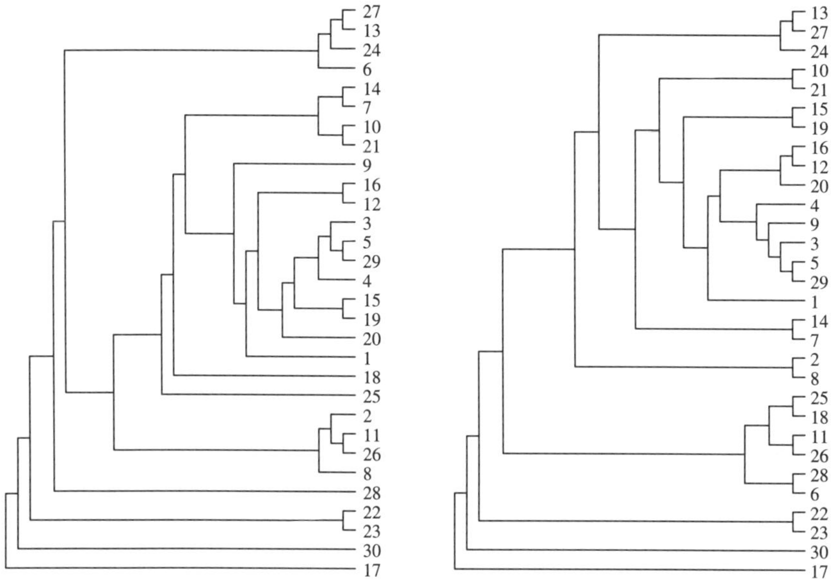

FIG. 10.11. The input to HYBRIDNUMBER. The tree resulting from the nuclear sequence is on the left, while the tree resulting from the chloroplast sequence is on the right.

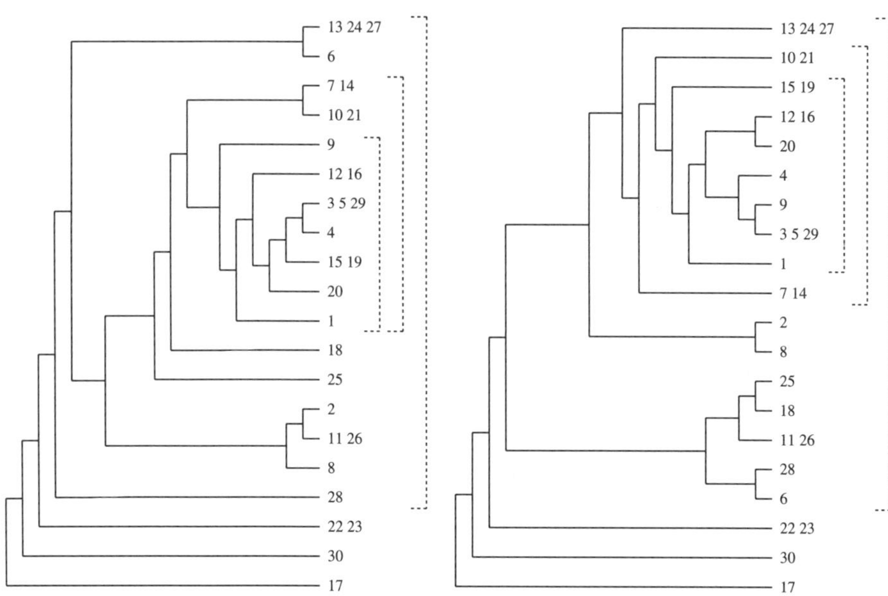

FIG. 10.12. The two phylogenetic trees resulting from repeated applications of Rule 1 to the two phylogenetic trees in Fig. 10.11.

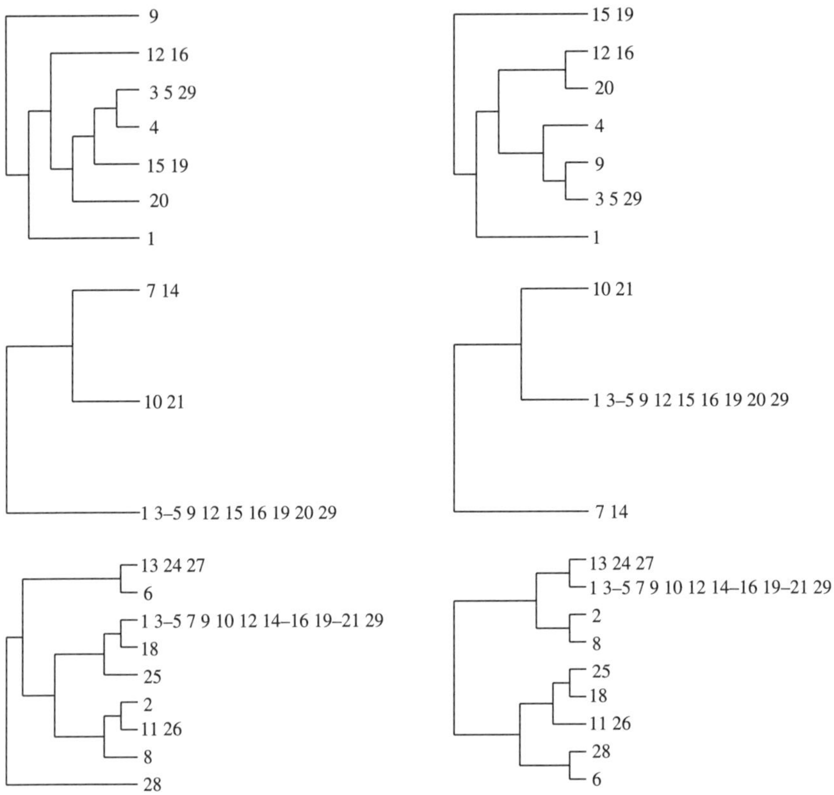

FIG. 10.13. The top pair of trees are the subtrees in Fig. 10.12 corresponding to the common cluster $\{1, 20, 15, 19, 4, 3, 5, 29, 12, 16, 9\}$. The bottom two pairs of trees are the resulting pairs of subtrees after two further iterations of HYBRIDNUMBER.

that are identical. We now exhaustively find the hybridization number of each of the three pairs of non-identical trees. The first pair has a hybridization number of 3, while the second and third pairs have hybridization numbers of 1 and 4, respectively. Adding the three numbers together results in the hybridization number of 8 for the phylogenetic trees shown in Fig. 10.11. The running time of an implementation of the algorithm HYBRIDNUMBER applied to the two trees in Fig. 10.11 is 19 seconds. Given that the trees contain 30 taxa and have a hybridization number of 8, this is remarkably quick.

We end this subsection with two further comments. Firstly, Nakhleh *et al.* [39] describe a polynomial-time heuristic for finding $h(\mathcal{T}, \mathcal{T}')$ that is based on an agreement-forest-type approach. In this heuristic, they obtain a certain agreement forest by repeatedly finding a maximum-agreement subtree of two trees to decompose $\mathcal{T}$ and $\mathcal{T}'$. For further details and the associated reconstruction algorithm, see [39]. Secondly, although we have not included the details here, it is straightforward to construct a hybridization network associated with HYBRIDNUMBER by combining our earlier algorithm HYBRIDNETWORK (Section 10.3.4) with the second part of Theorem 10.5. However, it is important to note that such a network is not necessarily unique. Typically, there will be a number of possibilities.

10.4.3 *Galled-trees*

Whenever one is confronted with an NP-hard problem, a natural consideration is to see if there exists a polynomial-time algorithm for special instances of the problem that are still meaningful. In this subsection, we describe one particular instance that has been very successful in this regard.

Ignoring the directions of the arcs, a *galled-tree* is a hybridization network in which every vertex is in at most one cycle. This means that, for every pair of cycles, their vertex sets (and thus arc sets) are disjoint. In keeping with the terminology in the literature, a cycle in a galled-tree is called a *gall.* First studied in [52], galled-trees have been subsequently studied both in the hybridization and recombination settings (see Section 10.5 for details on the latter setting). These include algorithmic studies [19, 20, 31, 32, 40, 48] and enumeration studies [45]. The original motivation for their study, whether correct or not, is that hybridization events are rare and so one may expect such events to be isolated, in which case conflicts in the initial collection of phylogenetic trees could be explained by a galled-tree.

Let $\mathcal{T}$ and $\mathcal{T}'$ be two rooted binary phylogenetic X-trees, and let $|X| = n$. Nakhleh *et al.* [40] describe an $O(mn)$ algorithm for deciding if there exists a galled-tree that displays $\mathcal{T}$ and $\mathcal{T}'$, and then constructs such a minimal network, where m is the hybridization number of this network. Note that there is a proviso on the network that they construct, in particular, it is minimal with respect to all other galled-trees that display $\mathcal{T}$ and $\mathcal{T}'$. However, this proviso is not necessary because of the following proposition.

Proposition 10.8 *Let $\mathcal{T}$ and $\mathcal{T}'$ be two rooted binary phylogenetic X-trees, and suppose that there is a galled-tree that displays $\mathcal{T}$ and $\mathcal{T}'$. Suppose that the smallest number of hybridization vertices in such a network is m. Then $h(\mathcal{T},\mathcal{T}') = m$.*

Before proving Proposition 10.8, we remark that an alternative, but equivalent, way to say Proposition 10.8 is that if there is a galled-tree that displays $\mathcal{T}$ and $\mathcal{T}'$, then there is such a galled-tree that minimizes the number of hybridization vertices over all networks that displays $\mathcal{T}$ and $\mathcal{T}'$. The algorithm in [40] is essentially equivalent to combining HYBRIDNUMBER and HYBRIDNETWORK, and so one could establish the proposition as a consequence of these algorithms. However, we prove it directly using Theorem 10.5.

Proof of Proposition 10.8 The proof is by induction on m. If $m = 0$, then $\mathcal{T}$ and $\mathcal{T}'$ are isomorphic, so $h(\mathcal{T},\mathcal{T}') = 0$ and the theorem holds. Now suppose that $m = k+1$ for some $k \geq 0$ and that the theorem holds whenever the smallest number of hybridization vertices in a galled-tree that displays the two input trees is at most k.

Let $\mathcal{H}$ be a galled-tree that displays $\mathcal{T}$ and $\mathcal{T}'$, and has the smallest number of hybridization vertices amongst all such networks. Because of the minimality condition, each hybridization vertex has indegree 2. For the purposes of the proof, we will refer to the unique vertex of a gall that is closer to the root than any other vertex of the gall as the *coalescent* vertex of the gall. Let w be the coalescent vertex of a gall Q in $\mathcal{H}$ such that there is no directed path in $\mathcal{H}$ from w to another vertex that is the coalescent vertex of a gall in $\mathcal{H}$. Before continuing, we make two observations:

(i) The subset W of X whose elements can be reached from w via a directed path is a cluster of both $\mathcal{T}$ and $\mathcal{T}'$.
(ii) The subtree of $\mathcal{T}$ induced by W can be obtained from the subnetwork of $\mathcal{H}$ that consists of all vertices and arcs that lie on a directed path from w by deleting one of the incoming arcs of the hybridization vertex in Q. Similarly, the subtree of $\mathcal{T}'$ induced by W can be obtained by deleting the other incoming arc of the hybridization vertex in Q.

Let $\mathcal{T}_w$ and $\mathcal{T}'_w$ be the rooted binary phylogenetic trees obtained from $\mathcal{T}$ and $\mathcal{T}'$, respectively, by replacing the subtrees $\mathcal{T}|W$ and $\mathcal{T}'|W$ with a single vertex labelled w, where $w \notin X$. By Theorem 10.5,

$$h(\mathcal{T},\mathcal{T}') = h(\mathcal{T}|W,\mathcal{T}'|W) + h(\mathcal{T}_w,\mathcal{T}'_w).$$

Since $\mathcal{T}|W$ is not isomorphic to $\mathcal{T}'|W$, we have that $h(\mathcal{T}|W,\mathcal{T}'|W) \geq 1$. But, by (ii), $h(\mathcal{T}|W,\mathcal{T}'|W) \leq 1$ and therefore $h(\mathcal{T}|W,\mathcal{T}'|W) = 1$. Consider $h(\mathcal{T}_w,\mathcal{T}'_w)$. Let $\mathcal{H}_w$ denote the galled-tree obtained from $\mathcal{H}$ by deleting each of the vertices that lie on a directed path from w except w itself. Since $\mathcal{H}$ displays $\mathcal{T}$ and $\mathcal{T}'$, it follows that $\mathcal{H}_w$ displays $\mathcal{T}_w$ and $\mathcal{T}'_w$. Now $\mathcal{H}_w$ has k galls. Suppose that there is a galled-tree that displays $\mathcal{T}_w$ and $\mathcal{T}'_w$, but has less galls than $\mathcal{H}_w$. Then one could use this network to obtain a galled-tree that displays $\mathcal{T}$ and $\mathcal{T}'$ by adjoining the subnetwork below w in $\mathcal{H}$ to w resulting in a galled-tree with less galls than $\mathcal{H}$; a contradiction to the minimality of $\mathcal{H}$. It now follows that amongst all galled-trees that display $\mathcal{T}_w$ and $\mathcal{T}'_w$, the galled-tree $\mathcal{H}_w$ has the smallest number of galls. By the induction assumption, this implies that $h(\mathcal{T}_w,\mathcal{T}'_w) = k$ and so

$$\begin{aligned} h(\mathcal{T},\mathcal{T}') &= h(\mathcal{T}|W,\mathcal{T}'|W) + h(\mathcal{T}_w,\mathcal{T}'_w) \\ &= k+1. \end{aligned}$$

This completes the proof of the proposition. □

Nakhleh *et al.* [40] propose a method for inferring hybridization networks that allows for errors in the estimation of the initial two gene trees. In brief, when methods such as maximum likelihood or maximum parsimony infer trees, there are a number of equally or close-to-equally good trees that could have been inferred. Thus the strict consensus of each such set of trees is perhaps a better representative of the original data set than one particular tree. However, this representative is typically unresolved, and so an interesting problem is the following. Given two rooted phylogenetic X-trees $\mathcal{T}_1$ and $\mathcal{T}_2$, determine if there is two rooted binary phylogenetic X-trees $\mathcal{T}_1'$ and $\mathcal{T}_2'$ such that $\mathcal{T}_i'$ is a refinement of $\mathcal{T}_i$ with the property that there is a galled-tree that displays $\mathcal{T}_1'$ and $\mathcal{T}_2'$. Moreover, if there is such a network, find $\mathcal{T}_1'$ and $\mathcal{T}_2'$ that minimizes the number of galls over all galled-trees that display $\mathcal{T}_1'$ and $\mathcal{T}_2'$. In [40], the authors provide a linear-time algorithm for when the galled-tree contains exactly one gall. Huynh *et al.* [31] significantly extend this result by providing a quadratic-time algorithm for this problem with no restrictions on the number of galls in the resulting galled-tree. Moreover, they also show that this algorithm easily extends to an efficient algorithm for an arbitrary number of input trees. For further details, we refer the reader to [31].

Controlling the way in which hybridization events occur in a network is a possible avenue for further polynomial-time algorithms. Indeed, recent positive results by Huson *et al.* [30] suggest that this control could be done in a number of successful ways.

10.5 Recombination networks

The *perfect phylogeny with recombination* is a problem that has a very similar flavour to that of MINIMUM HYBRIDIZATION. Indeed, the two problems are closely related. Instead of inputting a collection of trees, the input for this problem is a collection, B say, of binary sequences. However, the goal is essentially the same. Loosely speaking, this goal is to compute the minimum number of 'recombination' events to 'explain' B. Introduced by Hein [23, 24], there are now a number of papers on this problem, including [5, 17, 18, 19, 20, 48, 49, 50, 51, 52]. In this section, we describe this problem and its relationship with MINIMUM HYBRIDIZATION. This relationship will be used in Section 10.7.

An (n, m)-*recombination network* $\mathcal{N}$ is a rooted acyclic digraph with exactly n vertices of outdegree zero in which each vertex other than the root has either one or two incoming arcs, and each vertex of $\mathcal{N}$ is labelled with a binary sequence of length m. The sequence labelling the root is called the *root* or *ancestral* sequence. A vertex with two incoming arcs is called a *recombination* vertex. Each integer in $\{1, 2, \ldots, m\}$ is assigned to exactly one arc of $\mathcal{N}$ that is not directed towards a recombination vertex. Beginning with the root and its associated sequence, each of the binary sequences labelling the other vertices is based on the binary sequence of its parent and the incoming arc (in the case it is a non-recombination vertex) or its parents (in the case it is a recombination vertex). In particular, the sequences satisfy the following properties:

(i) If v is a non-recombination vertex with incoming arc e, then the sequence labelling v is obtained from the sequence labelling its parent by changing the i-th element (site) from 0 to 1 or 1 to 0 appropriately for each integer i assigned to e. If no integer is assigned to e, then the sequence labelling v is the same as its parent.

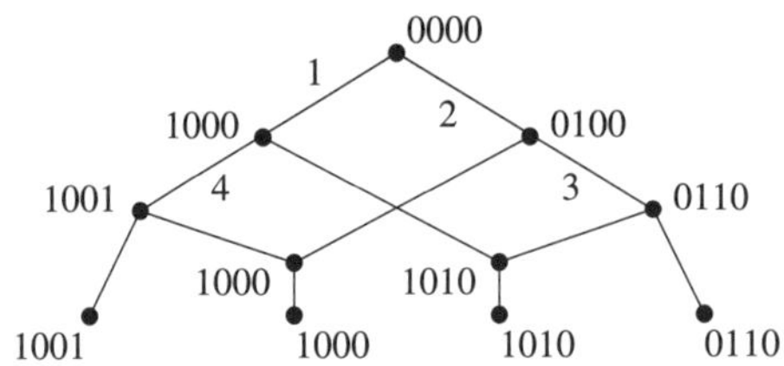

FIG. 10.14. A $(4, 4)$-recombination network in which the root sequence is the all-0 sequence.

(ii) If v is a recombination vertex, then, for some positive integer p strictly between 1 and m (that is, $1 < p < m$), the sequence labelling v is the concatenation of the first p elements of the sequence labelling one of its parents and the last $m - p$ elements of its other parent. To describe the corresponding recombination event one labels the incoming arcs either P or S depending upon which parent contributes the prefix part or the suffix part of the sequence, respectively, and also labels the recombination vertex with an ordered pair indicating the 'break-point'.

Biologically speaking, the mutations in (i) are called *point mutations* and, as each site in the sequence mutates exactly once, we are under the so-called *infinite sites model* of mutations. The recombination process in (ii) is called a *single-crossover recombination* as there is exactly one break-point in the resulting sequence. Even though this model of recombination is very simple, it is the basis of most applications of coalescent theory to recombining sequences [26].

As an example, a recombination network is shown in Fig. 10.14, where the root sequence is the all-0 sequence. For each recombination vertex in this example, the first two elements in the associated sequence come from its 'left' parent and the second two elements come from its 'right' parent. (We have omitted the labelling of the recombination vertices and their incoming arcs as described in (ii) above.) In the literature, a recombination network is commonly referred to as a 'phylogenetic network'.

Let B be a collection of n binary sequences of length m. An (n, m)-recombination network $\mathcal{N}$ *explains* B if the n vertices of outdegree zero are bijectively labelled with the elements of B. For example, the recombination network in Fig. 10.14 explains the collection $\{1001, 1000, 1010, 0110\}$ of binary sequences. Over all recombination networks that explain B, we are interested in finding one that has the minimum number of recombination vertices. The perfect phylogeny with recombination problem is formally stated as follows.

PERFECT PHYLOGENY WITH RECOMBINATION
Instance: A set B of n binary sequences of length m.

Goal: Find an (n, m)-recombination network $\mathcal{N}$ that explains B with minimum number of recombination vertices.
Measure: The number of recombination vertices in $\mathcal{N}$.

Depending upon whether the root sequence of the recombination network is specified or not specified in advance, the problem can be interpreted in one of two ways. If the root sequence is specified in advance, then, from a mathematical perspective, no generality is lost in always choosing the root sequence to be the all-0 sequence. We denote the minimum values for the two problems by $r(B)$ and $r^*(B)$, respectively, and note that $r^*(B) \leq r(B)$. The reason for the wording 'perfect phylogeny' is that the classical perfect phylogeny problem can be interpreted as the problem of deciding if there is a recombination network with no recombination vertices that explains B.

Recombination events are one of the primary influences on genetic variation amongst individuals of the same population. Recognizing how many and where in the sequence these events occur is expected to be a contributing factor in answering a number of important problems in genetics including those centred around genetic diseases. Thus the motivation for PERFECT PHYLOGENY WITH RECOMBINATION is similar to that for MINIMUM HYBRIDIZATION except that our input is now a collection of binary sequences. SNP (single nucleotide polymorphism) sequences satisfy this criteria and are now of great interest (for example, see [27]). Each sequence represents an individual of the same population and, in such a sequence, each site represents an allele of the species. In the case that the root sequence is specified in advance, a 0 denotes the ancestral allele, while a 1 denotes the derived (mutant) allele. Observe that $0 \to 1$ is the only allowable transition in this case.

There is a close relationship between MINIMUM HYBRIDIZATION and PERFECT PHYLOGENY WITH RECOMBINATION with the root sequence specified in advance. In particular, the former problem can be interpreted as a particular instance of the latter.

Using the construction in Wang *et al.* [52], let $\mathcal{T}$ and $\mathcal{T}'$ be two rooted binary phylogenetic X-trees and let $|X| = n$. Noting that $|E(\mathcal{T})| = |E(\mathcal{T}')| = 2(n-1)$, bijectively label the edges of $\mathcal{T}$ and $\mathcal{T}'$ with the elements of $\mathcal{C} = \{\chi_1, \chi_2, \ldots, \chi_{2(n-1)}\}$ and $\mathcal{C}' = \{\chi'_1, \chi'_2, \ldots, \chi'_{2(n-1)}\}$, respectively. Each of the elements in $\mathcal{C}$ and $\mathcal{C}'$ represent a site. Associated to each vertex v (resp. v') of $\mathcal{T}$ (resp. $\mathcal{T}'$) is the binary sequence of length $2(n-1)$ in which the i-th element is 1 if and only if χ_i (resp. χ'_i) labels an edge from the root of $\mathcal{T}$ (resp. $\mathcal{T}'$) to v (resp. v'). Now, for each $x \in X$, concatenate the sequences labelling x in $\mathcal{T}$ and $\mathcal{T}'$ with the sequence labelling x in $\mathcal{T}'$ following the sequence labelling x in $\mathcal{T}$. Let B be the resulting collection of n (concatenated) sequences of length $4(n-1)$. The following theorem due to Bordewich and Semple [12] provides the above mentioned close relationship.

Theorem 10.9 *Let $\mathcal{T}$ and $\mathcal{T}'$ be two rooted binary phylogenetic X-trees, and let B be the collection of binary sequences that is constructed from $\mathcal{T}$ and $\mathcal{T}'$ as* *above. Then*

$$h(\mathcal{T}, \mathcal{T}') = r(B).$$

The proof of Theorem 10.9 is constructive. In particular, if $\mathcal{H}$ is a minimum hybridization network that displays $\mathcal{T}$ and $\mathcal{T}'$, then there is a polynomial-time modification of $\mathcal{H}$ that results in a recombination network $\mathcal{N}$ that explains B with the all-0 sequence at the root and has $h(\mathcal{H})$ recombination vertices. On the other hand, if $\mathcal{N}$ is a recombination network explaining B with the all 0-sequence at the root and k recombination vertices, then $\mathcal{N}$ can be modified to produce a hybridization network that displays $\mathcal{T}$ and $\mathcal{T}'$ with k hybridization vertices. Again, this modification can be done in polynomial-time.

Remark In this section, we have restricted ourselves to single-crossover recombinations. However, we note here that more general recombinations called *multiple-crossover* recombinations have also been considered (for example, see [17, 18, 30]). Here, if v denotes the recombination vertex, then the sequence labelling v has the weaker property that, for all i, the i-th element in the sequence is the same as the i-th element in at least one of the parent sequences. By specifying, for all i, which parent the i-th element came from, the number of crossovers events is equal to the number of pairs $(j, j+1)$ in which the j-th element comes from one parent while the $(j+1)$-th element comes from the other parent. Extending the definition of an (n, m)-recombination network in the obvious way to allow for multiple-crossover events, the 'goal' of the optimization problem analogous to PERFECT PHYLOGENY WITH RECOMBINATION could be interpreted in one of two ways. Namely, minimize the number of recombination vertices in a network that explains an initial set B of binary sequences, or minimize the number of crossover events in a network that explains B. While the first interpretation has received a reasonable amount of attention, the second interpretation appears to have received little attention.

10.6 Hybridization networks in real time

An important biological requirement of hybridization networks is that hybridization events occur between contemporaneous taxa (past or present). Maddison [34] pointed out this requirement and, from a mathematical perspective, it has been considered in several papers since including [7, 38, 49, 51]. We begin this section by considering the problem of whether a given hybridization network is consistent with this requirement.

10.6.1 *Temporal representations*

Let $\mathcal{H}$ be a hybridization network with vertex set V, and let $\mathbb{N} = \{0, 1, 2, \ldots\}$. We say that $\mathcal{H}$ has a *temporal representation* if there is a map $f : V \to \mathbb{N}$ that satisfies the following two properties:

(i) If (u, v) is an arc of $\mathcal{H}$ with $d^-(v) = 1$, then $f(u) < f(v)$.
(ii) If (u, v) is an arc of $\mathcal{H}$ with $d^-(v) \geq 2$, then $f(u) = f(v)$.

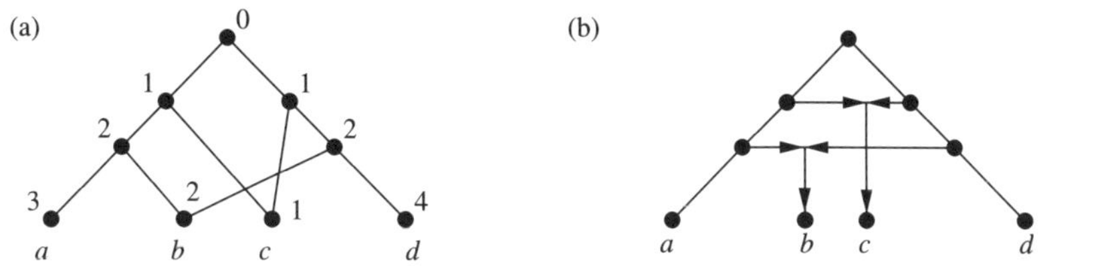

FIG. 10.15. (a) A temporal labelling of a hybridization network and (b) a 'real time' realization of this labelling.

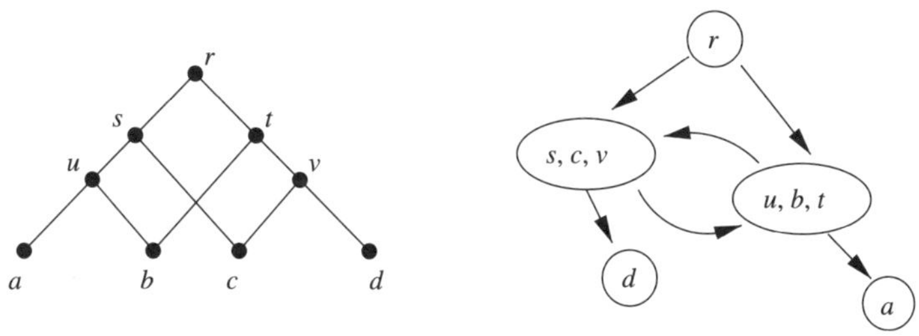

FIG. 10.16. (a) A hybridization network with no temporal representation and (b) its temporal digraph.

Such a map f is called a *temporal labelling* of $\mathcal{H}$. The purpose of (ii) is so that hybridization events occur with contemporaneous taxa. A temporal labelling of a hybridization network is shown in Fig. 10.15(a). A 'real time' realization of this labelled network is shown in Fig. 10.15(b).

All rooted phylogenetic trees have a temporal representation, but not all hybridization networks have such a representation. For example, the hybridization network in Fig. 10.16(a) has no temporal representation. The reason for this is that u and t, the parents of b, must coexist in time, while s and v, the parents of c, must also coexist in time. By considering the ancestor–descendant relationships of s and u, and t and v, this is not possible.

We next describe a simple polynomial-time algorithm for deciding whether a hybridization network has a temporal representation and, if so, constructs such a representation. Due to Baroni *et al.* [7], we begin by defining a particular digraph around which the algorithm is based. Let $\mathcal{H}$ be a hybridization network with vertex set V. Ignoring the direction of the arcs of $\mathcal{H}$, set

$$[v] = \{v\} \cup \{u \in V : \text{there is a path of hybridization arcs from } u \text{ to } v\},$$

where a *hybridization arc* is an arc that is directed into a hybridization vertex.

Note that we have partitioned V into equivalence classes, where $[v] = \{v\}$ precisely if v is not incident with a hybridization arc. Setting $[V] = \{[v] : v \in V\}$, we define the *temporal digraph* of $\mathcal{H}$ as the digraph whose vertex set is $[V]$ and where $[u]$ and $[v]$ are joined by an arc $([u], [v])$ if there is a vertex a in $[u]$ and a vertex b in $[v]$ such that (a, b) is an arc of $\mathcal{H}$ with $d^-(b) = 1$. For example, the digraph in Fig. 10.16(b) is the temporal digraph of the hybridization network in Fig. 10.16(a).

It turns out that $\mathcal{H}$ has a temporal representation if and only if its temporal digraph is acyclic and this is the basis of the following algorithm whose correctness is shown in [7].

Algorithm: TEMPREP($\mathcal{H}$)
Input: A hybridization network $\mathcal{H}$ with vertex set V.
Output: A temporal labelling of $\mathcal{H}$ or the statement $\mathcal{H}$ *has no temporal labelling.*

1. Construct the temporal digraph $D_{\mathcal{H}}$ of $\mathcal{H}$.
2. Find an acyclic ordering, $V_0, V_1, \ldots, V_k$ say, of $D_{\mathcal{H}}$. If there is no such ordering, then return $\mathcal{H}$ *has no temporal representation.*
3. Define $f : V \rightarrow \mathbb{N}$ by setting $f(v) = i$ for all $v \in V$, where $[v] \in V_i$.
4. Return the map f.

If a map f is returned by the algorithm, then f is a temporal labelling of $\mathcal{H}$. It is important to note that a temporal labelling of a hybridization network is no more than an ordering of when past or present taxa appeared. Consequently, it is the ordering on the vertices of V that is important and not the actual values.

If one is interested in obtaining, up to isomorphism, all temporal labellings of $\mathcal{H}$, then the above algorithm can be easily modified to output a list of all such labellings, where a new labelling is outputted in polynomial-time and where two labellings are non-isomorphic if the relative orderings of the vertices are not the same. Essentially, one selects non-empty subsets of vertices that have indegree zero instead of a single vertex in the process of finding an acyclic ordering. All such orderings result in a distinct temporal labelling and all such labellings can be obtained this way. For further details, see [7].

We end this subsection with the following remark. If a hybridization network $\mathcal{H}$ does not have a temporal representation, then Moret *et al.* [38] observed that, by allowing for missing taxa, one could resolve this issue without adding to the hybridization number of $\mathcal{H}$. For example, consider the hybridization network in Fig. 10.16(a). By creating two new vertices that subdivide the arcs (t, b) and (s, c), and joining pendant arcs to each of these new vertices with new taxa, the resulting hybridization network has a temporal representation. The role of such taxa is to carry a gene or combination of genes from the past into some time when it can passed on into the new hybrid species. Of course, whether such taxa exist or existed is a separate question.

10.6.2 *Time-ordered rooted subtree prune and regraft operations*

Realizing the importance that time places on possible scenarios for evolutionary histories, Song and Hein [49, 51] (also see [26]) considered a more restrictive notion of the rooted subtree prune and regraft operation. This restriction allows one to attack the problem of PERFECT PHYLOGENY WITH RECOMBINATION in which the root sequence is not specified in advance using rooted subtree prune and regraft operations.

Let $\mathcal{T}$ be a rooted binary phylogenetic tree and let $\mathring{V} = \{v_1, v_2, \ldots, v_{n-2}\}$ be the set of interior vertices of $\mathcal{T}$. A *total ordering* on $\mathring{V}$ is a binary relation $<_\mathcal{T}$ given by $v_i <_\mathcal{T} v_j$ if the hypothetical ancestor or speciation event represented by v_i predates the hypothetical ancestor or speciation event represented by v_j. In mathematics, total orderings are also called *linear orderings*. We say that $\mathcal{T}$ is *ordered* if $\mathring{V}$ is totally ordered. By default, such an ordering must preserve the ancestor–descendant relationships given by the topology of $\mathcal{T}$.

In performing a rooted subtree prune and regraft operation on an ordered tree $\mathcal{T}$ one must preserve the ordering on $\mathring{V}$. In particular, referring to the notation in the definition of a rSPR operation in Section 10.3, for all $v_i, v_j \in \mathring{V} - \{u\}$, we have that $v_i <_{\mathcal{T}'} v_j$ precisely if $v_i <_\mathcal{T} v_j$, where u is the 'parent' vertex of the root of the subtree being pruned, $\mathcal{T}$ is the initial tree, and $\mathcal{T}'$ is the tree resulting from the rSPR operation. Given two ordered rooted binary phylogenetic X-trees, there is a sequence of (ordered) rSPR operations that transforms one tree into the other. For further combinatorial results on this operation and the ordinary rSPR operation, see Song [46, 47].

Now let B be a collection of binary sequences of equal length m. For each i, the i-th sites in the sequences induce a character χ_i. Under the infinite-sites model of mutation, let $\mathcal{P}_i$ be the collection of ordered rooted binary phylogenetic X-trees that display χ_i, that is $\mathcal{P}_i$ is the collection of all such trees for which there exists an edge whose deletion induces the bipartition of X induced by the character states in χ_i. Consider the problem of minimizing the following sum:

$$\sum_{i=1}^{m-1} d_{\text{rSPR}}(\mathcal{T}_i, \mathcal{T}_{i+1}), \tag{10.2}$$

where $\mathcal{T}_i \in \mathcal{P}_i$ for all i and $d_{\text{rSPR}}(\mathcal{T}_i, \mathcal{T}_{i+1})$ denotes the minimum number of (ordered) rSPR operations to transform $\mathcal{T}_i$ into $\mathcal{T}_{i+1}$. It turns out that the minimum value of this sum is equal to $r^*(B)$, the optimal value of PERFECT PHYLOGENY WITH RECOMBINATION in which the root sequence is not specified in advance (Yun Song, private communication). Thus $r^*(B)$ can be written in terms of the rSPR distance on ordered rooted binary phylogenetic trees. Moreover, a lower bound for $r^*(B)$ can be obtained by interpreting the terms in the sum in (10.2) as the ordinary rSPR distance between two rooted binary phylogenetic trees, where the total ordering on the interior vertices is ignored.

The number of ordered rooted binary phylogenetic trees grows significantly faster than the number of (ordinary) rooted binary phylogenetic trees, and so as it currently stands the above approach to computing $r^*(B)$ exactly is limiting in practice. Nevertheless, by studying a particular data set for which previous lower bounds have been calculated, Song and Hein have shown it can work. For further details, see [49, 51] and note that Song and Hein use the terminology 'ancestral recombination graph' instead of recombination network.

10.7 Computational complexity

In this section, we discuss some of the computational issues associated with the three main problems that we have discussed in this chapter, namely MINIMUM HYBRIDIZATION, MINIMUM rSPR, and PERFECT PHYLOGENY WITH RECOMBINATION. Throughout this section, the interpretation of the last of these problems will always be the one in which the root sequence is specified in advance.

The following theorem, which we have alluded to several times in this chapter, is due to Bordewich and Semple [11, 12].

Theorem 10.10 *Each of the optimization problems* MINIMUM HYBRIDIZATION, MINIMUM rSPR, *and* PERFECT PHYLOGENY WITH RECOMBINATION *is* NP*-hard.*

The proofs of the NP-hardness of MINIMUM HYBRIDIZATION and MINIMUM rSPR make use of their characterizations in terms of agreement forests and use ideas originating from Hein *et al.* [25]. The NP-hardness of PERFECT PHYLOGENY WITH RECOMBINATION follows from Theorem 10.9 and the polynomial-time constructions mentioned after it. To avoid repetition, these comments are also valid for Theorem 10.11.

Despite the negativity of Theorem 10.10, there are some positive results for these problems. Fixed-parameter algorithms are a practical way to find optimal solutions of NP-hard problems if the parameter measuring the hardness of the problem is small. For MINIMUM rSPR, Bordewich and Semple [11] showed that there is such an algorithm where the rSPR distance itself is the parameter. In particular, instead of computing the rSPR distance between two rooted binary phylogenetic X-trees by an exhaustive search resulting in an algorithm that takes time $O((2n)^{2k})$ where $n = |X|$ and $k = d_{\mathrm{rSPR}}(\mathcal{T}, \mathcal{T}')$, they showed that there is a parameterized algorithm for computing this distance in $O(f(k) \cdot p(n))$ where $f(k)$ is some computable function depending on k and p is a polynomial in n. The important point of this running time is that n and k are now separated which means that, provided k is small, computing $d_{\mathrm{rSPR}}(\mathcal{T}, \mathcal{T}')$ may be efficiently possible even when n is large. The important part of the analysis is Theorem 10.4.

Translating the setting in [22], Hallet and Lagergren give a fixed-parameter algorithm for a problem that is a restriction of MINIMUM HYBRIDIZATION (also see [1] for a description of the algorithm). Parameterized by the hybridization number of the two trees, Bordewich and Semple [13] recently gave a fixed-parameter algorithm for MINIMUM HYBRIDIZATION in general. For further details of this last algorithm and an analysis of how well it works in practice, we refer the interested reader to [13] and [10], respectively. For those wanting to find out more about fixed-parameter algorithms, we refer the reader to [15] and [29]. The latter is an easy-to-read introduction to fixed-parameter algorithms and describe three techniques for developing such algorithms.

For computationally hard problems, polynomial-time approximation algorithms can efficiently find feasible solutions that are sometimes arbitrarily close to the optimal solution. In particular, for a minimization problem, an r-approximation algorithm means that, for all instances, the ratio of the size of the feasible solution outputted by the algorithm and the size of an optimal solution is at most r. The existence of polynomial-time approximation algorithms varies greatly amongst NP-hard problems. For example, regardless of the choice of r, there is no such algorithm for the general travelling salesman problem unless $P = NP$, while for some problems π, no matter how close r is to 1, there is always such an algorithm. In this latter case, we say that π exhibits a *polynomial-time approximation scheme* (PTAS). Theorem 10.11 is due to Bordewich and Semple [12].

Theorem 10.11 *Each of the optimization problems* MINIMUM HYBRIDIZATION, MINIMUM RSPR, *and* PERFECT PHYLOGENY WITH RECOMBINATION *is* APX-*hard. In particular, for each of these problems there is no polynomial-time approximation scheme unless* $P = NP$.

For each of our optimization problems, the implication of Theorem 10.11 is that, unless $P = NP$, there is some fixed constant r strictly bigger than 1 for which there is no polynomial-time r-approximation algorithm. It is shown in [12] that, for each of these problems, r is at least $\frac{2113}{2112}$.

Two polynomial-time approximation algorithms for finding the 'TBR distance' between two unrooted phylogenetic trees have appeared in the literature [25, 42]. In many ways, this distance is the unrooted analogue of the rSPR distance. Both are stated as 3-approximation algorithms, however, each of these algorithms have been subsequently shown to be incorrect in some way. Nevertheless, using these approaches, Bonet *et al.* [9] describe a polynomial-time 5-approximation algorithm for MINIMUM RSPR. Intuitively, this algorithm builds an agreement forest locally. Currently, there appears to be no such algorithm for MINIMUM HYBRIDIZATION. One might hope that the algorithm in [9] extends to MINIMUM HYBRIDIZATION, but, due to the additional global condition on an acyclic-agreement forest, it seems unlikely that such an approach will work. For an excellent reference on approximation algorithms, see [4].

10.8 Concluding remarks

The understanding and analysis of reticulation in evolution is playing a prominent role in modern-day phylogenetics. In this chapter, we considered one particular, but central, aspect; namely the problem of finding the smallest number of reticulation events that are required to explain the evolution of a collection of species under consideration subject to some initial input. For us, the input was a collection of rooted phylogenetic trees. The approach we have taken here is analytical so as to provide a theoretical foundation for algorithmic solutions to the problem. Furthermore, our main interest has been on a general solution rather than one that is restricted in some way. Unfortunately, despite the fixed-parameter algorithms for MINIMUM RSPR and MINIMUM HYBRIDIZATION, and the divide-and-conquer algorithm for MINIMUM HYBRIDIZATION described in this chapter, we are always going to be limited in finding exact solutions because of the NP-hardness of these problems. This turns our attention to future work.

A number of papers have considered efficient algorithms for computing lower bounds for Perfect Phylogeny with Recombination (for example, see [5, 21, 28, 37, 50]). While one could use the constructions outlined after Theorem 10.9 and these results, it appears that little attention has been given to finding such algorithms directly for Minimum rSPR and Minimum Hybridization. Given the incorrectness of related approximations, a mathematically challenging task is to improve the 5-approximation algorithm for Minimum rSPR. Whether Minimum Hybridization even has such an algorithm, regardless of the size of the ratio, is an interesting question.

In this chapter, we have only considered combinatorial questions. While a combinatorial understanding of reticulation is far from complete, it is statistical questions that will eventually need to be addressed. For example, how can one use differing bootstrap support values for conflicting phylogenies to quantify and distinguish between genuine reticulation and other biological processes that give rise to conflicts such as lineage sorting? Combinatorial considerations are often the first steps towards statistical-based approaches in phylogenetics and so it is highly likely that combinatorial insights into hybridization networks will aid the development of such approaches to reticulation.

Acknowledgements

Many thanks to Peter Lockhart, Katherine St. John, and Yun Song for a number of helpful discussions during the writing of this chapter, and Simone Linz for providing the figures for the grass data set example in Section 10.4.2. This work was supported by the New Zealand Marsden Fund (UOC310).

References

[1] Addario-Berry, L., Hallett, M., and Lagergren, J. (2003). Towards identifying lateral gene transfer events. In: *Proceedings of the Pacific Symposium on Biocomputing* (PSB 2003) (ed. R. S. Altman et al.), pp. 279–290.

[2] Allan, H. H. (1961). *Flora of New Zealand, Volume I, Indigenous tracheophyta: Psilopsida, Lycopsida, Filicopsida, Gymnospermae, Dicotyledones.* Government Printer, Wellington, World Scientific, Singapore.

[3] Allen, B. L. and Steel, M. (2001). Subtree transfer operations and their induced metrics on evolutionary trees. *Annals of Combinatorics*, **5**, 1–13.

[4] Ausiello, G., Crescenzi, P., Gambosi, G., Kann, V., Marchetti-Spaccamela, A., and Protasi, M. (1999). *Complexity and Approximation.* Springer, Berlin.

[5] Bafna, V. and Bansal, V. (2004). The number of recombination events in a sample history: conflict graph and lower bounds. *IEEE/ACM Transactions on Computational Biology and Bioinformatics*, **1**, 78–90.

[6] Baroni, M., Semple, C., and Steel, M. (2004). A framework for representing reticulate evolution. *Annals of Combinatorics*, **8**, 391–408.

[7] Baroni, M., Semple, C., and Steel, M. (2006). Hybrids in real time. *Systematic Biology*, **55**, 46–56.

[8] Baroni, M., Grünewald, S., Moulton, V., and Semple, C. (2005). Bounding the number of hybridization events for a consistent evolutionary history. *Mathematical Biology*, **51**, 171–182.

[9] Bonet, M. K., St. John, K., Mahindru, R., and Amenta, N. (2006). Approximating subtree distances between phylogenies. *Journal of Computational Biology*, **13**, 1419–1434.

[10] Bordewich, M., Linz, S., St. John, K., and Semple, C. A reduction algorithm for computing the hybridization number of two trees. *Evolutionary Bioinformatics*, in press.

[11] Bordewich, M. and Semple, C. (2004). On the computational complexity of the rooted subtree prune and regraft distance. *Annals of Combinatorics*, **8**, 409–423.

[12] Bordewich, M. and Semple, C. Computing the minimum number of hybridisation events for a consistent evolutionary history. *Discrete Applied Mathematics*, **155**, 806–830.

[13] Bordewich, M. and Semple, C. Computing the hybridization number of two phylogenetic trees is fixed-parameter tractable. *IEEE/ACM Transactions on Computational Biology and Bioinformatics*, in press.

[14] Doolittle, W. F. (1999). Phylogenetic classification and the universal tree. *Science*, **284**, 2124–2128.

[15] Downey, R. and Fellows, M. (1998). *Parameterized Complexity*. Springer, New York.

[16] Ellstrand, N. C., Whitkus, R., and Rieseberg, L. H. (1996). Distribution of spontaneous plant hybrids. *Proceedings of the National Academy of Sciences*, **93**, 5090–5093.

[17] Gusfield, D. (2005). Optimal, efficient reconstruction of root-unknown phylogenetic networks with constrained and structured recombination. *Journal of Computer and System Sciences*, **70**, 381–398.

[18] Gusfield, D. and Bansal, V. (2005). A fundamental decomposition theory for phylogenetic networks and incompatible characters. In: *Proceedings of the Ninth Annual International Conference on Research in Computational Molecular Biology* (RECOMB 2005) (ed. S. Miyano *et al.*), Lecture Notes in Bioinformatics, Vol. 3500, pp. 217–232. Springer, Berlin.

[19] Gusfield, D., Eddhu, S., and Langley, C. (2004). Optimal, efficient reconstruction of phylogenetic networks with constrained recombination. *Journal of Bioinformatics and Computational Biology*, **2**, 173–213.

[20] Gusfield, D., Eddhu, S., and Langley, C. (2004). The fine structure of galls in phylogenetic networks. *INFORMS Journal on Computing*, **16**, 459–469.

[21] Gusfield, D., Hickerson, D., and Eddhu, S. An efficiently-computed lower bound on the number of recombinations in phylogenetic networks: theory and empirical study. *Discrete Applied Mathematics*, **155**, 806–830.

[22] Hallett, M. and Lagergren, J. (2001). Efficient algorithms for lateral gene transfer problems. In: *Proceedings of the Fifth Annual International Conference on Research in Computational Molecular Biology* (RECOMB 2001), pp. 149–156. ACM Press, New York.

[23] Hein, J. (1990). Reconstructing evolution of sequences subject to recombination using parsimony. *Mathematical Biosciences*, **98**, 185-200.

[24] Hein, J. (1993). A heuristic method to reconstruct the history of sequences subject to recombination. *Journal of Molecular Evolution*, **36**, 396-405.

[25] Hein, J., Jing, T., Wang, L., and Zhang, K. (1996). On the complexity of comparing evolutionary trees. *Discrete Applied Mathematics*, **71**, 153–169.

[26] Hein, J., Schierup, M., and Wiuf, C. (2005). *Gene Genealogies, Variation and Evolution: A Primer in Coalescent Theory*. Oxford University Press, Oxford.

[27] Hinds, D, Stuve, L., Nilsen, G., Halperin, E., Eskin, E., Gallinger, D., Frazer, K., and Cox, D. (2005) Whole-genome patterns of common DNA variation in three human populations. *Science*, **307**, 1072–1079.

[28] Hudson, R. and Kaplan, N. (1985). Statistical properties of the number of recombination events in the history of a sample of DNA sequences. *Genetics*, **111**, 147–164.

[29] Hüffner, F., Niedermeier, R., and Wernick, S. Techniques for practical fixed-parameter algorithms, submitted.

[30] Huson, D. H., Klöpper, T., Lockhart, P. J., and Steel, M. (2005). Reconstruction of reticulate networks from gene trees. In: *Proceedings of the Ninth Annual International Conference on Research in Computational Molecular Biology* (RECOMB 2005) (ed. S. Miyano *et al.*), Lecture Notes in Bioinformatics, Vol. 3500, pp. 233–249. Springer, Berlin.

[31] Huydn, T. N. D., Jansson, J., Nguyen, N. B., and Sung, W. -K. (2005). Constructing a smallest refinining galled phylogenetic network. In: *Proceedings of the Ninth Annual International Conference on Research in Computational Molecular Biology* (RECOMB 2005) (ed. S. Miyano *et al.*), Lecture Notes in Bioinformatics, Vol. 3500, pp. 265–280. Springer, Berlin.

[32] Jansson, J., Nguyen, N. B., and Sung, W. -K. (2006). Algorithms for combining rooted triples into a galled phylogenetic network. *SIAM Journal on Computing*, **35**, 1098–1121.

[33] Linz, S. Reticulation in evolution. Unpublished PhD thesis, Heinrich-Heine Universität, in preparation.

[34] Maddison, W. (1997). Gene trees in species trees. *Systematic Biology*, **46**, 523–536.

[35] Mallet, J. (2005). Hybridization as an invasion of the genome. *Trends in Ecology and Evolution*, **20**, 229–237.

[36] McBreen, K. and Lockhart, P. J. (2007). Recostructing reticulate evolutionary histories of plants. *Trends in Plant Science*, **11**, 398–404.

[37] Myers, S. and Griffiths, R. (2003). Bounds on the minimum number of recombination events in a sample history. *Genetics*, **163**, 375–394.

[38] Moret, B. M. E., Nakhleh, L., Warnow, T., Linder, C. R., Tholse, A., Padolina, A., Sun, J., and Timme, R. (2004). Phylogenetic networks: modeling, reconstructibility, and accuracy. *IEEE/ACM Transactions on Computational Biology and Bioinformatics*, **1**, 1–11.

[39] Nakhleh, L., Ruths, D., and Wang, L. S. (2005). RIATA-HGT: a fast and accurate heuristic for reconstructing horizontal gene transfer. In: *Proceedings of the Eleventh International Computing and Combinatorics Conference* (COCOON 05) (ed. L. Wang), Lecture Notes in Computer Science, Vol. 3595, pp. 84–93. Springer, Berlin.

[40] Nakhleh, L., Warnow, T., Linder, C. R., and St. John, K. (2005). Reconstructing reticulate evolution in species—theory and practice. *Journal of Computational Biology*, **12**, 796–811.

[41] Olsen, G. J., Matsuda, H., Hagstrom, R., and Overbeek, R. (1994). fastDNAmL: a tool for construction of phylogenetic trees of DNA sequences using maximum likelihood. *Computing Applications in the Biosciences*, **10**, 41–48.

[42] Rodrigues, E. M., Sagot, M. -F., and Wakabayashi, Y. (2001). Some approximation results for the maximum agreement forest problem. In: *Approximation, Randomization and Combinatorial Optimization: Algorithms and Techniques* (APPROX and RANDOM) (ed. M. Goemans *et al.*), Lecture Notes in Computer Science, Vol. 2129, pp. 159–169. Springer, Berlin.

[43] Schmidt, H. A. (2003). Phylogenetic trees from large data sets. Unpublished PhD thesis, Heinrich-Heine Universität.

[44] Semple, C. and Steel, M. (2003). *Phylogenetics*. Oxford University Press.

[45] Semple, C. and Steel, M. (2006). Unicyclic networks: compatibility and enumeration. *IEEE/ACM Transactions on Computational Biology and Bioinformatics*, **3**, 84–91.

[46] Song, Y. S. (2003). On the combinatorics of rooted binary phylogenetic trees. *Annals of Combinatorics*, **7**, 365–379.

[47] Song, Y. S. (2006). Properties of subtree-prune-and-regraft operations on totally-ordered phylogenetic trees. *Annals of Combinatorics*, **10**, 147–163.

[48] Song, Y. S. (2006). A concise necessary and sufficient condition for the existence of a galled-tree. *IEEE/ACM Transactions on Computational Biology and Bioinformatics*, **3**, 186–191.

[49] Song, Y. S. and Hein, J. (2003). Parsimonious reconstruction of sequence evolution and haplotyde blocks: finding the minimum number of recombination events. In: *Algorithms in Bioinformatics* (WABI 2003) (ed. G. Benson and R. Page), Lecture Notes in Bioinformatics, Vol. 2812, pp. 287–302. Springer, Berlin.

[50] Song, Y. S. and Hein, J. (2004). On the minimum number of recombination events in the evolutionary history of DNA sequences. *Journal of Mathematical Biology*, **48**, 160–186.

[51] Song, Y. S. and Hein, J. (2005). Constructing minimal ancestral recombination graphs. *Journal of Computational Biology*, **12**, 147–169.

[52] Wang, L., Zhang, K., and Zhang, L. (2001). Perfect phylogenetic networks with recombination. *Journal of Computational Biology*, **8**, 69–78.

INDEX